Handbook to IEEE Standard 45™:

A Guide to Electrical Installations on Shipboard

Mohammed M. Islam

Published by
Standards Information Network
IEEE Press

Trademarks and Disclaimers

IEEE believes the information in this publication is accurate as of its publication date; such information is subject to change without notice. IEEE is not responsible for any inadvertent errors.

IEEE
3 Park Avenue, New York, NY 10016-5997, USA

Yvette Ho Sang, Manager, Standards Publishing
Jennifer Longman, Managing Editor
Mike Fisher, Project Editor
Linda Sibilia, Cover Designer

Review Policy

The information contained in IEEE Press/Standards Information Network publications is reviewed and evaluated by peer reviewers of relevant IEEE Technical Societies, Standards Committees and/or Working Groups, and/or relevant technical organizations. The authors addressed all of the reviewers' comments to the satisfaction of both the IEEE Standards Information Network and those who served as peer reviewers for this document.

The quality of the presentation of information contained in this publication reflects not only the obvious efforts of the authors, but also the work of these peer reviewers. The IEEE Press acknowledges with appreciation their dedication and contribution of time and effort on behalf of the IEEE.

To order IEEE Press Publications, call 1-800-678-IEEE.

Print: ISBN 0-7381-4101-1   SP1135

See other IEEE standards and standards-related product listings at:

http://standards.ieee.org/

Trademarks

IEEE Standards designations are trademarks of the Institute of Electrical and Electronics Engineers, Incorporated (www.ieee.org/)

National Electrical Code and NEC are both registered trademarks of the National Fire Protection Association, Inc. (www.nfpa.org/)

Acknowledgement

This Handbook maintains the format of IEEE Std 45™-2002 for easy reference and understanding. It also provides the necessary background to justify the recommendations and provides simple formulae, tables, and diagrams for understanding recommended design features.

This Handbook is a collection of many years of shipbuilding design development experience, as well as the experience of the IEEE Std 45 developmental process. The author wishes to thank all the individuals who have encouraged him and contributed in the preparation of this Handbook. The author also wishes to thank all IEEE Std 45 Working Group members for sharing their technical know-how and expertise over the years, as well as the technical experts in this field whose works may have been quoted in this Handbook.

Special thanks to Joe English, Director of Northrop Grumman Ship Systems, Electrical Engineering, for his encouragement in writing this Handbook.

Special thanks to fellow engineers who have advised and clarified many technical issues over the years.

Dedicated to my parents.

- Moni Islam, July 2004

Author

Mohammed M. (Moni) Islam is R&D Supervisor of Applied
Science at Northrop Grumman Ship Systems. He has thirty-four
years of diversified shipboard electrical engineering experience and
has played significant roles in every part of new shipbuilding and
ship modernization engineering. Mr. Islam also currently serves as
the IEEE-45 central committee Vice-Chair and is a member of
IEEE-1580 working group. He has been involved in the "All
Electric Ship" R&D programs for many years and was the principal
investigator of the Ship Smart-System Design (S3D) feasibility
study, an ONR funded research and development project. He
received his Bachelor of Marine Engineering Technology from the
Merchant Marine Academy of Bangladesh in 1969, and Bachelor of Electrical Engineering Degree
with Honors from the State University of New York, Fort Schuyler Maritime College, in 1975.

Preface

IEEE Std 45™-2002 is a recommended practice for electrical installations on shipboard.
The standard is neither a regulatory body requirement nor the rules for building ships. It provides vital
guidelines for the safe operation of electrical equipment, for the safe operation of the ship, and for the
safety of shipboard personnel. These guidelines emphasize safety and security of electrical and electronic
equipment installation, equipment selection, and system coordination. The responsibility for
implementing these recommendations belongs to everyone dealing with the shipboard electrical
equipment and electrical system, such as the electrical engineers, the electrical designers, the electrical
cable pullers, the electrical equipment installers, the shipboard equipment and system testers, and the
troubleshooters.

At any voltage level, electricity can be deadly. Traditionally, shipboard electrical voltage ratings have
been 12 V, 24 V, 110 V, and 460 V for grounded and ungrounded installations. Until recently, the 460 V
level was high for shipboard installation. In recent years the voltage level has risen to 4100 V, 6600 V,
11 000 V, and 13 800 V. The power requirement has increased from a few megawatts to hundreds of
megawatts. Power generation and distribution at different voltages and at hundreds of megawatts have
become a big challenge. The IEEE Std 45 recommendations are a supplement to American Bureau of
Shipping (ABS) rules and United States Coast Guard (USCG) regulations for commercial ships. The
advent and proliferation of information technology has made available enormous amounts of technical
information related to shipbuilding innovations, rules, regulations, and standards. This access has
impacted the endeavor to standardize international rules and regulations surrounding shipbuilding.
Information technology has helped tremendously to make necessary information available at the click of a
mouse. The responsibility of gaining knowledge of available shipbuilding rules, regulations, and
recommendations around the globe and adapting the most appropriate ones must be done at a very fast
pace. The adaptation of the very process of technical innovation is also a universal challenge of building a
bridge from present to future shipbuilding in order to meet tomorrow's demand.

The concept of IEEE Std 45 arose with the same objective as the National Electric Code® (NEC®).
Acceptable standards are needed because no two persons will view something in the same way, interpret it
in the same way, and implement it in the same way. These standards are critical in applying technology,
which is a time domain domino scenario by the very nature of innovation. As we build for the future, we
have to live with the present. We must write down the most probabilistic aspect of an idea and agree to
follow it. The accepted norm of today may not be the norm of tomorrow; however, it is appropriate for
today because it works to an accepted level and meets safety requirements of shipboard installation.

IEEE Std 45 has been accepted as the recommended practice, not only in the United States, but also
globally. It was revised and published in October 2002. Many industry experts have contributed many
years of experience in the shipboard electrical engineering field. This task, however, has been presented
with a significant challenge due to global cooperation initiative, namely harmonization and globalization.
IEEE Std 45-2002 is in compliance with the NEC, the standards developed by the National Electrical
Manufacturers' Association (NEMA), the Underwriters Laboratory (UL), and the American Association
of Testing and Material (ASTM), as well as ABS Rules, the Code of Federal Regulations (CFR) of the
United States Department of Transportation, and various military specifications. The very process of
equipment specification, manufacturing, installation, and testing has attained solid ground by the repeated
revision of existing standards and the addition of new ones. International Electrotechnical Commission
(IEC) standards are also applicable for shipboard installation. The United States is signatory to the IEC
standards through the United States National Committee of the IEC, administered by the American
National Standard Institute (ANSI). IEC standards differ from US standards in numerous ways, such as

voltage levels, units of measurement, equipment rating, ambient rating, enclosure type, and equipment location classification. One must understand differences to ensure applicability and interchangeability, and to combine use of US standard equipment with IEC standard equipment. Most of the US standards committees agreed to adopt, supplement, or change US standards to comply with IEC standards. The IEEE Std 45-2002 committee has also agreed to adopt IEC standards by directly replacing, or modifying the existing standards. These changes must be clearly understood in order to ensure that the safety and security of life and equipment are not compromised.

This Handbook provides a detailed background of the changes in IEEE Std 45-2002 and the reasoning behind the changes as well as explanation and adoption of other national and international standards. This Handbook provides necessary technical details in a simplified form to enhance understanding of the requirements for technical and non-technical people in the maritime industry.

The maritime electrical industry has been experiencing tremendous changes. The "all-electric ship" and "electric propulsion" are paving the future of shipboard electrical power generation, distribution, and management from a few megawatts to hundreds of megawatts. Innovations such as electric propulsion, podded propulsion, and the Z drive present new challenges to the existing standards. By the very nature of a standard's publication period, it is not possible to maintain pace with the changes in the shipbuilding industry.

Table of Contents

List of Figures

List of Tables

1   Overview

IEEE Std 45™-2002 is a standard (a set of recommendations and specifications) relating to shipboard electrical installations. The purpose of this Handbook is to enhance understanding of that standard, describing its application to common steps in shipboard electrical design and relation to other standards, so that ship builders can design and implement a shipboard electrical system in accordance with this standard as well as the other related standards. In addition, ship owners can use this Handbook to help them understand the requirements in the 1998 edition and the changes made in the 2002 edition. Ship owners should understand the regulatory body requirements, such as from the American Bureau of Shipping (ABS), United States Coast Guard (USCG), Det Norske Veritas (DNV), and IEEE Std 45, to prepare better outline specifications that will guide the ship designers. The ship design engineers need to understand the requirements and the changes to the 1998 edition of IEEE Std 45 to design shipboard electrical systems. A ship's electrical power generation has increased to hundreds of megawatts demanding voltage requirements from 460 V to 4160 V, 6600 V, 11 000 V, 13 800 V, 25 000 V, and 35 000 V. This Handbook addresses these changes by providing additional explanation and necessary information. The ship builder needs to understand the changes so that appropriate measures can be taken to build ships in compliance with these recommendations and requirements. Ship crews must also understand these requirements and get proper training to operate shipboard systems.

IEEE Std 45-2002 has made tremendous strides in adapting international standardization guidelines such as the incorporation of SI units, detailed review of the NEMA enclosure and IEC type IP enclosure, and mixing and matching of US standard equipment and IEC standard equipment. This Handbook explains those changes for better understanding of the recommendations.

1.1   Scope

The scope of the Handbook for IEEE Std 45-2002 is to explain the IEEE Std 45 recommendations in line with regulatory body requirements and explain major changes in IEEE Std 45-2002. In this Handbook, the recommendations are further amplified with explanations, simple drawings, and examples. In addition, this Handbook describes related standards and requirements that come into play during the real-world design of shipboard electrical systems; the interaction between IEEE, ABS and governmental standards is addressed in detail.

1.2   Disclaimer

IEEE Std 45-2002 is the recommended practice for shipboard electrical installation. IEEE 45 was prepared by a voluntary consensus body to provide assistance and guidance to the requirements of national and international regulatory agencies governing electrical engineering requirements for shipboard installation. This Handbook is an effort to assist in understanding those recommendations. If there is any conflict with the recommendations of IEEE Std 45™-1983, IEEE Std 45™-1998, or IEEE Std 45-2002, the recommendations of IEEE Std 45-2002 will prevail. If there is any conflict with regulatory body requirements, regulatory body requirements will apply.

1.3   History of IEEE Std 45

IEEE Std 45 dates back to 1920 and has traditionally been a standard for the construction of commercial ships. In recent history, the commercial shipbuilding industry in the United States has experienced a decline; at the same time, however, we have seen a resurgence of military shipbuilding. The military are increasingly looking for Commercial Off the Shelf (COTS) componentry in order to reduce their costs. At the same time, their standards remain very high. Accordingly, standards such as IEEE Std 45 are seeing increasing use in the military shipbuilding arena.

IEEE Std 45 editions were issued in 1920, 1927, 1930, 1938, 1940, 1945, 1948, 1951, 1955, 1958, 1962, 1967, 1971, 1977, 1983, 1998, and 2002.

The first sponsor of IEEE Std 45 was the Marine Committee (MC) of the IEEE Industry Application Society (IAS), which subsequently became the Marine Transportation Committee (MTC). This sponsorship continued until the 2002 revision, when the International Marine Industry Committee (IMIC) of the IEEE Industry Application Society (IAS) became the sponsor.

In 2003, the IEEE 45 sponsor was changed from the International Marine Industry Committee (IMIC) of the IEEE Industry Application Society (IAS) to the Petroleum and Chemical Industrial Committee (PCIC) of the IEEE Industry Application Society (IAS).

1.4 Overall approach of this Handbook

This book takes a *task*-based approach to the subject matter of IEEE Std 45. The task is the design of a shipboard electrical system. Each chapter addresses a component of that design process in a logical, step-by-step fashion. IEEE Std 45-2002 addresses many of these steps, while some other steps are addressed only slightly. The designer of electrical systems must be familiar with many standards, of which IEEE Std 45 is but one. This Handbook therefore touches on many of the standards beyond IEEE Std 45, to provide a complete handbook reference to shipboard electrical systems.

Commercial ship design and development is done in numerous multi-disciplined phases such as Hull, Machinery, and Electrical (HM&E). This allows understanding of the performance requirements and operational needs of the ship, which constitutes baseline requirements for the design solution. We will address electrical design development issues here in support of IEEE Std 45. The development phases are: Concept Studies, System Definition, Preliminary Design, Detail Design, Equipment Procurement, Equipment Installation, and testing. Each phase is briefly summarized below.

a. **Concept study** – The initial detailed operational concept; for instance, cruise ship for transatlantic voyage, million-barrel tanker for Alaskan Crude Oil, direct diesel drive or electric propulsion drive. Based on the initial concept, the requirements of the electrical system are explored.

- For direct drive ship, the electrical requirement usually is to support ship service power generation and distribution. The ship service power generation usually consists of multiple ship service generators for redundancy requirement and emergency generator for electrical blackout prevention. The emergency generator usually get automatically started under zero voltage detection in the switchboard and automatically takes over the emergency loads, such as fire pump, emergency lighting, navigation equipment.
- For electric drive ship, the electrical requirement is to support ship's propulsion power in addition to ship service and emergency power requirement the electrical power generation and distribution can be ten to twenty times the electrical power requirement of non-electric drive ship. The high power requirement often demands a higher voltage requirement. The duty p rofile of the propulsion drive motors is different from other ship service motors due to the propulsion duty requirement.
- The ship service power generation and distribution usually is 460 V, the electric propulsion drive power generation and distribution can be 6600 V or higher.
- The distribution switchboards are medium voltage switchboard, 450 V switchboard and 120 V switchboard and panel boards.

b. **System definition**– The system definition usually provides the system requirement to support the performance requirement such as the tanker propulsion will be diesel driven and the cargo pumps will be electric driven or hydraulically driven to support the unloading requirement. Another requirement might be the propulsion and cargo pumps will be electric driven. These are the baseline requirement usually established by the ship owner.

c. **Preliminary design**—Concept study and system definition lead to the preliminary design phase. The preliminary design is performed to support the baseline requirement of the ship. Preliminary electrical one-line diagram and electrical load analysis are performed showing number and ratings of generators, such as ship service generators, propulsion generators, and emergency generator, as applicable. The preliminary design is performed in compliance with established rules and regulations such American Bureau of Shipping (ABS), United States Coast Guard, IEEE Std 45, Lloyd's, DNV etc. as applicable. The preliminary design provides information supporting redundancy requirements as well as emergency requirements.

d. **Detailed design**—The detailed design phase is usually after the contract award. In this phase every system development is required for equipment procurement, ship construction drawings, equipment installation and testing. The following systems comprise an electrical system ship construction drawing development:

- Electrical power system calculations, such as electrical one-line diagram, load analysis, voltage drop calculation, fault current calculation, power system deck plans
- Electrical wire ways development
- Lighting system – ship service and emergency
- Internal communications
- External communications
- Centralized control system

These detailed drawings are used to procure equipment, make foundations, and install equipment

e. **Testing**—On completion of equipment installation, appropriate service to the equipment is provided as preparation of equipment testing and system testing. The testing is performed in two phases, dockside testing and at sea testing. Completion of at sea testing is prerequisite for the delivery of ship.

Traditionally, shipboard electric system design is supported by numerous electrical system network analyses in accordance with applicable regulatory body requirements. These electrical system analyses are namely electrical system load analysis, voltage drop calculations, electrical load flow calculations, electrical short-circuit analysis, and circuit breaker coordination and selectivity study.

The advent of all-electric ships and advances in components, such as the development of high-efficiency, power-dense, lightweight motors, transformers, drives, and other devices require careful and, in many cases, entirely unique integration into the ship system design and development. These imperatives motivate the use of tools for the system wide analysis of shipboard electric power systems. In order to perform these analyses, digital simulation of electrical systems is essential. To establish the physical laws governing power-system components and their interactions data are collected in large numbers and analyzed to predict the behavior of a system. Digital modeling and simulation are fast, accurate, and convenient approaches for predicting power-system behavior.

The new technology of electric-driven propulsion power systems requires additional system analyses such as power system dynamic analysis, harmonic analysis, network transient analysis, electric motor drive performance analysis, and extensive modeling simulation of power generation and distribution. These analyses are usually done by the use of computer-aided engineering and design tools.

Some of the most crucial elements of shipboard electrical system design are outlined below, along with their place in this Handbook.

1.4.1 Electrical load flow calculations (Handbook, Section 6)

The aim of the load flow calculations is to determine whether the thermal and mechanical stresses on equipment are below the maximum design values. Load flow calculations are performed to find steady-state values of loads in the generation and distribution network. They are normally done prior to the short-circuit calculations in order to obtain the steady-state power requirement of the system.

1.4.2 Voltage deviation calculations—steady state and transient (Handbook, Section 7)

The steady state voltage deviation requirement is +5% -6% to -10% for 460 V and around 3% for 120 V distribution. Refer to Section 7 for details.

When starting heavy ship service consumers such as motors or energizing large transformers, the startup transient current may be several times larger than the nominal full load current. For a motor typical transient load 5 to 8 times the full load current, and for a transformer, up to 10 to 12 times the full load current. The high startup current will give a voltage disturbance, or voltage drop in the network. The regulatory requirement is to limit the voltage transient at the range of -15% and $+20\%$. There are analytical methods for calculating such voltage transients as well as the steady state voltage drops, although the most accurate result is found by numerical simulations. In order not to exceed the required voltage variations, there may be a need to adjust the characteristics of the generator or the large consumers, or to introduce a means to reduce the startup transients. For motors, such means could be soft-starting, star-delta starters, or autotransformer starters or solid-state adjustable speed drive. For transformers starting transient pre-magnetizing could be evaluated or the incoming circuit breaker must be selected as such the breaker is capable of sustaining the transformer magnetizing period.

1.4.2.1 Network transient analysis—modeling and simulations

In addition to the fault calculations and voltage drop calculations, there is a need for thorough analysis of the transient behavior of the network during and after clearing a fault. Typically, such analysis includes voltage and frequency stability (i.e., will the voltage and frequency of the generators be reestablished after the fault is cleared?) and re-acceleration of motor loads (i.e., will essential motors be able to accelerate without tripping after the fault is cleared?). Such analyses are extensive and require accurate modeling of the network and regulators for voltage and frequency in order to be reliable.

1.4.3 Harmonic distortion (Handbook, Section 4.2)

The harmonic distortion calculation and harmonic management are significant for power electronic driven starting and control systems such as ship service auxiliary motors, and variable frequency driven electric propulsion systems.

Power electronic driven systems such as frequency converters are inherently nonlinear, and the currents to the drive are not sinusoidal but distorted by harmonic components. Assuming that the converter and transformer are symmetrically designed and the output stage of the converter is de-coupled from rectifier current, only the characteristic harmonic components are present in the input currents to the line supply of the frequency converter. For a six-pulse converter drive, these harmonics are

Harmonics (H)th order = $6 \times (n \pm 1)$ where n is integer such as 1, 2, 3, 4, 5....

$H = 5^{th}, 7^{th}, 11^{th}, 13^{th}, \ldots\ldots\ldots$

In a 12-pulse converter, multiples of sixth harmonics, which are present in the secondary and tertiary windings of the feeding transformer, will cancel in the primary windings. Hence the remaining harmonic current components will be of the order

Harmonics (H)th order = $12 \times (n \pm 1)$ where n is integer such as 1, 2, 3, 4, 5....

H = 11th, 13th, 23rd, 25th, ...

The total harmonic distortion (THD) is a measure of the total content of harmonic components in a measured current, THD(i), or voltage, THD(v):

$$THD(i) = 100\% \times \frac{\sqrt{\sum_{h=2}^{\infty} i_{(h)}^{2}}}{i_{(1)}}, \text{ and } THD(v) = 100\% \times \frac{\sqrt{\sum_{h=2}^{\infty} v_{(h)}^{2}}}{v_{(1)}}$$

1.5   Cross references to other standards

There are many standards related to shipbuilding. IEEE Std 45-2002 is but one of them, and the ship designer must take all of them into account to varying degrees. Accordingly, a major focus of this Handbook is to provide extensive cross-references between IEEE standards and other shipbuilding standards, such as IEC, ABS, USCG, DVN, NEMA, and ANSI.

The sections in this Handbook have been arranged for better understanding of intersystem and intrasystem functionality of shipboard electrical systems, particularly the power generation, distribution, controls, and communications. The reader may be familiar with IEEE Std 45-1998; a cross reference from that standard to the 2002 standard is supplied in Appendix H. Additional details of the changes are provided in appropriate sections in this Handbook. Note that Clause 8 in IEEE Std 45-1998 cable manufacturing was removed from IEEE Std 45-2002 and published in a stand-alone standard, IEEE Std 1580™-2001.

Other relevant standards include the Ship Work Breakdown Sequence (SWBS), promulgated by the US Coast Guard (USCG) and the American Bureau of Shipping (ABS) rules.

1.5.1   Ship Work Breakdown Sequence (SWBS)

The United States Navy uses unique system sequence numbers called ship work breakdown sequence (SWBS). SWBS provides a structured sequence number for shipboard system and sub-system, such as SWBS 304 for electric cables. Refer to the United States Navy, General Specifications for Ships, 1995 edition for shipboard SWBS numbers. The SWBS numbers are sometimes used in commercial shipbuilding as well. The SWBS numbering for electrical engineering sub-systems is provided in Appendix H.

1.5.2   American Bureau of Shipping (ABS)

ABS rules are quoted in this Handbook as have been quoted in the IEEE Std 45. The ABS rules are published in multiple parts: Part 4 is for machinery. In part 4, chapter 8 is for electrical and chapter 9 is for automation.

Part 4, Chapter 8 for electrical system is organized as follows:

4-8-1	Electrical System – General
4-8-2	System Design
4-8-3	Electrical equipment
4-8-4	Shipboard installation and test
4-8-5	Special Systems

Each group again subdivides by paragraph numbers and subdivisions.

Part 4, Chapter 9 for Remote Propulsion Control and Automation, which is organized as follows:

4-9-1	General Provisions
4-9-2	Remote Propulsion Control
4-9-3	ACC Notation
4-9-4	ACCU Notation
4-9-5	Installation, Test and Trials
4-9-6	Computerized System
4-9-7	Equipment

Paragraph numbers and subdivisions further subdivide each group. ABS adapted this numbering system from their 2000 publication.

1.5.3   Safety of Life At Sea (SOLAS)

The International Convention for the Safety of Life at Sea (SOLAS) was adopted on November 1974 by the International Conference on Safety of Life at Sea, convened by the International Maritime Organization (IMO). SOLAS Regulation Part 1 contains eleven chapters:

Chapter I	General Provisions
Chapter II-1	Construction-Subdivision, machinery and electrical installations
Chapter II-2	Construction – Fire protection, fire detection, and fire extinction
Chapter III	Life-saving appliances and arrangements
Chapter IV	Radiocommunications
Chapter V	Safety of Navigations
Chapter VI	Carriage of cargo
Chapter VII	Carriage of dangerous goods
Chapter VIII	Nuclear ships
Chapter IX	Management for safe operation of ships
Chapter X	Safety measure for High Speed craft
Chapter XI	Special measure for maritime safety

1.5.4   Shipboard electrical equipment construction, testing and certification

Marine electrical equipment should be constructed, tested and certified in accordance with the requirements of applicable standards. The equipment testing, certification and labeling should be done by third party with authority having jurisdiction in the field. For an example, the NEMA switchboard for marine application should have UL labeling. However for UL certification, the busbar and the circuit breakers are individually tested and labeled. The shipboard generator should be tested in accordance with applicable NEMA requirements; ABS will go through the testing process, and witness the test and label the generator as compliant with the testing requirements. (For details refer to IEEE Std 45-2002, 1.6 and 1.7)

For ABS class ships, ABS requirements must be met. For ACP (Alternate Compliance Program, promulgated by ABS and USCG together) class ships, both ABS requirements and ACP requirements must be met. IEEE 45 provides recommendations in addition to regulatory body rules and regulations, which are supplemental recommendations for consideration of better operation, performance, etc.

1.6   Notes on the text in this Handbook

The Sections in this Handbook do not follow the sections in IEEE Std 45-2002 one-for-one. Rather, this book takes a *task-based* approach—the task being the design of a shipboard electrical system from the ground up. In each section, the general goal of the task is first described, and then the recommendations of IEEE Std 45-2002 are described. Interactions with other standards such as ABS and SWBS are also described.

The sections will refer to the relevant section of IEEE Std 45-2002 for further detail. In addition, the relevant SWBS and ABS numbers are also called out in the section headers.

This Handbook refers to the IEEE and other standards in varying ways—by quoting, paraphrasing or simply describing. In order to be clear about when we are quoting directly and when we are paraphrasing, the following conventions have been adopted to inform you of whether the Handbook is directly quoting, paraphrasing, or summarizing other standards.

a. A border with a **solid line** around the text indicates quoted material. Appropriate reference information is provided acknowledging the source of the quotation.

b. A border with a **dotted line** around the text indicating extract only of quoted material. The extract text may have some changes to simplify the reference material and/or to limit the quoted text space. Appropriate references information is provided, acknowledging source of the quotation.

c. A border with a **wiggly line** around the text indicates that the encased is extract from quoted material, however may not be the exact quotation. Appropriate reference number may not be shown, as the material may not match the quotation. This is to indicate that the material source is a quotation.

Finally, some sections in this book (particularly the latter sections) describe standards and requirements that are not part of IEEE Std 45-2002. This is done in the interest of completeness. Many of these areas are addressed only scantily by IEEE Std 45 because they have been adequately described by other regulatory bodies. Nevertheless, the task of shipboard electrical design requires a good understanding of these areas, and so they have been included in this Handbook.

2   Ship service and propulsion generator (SWBS 235)

2.1   General

The ship service power is generated by ship service generators, propulsion generators, emergency generators, and uninterruptible power sources. This section is dedicated to the ship service generator and propulsion generator. In case of a prime mover-driven propulsion system, the ship service electric power is generated by ship service generators. In case of an electrical propulsion system, ship's electrical propulsion power as well as ship service are generated by the propulsion generator, often supplemented by a smaller ship service generator. The regulatory body requirement is to provide redundant generators, ensuring availability of electrical power under all operating conditions. It is also a requirement to run the generators under the most fuel-efficient condition. The shipboard electrical loading scenarios typically are: at shore, at anchor, maneuvering, cruising and other specific service-related loading. The generator prime movers must be provided with the required starting system, governor system and loading characteristics, ensuring compatibility of the intended service. Number of generators and their ratings are to be properly calculated so that the electrical power generation is sufficient for all operating conditions of the ship. In order to ensure proper control and monitoring of these generators, IEEE Std 45 recommends a number of instruments in the switchboard as a minimum requirement. The propulsion generator ratings are usually much higher than ship service generators, as these generators are required to provide power to ship service loads as well as propulsion loads.

2.1.1   IEEE Std 45-2002 Standard Voltages

Per table 3 of the IEEE standard, the following voltages are recognized as standard.

IEEE Std 45-2002 Table 3—Standard voltages

Standard	AC(V)	DC(V)
Power utilization	115-200-220-230-350- 440-460-575-660-2300-3150-4000-6000- 10600-13200	115 and 230
Power generation	120-208-230-240-380- 450-480-600-690-2400-3300-4160-6600- 11000-13800	120 and 240

The typical usage and frequencies associated with these voltages are briefly summarized in Table 3 below.

AC voltage generation 60 Hz and 50 Hz (V)	AC voltage distribution 60 Hz and 50 Hz (V)	Remarks
120	115	Mostly US applications
208	200	
230	220	Mostly IEC application and some US commercial applications
240	230	
380	350	
400	380	50 Hz
450	440	Mostly military applications
480	460	Mostly commercial applications
600	575	
690	660	Mostly IEC applications
2400	2300	50 Hz
3300	3150	60 Hz and 50 Hz

AC voltage generation 60 Hz and 50 Hz (V)	AC voltage distribution 60 Hz and 50 Hz (V)	Remarks
4160	4000	60 Hz and 50 Hz
6600	6000	50 Hz
11 000	10 600	60 Hz and 50 Hz
13 800	13 200	60 Hz and 50 Hz

Table 2-1: US and IEC Shipboard Power Generation and Distribution Levels at 50 Hz and 60 Hz

2.1.2 Generators

Modern shipbuilding, both military and commercial, employs mostly AC power generation and AC distribution. The generators are synchronous machines, with a magnetizing winding on the rotor carrying a DC current, and a three-phase stator winding where the magnetic field from the rotor current induces a three-phase sinusoidal voltage when the rotor is rotated by the prime mover. The frequency f [Hz] of the induced voltages is proportional to the rotational speed n [RPM] and the pole number p in the synchronous machine:

$$f = \frac{p}{2} \cdot \frac{n}{60}$$

For 60 Hz systems a two-pole generator gives 3600 rpm, a four-pole at 1800 rpm, and a six-pole at 1200 rpm, etc (US standard application). For 50 Hz IEC applications, 3000 rpm, 1500 rpm, 1000 rpm for two-, four-, and six-pole machines. A large medium speed engine normally considered at 514 rpm, 14 poles and 720 rpm at 10 pole for 60 Hz network or for 50 Hz system 600 rpm, 10 poles or 750 rpm, 8 pole.

In synchronous generator design, the DC current is transferred to the magnetizing windings on the rotor by brushes and slip-rings. The new generators are equipped with a brushless excitation system. The brushless exciter generator is a synchronous machine with DC magnetization of the stator and rotating three-phase windings and a rotating diode rectifier. Figure 2-1 shows typical exciters schematics.

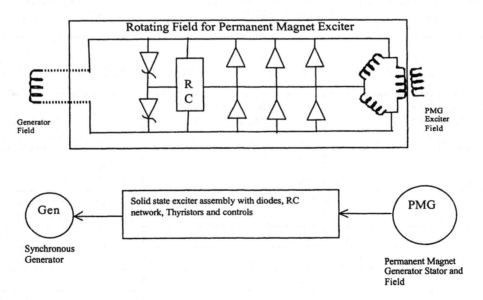

Figure 2-1: Typical Brushless Permanent Magnet Generator (PMG) Exciter for Synchronous Generator
(Example Only)

The brushless exciter is typically mounted on the generator rotor shaft. The solid state consists of rotating diodes, thyristors and RC snubber circuit. The permanent magnet generator (PMG) produces the excitation current for the generator. The synchronous generator excitation is controlled by an automatic voltage regulator (AVR) system, which senses the terminal voltage of the generator and compares it with a reference value. According to most applicable regulations, the stationary voltage variation on the generator terminals should not exceed ± 2.5% of nominal voltage. For electrical generation system stability, transient load variation should exceed the voltage variation of

−15% to +20% of the nominal voltage. In order to maintain the transient voltage requirement, the AVR is normally also equipped with a feed-forward control function based on measuring the stator current.

In addition to the magnetizing winding, the synchronous generator rotor is also equipped with a damper winding which consists of axial copper bars threaded through the outer periphery of the rotor poles, and short circuited by a copper ring in both ends. The main purpose of the damper winding is to introduce an electromagnetic damping to the stator and rotor dynamics. A synchronous machine without damper winding is inherently without damping and would give large oscillations in frequency and load sharing for any variation in the load.

The determination of power rating for generators and prime movers requires careful calculation and analysis. The regulatory body requirement is to prepare comprehensive electrical load analysis for each electrical power consuming load such as propulsion, propulsion auxiliaries, ship service auxiliaries, heating ventilation and air-conditioning (HVAC), normal lighting, emergency lighting on board ship, and the load classification such as emergency service, vital service, non-vital service etc.

2.2 IEEE Std 45-2002, ABS-2002 and IEC for generator size and rating selection

Ship service generator size, rating and quantity requirements are very well defined in IEEE Std 45, ABS and USCG with some slight differences. IEEE Std 45 requirements are in Section 7, ABS-2002 requirements are in part 4, Chapter 8, Section 3, and USCG requirements are in 46 CFR, Regulations subparts 111-10 and 111-12. The propulsion generator size, rating, and quantity requirements are somewhat different from ship service generators. It is very important to understand the operational requirements prior to adapting a set of rules. Due to the size of the propulsion generator the quantity of generators and redundancy requirement may be different from small size ship service generators. The propulsion generator requirements are give in IEEE Std 45, Section 31. The most relevant generator size, rating, and quantity requirements are quoted below.

2.2.1 IEEE Std 45-2002, 7.4.2, Selection and sizing of generators (extract)

In determining the number and capacities of generating sets to be provided for a vessel, careful consideration should be given to the normal and maximum load demand (i.e., load analysis) as well as for the safe and efficient operation of the vessel when at sea and at port. The vessel must have at least two generating sources. For ships, the number and rating of the main generating sets should be sufficient to provide one spare generating set (one set not in operation) at all times to service the essential and habitable loads. For MODUs, with the largest generator off-line, the combined capacity of the remaining generators must be sufficient to provide normal (non-drilling) load demands.

In selecting the capacity of an AC generating plant, particular attention should be given to the starting current of AC motors supplied by the system. With one generator held in reserve and with the remaining generator set(s) carrying the minimum load necessary for the safe operation of the ship, the voltage dip resulting from the starting current of the largest motor on the system should not cause any motor already running to stall or control equipment to drop out. It is recommended that this analysis be performed when total horsepower of the motor capable of being started simultaneously exceeds 20% of the generator nameplate kVA rating. The generator prime-mover rating may also need to be increased to be able to accelerate motor(s) to rated speed. Techniques such as soft starting (i.e., reduced voltage autotransformer starters, electronic soft starters, and variable frequency drives) may be utilized to reduce the required capacity of generators when motor starting is of concern.

Note: For SOLAS (Safety Of Life At Sea) requirement refer to SOLAS Chapter II-1, Regulation 41

2.2.2 IEEE Std 45-2002, 31.3.1, General (electric propulsion prime movers) (extract)

The design of an integrated electric power system should consider the power required to support ship service loads and propulsion loads under a variety of operating conditions, with optimum usage of the installed and running generator sets.

In order to prevent excessive torsional stresses and vibrations, careful consideration should be given to coordination of the mass constants, elasticity constants, and electrical characteristics of the system. The entire system includes prime movers, generators, converters, exciters, motors, foundations, slip-couplings, gearing, shafting, and propellers….

Systems having two or more propulsion generators, two or more propulsion drives, or two or more motors on one propeller shaft should be so arranged that any unit may be taken out of service and disconnected electrically, without affecting the other unit.

2.2.3 IEEE Std 45-2002, 31.4, Prime movers for integrated power and propulsion plants

Prime movers, such as diesel engines, gas turbines, or steam turbines, for the generators in integrated electric power systems shall be capable of starting under dead ship conditions [see dead ship definition by ABS below] in accordance with requirements of the authority having jurisdiction. Where the speed control of the propeller requires speed variation of the prime mover, the governor should be provided with means for local manual control as well as for remote control.

The prime mover rated power, in conjunction with its overload and the large block load acceptance capabilities, should be adequate to supply the power needed during transitional changes in operating conditions of the electrical equipment due to maneuvering, sea, and weather conditions. Special attention should be paid to the correct application of diesel engines equipped with exhaust gas-driven turbochargers to ensure that sudden load application does not result in a momentary speed reduction in excess of limits specified in Table 4 [see 4.3.2 in this Handbook].

When maneuvering from full propeller speed ahead to full propeller speed astern with the ship making full way ahead, the prime mover should be capable of absorbing a proportion of the regenerated power without tripping from over-speed when the propulsion converter is of a regenerative type. Determination of the regenerated power capability of the prime mover should be coordinated with the propulsion drive system. The setting of the over-speed trip device should automatically shut down the unit when the speed exceeds the designed maximum service speed by more than 15%. The amount of the regenerated power to be absorbed should be agreed to by the electrical and mechanical machinery manufacturers to prevent over-speeding.

Electronic governors controlling the speed of a propulsion unit should have a backup mechanical fly-ball governor actuator. The mechanical governor should automatically assume control of the engine in the event of electronic governor failure. Alternatively, consideration would be given to a system, in which the electronic governors would have two power supplies, one of which should be a battery. Upon failure of the normal supply, the governor should be automatically transferred to the alternative battery power supply. An audible and visual alarm should be provided in the main machinery control area to indicate that the governor has transferred to the battery supply. The alternative battery supply should be arranged for trickle charge to ensure that the battery is always in a fully charged state. An audible and visual alarm should be provided to indicate the loss of power to the trickle charging circuit. Each governor should be protected separately so that a failure in one governor will not cause failure in other governors. The normal electronic governor power supply should be derived from the generator output power or the excitation permanent magnet alternator. The prime mover should also have a separate over-speed device to prevent runaway upon governor failure.

2.2.4 ABS-2002 Section 4-8-2-3.1.3 Generator Engine Starting from Dead Ship Condition (extract)

Dead ship (blackout) condition is the condition under which the main propulsion plant, boiler and auxiliaries are not in operation due to the unavailability of power from the main power source. See ABS 4-8-2-4-1-1/7.7. In restoring the propulsion, no stored energy for starting the propulsion plant, the main source of electrical power and other essential auxiliaries is to be assumed available. It is assumed that means are available to start the emergency generator at all times. The emergency source of electrical power may be used to restore the propulsion, provided its capacity either alone or combined with other source of electrical power is sufficient to provide at the same time those services required to be supplied by ABS 4-8-2/5.5.1 to 4-8-2/5.5.8.

The emergency generator and other means needed to restore the propulsion are to have a capacity such that the necessary propulsion starting energy is available within 30 min of dead ship (blackout) condition as defined above. Emergency generator stored starting energy is not to be directly used for starting the propulsion plant, the main source of electrical power and/or other essential auxiliaries (emergency generator excluded). For steam ships, the 30 min time limit is to be taken as the time from dead ship (blackout) condition to light-off of the first boiler.

Note: For SOLAS requirement refer to SOLAS Chapter II-1, Regulation 42.3.4 and 43.3.4

2.2.5 IEEE Std 45-2002, 7.3.4, Governors (Propulsion and Ship Service Generator Engines)

The prime-mover governor performance is critical to satisfactory electric power generation in terms of constant frequency, response to load changes, and the ability to operate in parallel with other generators.

The steady state speed variation should not exceed 5% (e.g., 3 Hz for a 60 Hz machine) of rated speed at any load condition.

Each prime mover should be under control of a governor capable of limiting the speed, when full load is suddenly removed, to a maximum of 110% of the rated speed. It is recommended that the speed variation be limited to 5% or less of the over-speed trip setting.

The prime mover and regulating governor should also limit the momentary speed variation to the values indicated in this sub-clause. The speed should return within 1% of the final steady state speed in a maximum of 5 seconds or as set by the limits specified in Table 7 IEEE Std 45-2002.

For emergency generators, the prime mover and regulating governor shall be capable of assuming the sum total of all emergency loads upon closure to the emergency bus. The response time and speed deviation shall be within the tolerances indicated in the Table 7 IEEE Std 45-2002.

Table 7— Response time and speed deviation requirements

Load (%)	Response time (s)	Speed deviation (%)
0 to 50, 50 to 0	5.0	10
50 to 100, 100 to 50	5.0	10

Generator sets should be capable of operating successfully in parallel when defined as follows: If at any load between 50% and 100% of the sum of the rated loads on all generators, the load (kW) on the largest generator does not differ from the other by more than ± 15% of the rated output or +25% of the rated output of any individual generator, whichever is less, from its proportionate share. The starting point for the determination of the successful load distribution requirements is to be at 75% load with each generator carrying its proportionate load.

2.2.6 IEEE Std 45-2002, 7.4.7, Voltage Regulation (Propulsion and Ship Service Generator)

At least one voltage regulator should be provided for each generator. Voltage regulation should be automatic and should function under steady state load conditions between 0% and 100% load at all power factors that can occur in normal use. Voltage regulators should be capable of maintaining the voltage within the range of 97.5% to 102.5% of the rated voltage. A means of adjustment should be provided for the voltage regulator circuit. Voltage regulators should be capable of withstanding shipboard conditions and should be designed to be unaffected by normal machinery space vibration.

Solid-state voltage regulators are recommended for high reliability, long life, fast response, and stable regulation. Regulator systems should be protected from under-frequency conditions. It is recommended that voltage regulators for machines rated in excess of 150 kW be provided with under-frequency and over-voltage sensors for protection of the voltage regulators.

Under motor starting or short-circuit conditions, the generator and voltage regulator together with the prime mover and excitation system should be capable of maintaining short-circuit current of such magnitude and duration as required to properly actuate the associated electrical protective devices. This shall be achieved with a value of than not less than 300% of generator full-load current for a duration of 2 seconds, or of such additional magnitude and duration as required to properly actuate the associated protective devices.

For single-generator operation (no reactive droop compensation), the steady state voltage for any increasing or decreasing load between zero and full load at rated power factor under steady state operation should not vary at any point more than ± 2.5% of rated generator voltage. For multiple units in parallel, a means should be provided to automatically and proportionately divide the reactive power between the units in operation.

Under transient conditions, when the generator is driven at rated speed at its rated voltage, and is subjected to a sudden change of symmetrical load within the limits of specified current and power factor, the voltage should not fall below 80% nor exceed 120% of the rated voltage. The voltage should then be restored to within ± 2.5% of the rated voltage in not more than 1.5 s.

In the absence of precise information concerning the maximum values of the sudden loads, the following conditions should be assumed: 150% of rated current with a power factor of between 0.4 lagging and zero to be applied with the generator running at no-load, and then removed after steady state conditions have been reached.

For two or more generators with reactive droop compensation, the reactive droop compensation should be adjusted for a voltage droop of no more than 4% of rated voltage for a generator. The system performance should then be such that the average curve drawn through a plot of the steady state voltage vs. load for any increasing or decreasing load between zero and full load at rated power factor droops no more than 4% of rated voltage. No recorded point varies more than ± 1% of rated generator voltage from the average curve.

Isochronous operation of a single generator operating alone is acceptable. However, where two or more generators are arranged to operate in parallel, it is recommended that isochronous kilowatt load sharing governors and voltage regulation with reactive differential compensation capabilities be provided. Care should be taken if operating machines in parallel to ensure that the system minimum load does not decrease and cause a reverse power condition.

If voltage regulators for two or more generators are installed in the switchboard and located in the same section, a physical barrier should be installed to isolate the regulators and their auxiliary devices.

Where power electronic devices (such as variable frequency drives, soft starters, and switching power supplies) create measurable waveform distortions (harmonics), means should be taken to avoid malfunction of the voltage regulator, e.g., by conditioning of measurement inputs by means of effective passive filters.

Power supplies and voltage sensing leads for voltage regulators should be taken from the "generator side" of the generator circuit breaker. Normally, voltage-sensing leads should not be protected by an over-current protection device. If short-circuit protection is provided for the voltage sensing leads, this short-circuit protection should be set at no less than 500% of the transformer rating or interconnecting wiring ampacity, whichever is less. It is recommended that a means be provided to disconnect the voltage regulator from its source of power.

2.3 Additional details of sizing ship service generators

Detailed electrical load analysis should be made to select the size of ship service generators. The radial electrical system generator sizing requirement is different from the ring distribution. The radial system in general consists of ship service generators and emergency generator. The ring bus distribution system often consists of multiple generators and may not have a dedicated emergency generator. The ship service generator sizing requirement is as follows (For additional details refer to Section 6, Electrical load analysis for determination of generator rating):

a) Ship Service Generator for radial distribution: (Refer to Figure 5-1for typical radial bus power distribution)
— The generator size must be equal to or bigger than the maximum steady state worst-case load requirement of the system.
— The generator must be able to sustain the largest direct on line (DOL) motor starting load (Starting current and transient voltage dip) of the system. If the DOL starting requirement leads to bigger size generator and engine, detail calculation must be performed to replace the DOL starting system with Wye-Delta starting, with open transition or closed transition, auto-transformer starting, soft starting, or variable frequency drive. The calculation must be accompanied by trade study for service requirement, space requirement and cost impact. Always remember, the radial distribution system may not have additional generator to support the loading requirements.
— The generator must be able to sustain the harmonic distortion effect of the system contributed by non-linear loads in the system

b) Ship Service Generator for ring bus distribution system (Refer to Figure 5-8 for typical ring bus power distribution)

— The ring bus distribution is usually set-up with multiple generators in the system so that generators can be connected in parallel to support steady state loading under automatic or manual power management system. In this configuration, the sizing of generators is not as complex as the radial configuration. However, in radial configuration, the there is no dedicated emergency generator, where the redundant power feeder requirement can get complicated.

2.4 Engine governor characteristics

Governors for engine generator sets used in parallel operation should be of the electronic load sharing type. Such governors are specifically designed for isochronous operation at any load from zero to 100% load. These governors cause their respective prime movers or engines to share load proportional to their horsepower or kilowatt rating and as a function of direct measurement of the electrical load (in kilowatt) on a specific engine governor set. Proportional division is accomplished via feedback through a load share loop, which connects all governors to all generator sets on-line at any given time. Such governors provide the following functions:

a) Rapid response to load changes
b) Stable system operation
c) Paralleling dissimilar sized engine generator sets
d) Speed regulation of .25%

2.5 Generator voltage regulator characteristics

It is the function of the voltage regulator to control the excitation of the generator so as to maintain constant generator terminal voltage, within defined limit e.g. +6%, to -10%. Since most engine generator sets used in today's power systems are brushless type generators, relatively inexpensive solid-state voltage regulator can be furnished to provide steady state regulation of 1 to 2% under any load condition from zero to 100% load. These solid-state controls are also capable of rapid response to load changes and of boost excitation to provide current magnitude capable of achieving selective coordination with the over-current protective devices in the power system. As with the governors, voltage regulators are also available with droop characteristics. The purpose of a droop characteristic in a voltage regulator is to enable the generator to share reactive component of the load in proportion to the kilovolt ampere rating of the generator.

The amount of reactive load is also an important design factor. For example, if an engine is delivering 800 kW at .8pf, then the engine is delivering 1000 kVA. The reactive power, kilovar, equals the square root of kilovoltampere squared minus kilowatt squared. So for this example, the generator would be delivering 600 kvar reactive power. When generator sets are operating in parallel, allowing voltage to drop or droop, as the current out of the machine increases, it causes the generators to share load almost proportionately. Why "almost proportionately"? It stems from dissimilarities in current-transformer-to-full-load-rating ratio discrepancies. Because the electrical loads driven by to-days power systems require precise frequency and voltage levels, droop compensation in the voltage levels, droop compensation in the voltage control for kilovar sharing is as objectionable as speed droop is for kilowatt sharing between the engines. Modern voltage regulators for both brushless and brush-type generators are readily available and provide constant voltage from no load to full load while achieving proportionate kilovar sharing between the generators.

In most cases, the same regulator that would be used for droop compensation can also be used for cross current compensation. The only difference lies in the connections of the regulator in the power system. Cross current compensation for proportionate reactive load division (kilovar sharing) is highly desirable design feature for paralleling systems employing automatic unattended operation. The designer should also specify cross-current compensation for the voltage regulation of these systems.

2.6   Droop characteristics – generator set

When selecting engine generator sets for parallel operation, the size of the sets is determined by analyzing the voltage and frequency requirements in term of the load, transient response, stability, and droop.

Droop is a function of t he difference between no-load and full-load operation. It is the percent difference in the values based on the no-load value. For instance a machine with 3% droop and a no-load frequency of 61.8 Hz, the full load frequency will be 97% of the no-load frequency of 60 Hz. The frequency droop for this set -up is therefore 1.8 Hz. Similarly, if the voltage droop is 5% on a nominal system voltage of 480 V AC, the voltage droop is 24 V AC.

However, today's engine generator sets can be furnished with relatively simple and reliable electronic governors and voltage regulators, which make these sets suitable for unattended automatic paralleling and load sharing. This is a basic key to paralleling engine generators for emergency power systems. The electronic governor provides isochronous operation, and automatic proportionate load division which make possible the automatic paralleling of dissimilar size sets. The electronic governor provides a more adaptable engine generator set because it will permit paralleling at any time without necessitating adjustment or requiring droop. Similarly, voltage regulators are available to achieve automatic reactive load division to provide constant voltage systems. These devices help to make automatic unattended emergency power paralleling system highly practical.

2.7   Typical generator prime mover set-ups

Figures 2-2, 2-3, and 2-4 show typical generator prime mover set -ups. These drawings are presented with minimum explanation and not intended to show a complete or proven design.

2.7.1   Ship service electric generator for 460 V generation and 450 V distribution (Handbook Figure 2-2)

Figure 2-2 shows the ship service power generation with two ship service diesel generators supplying power to the 450 V AC, three phase, 60 Hz ship service switchboard. Two generators are installed for redundancy consideration.

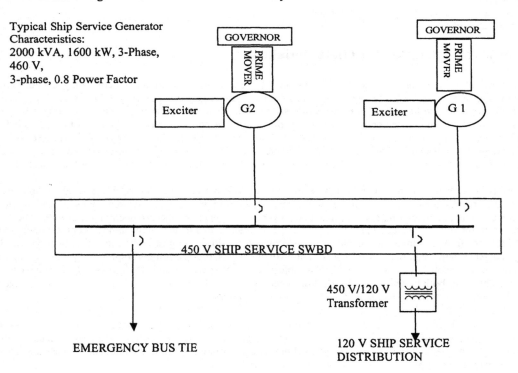

Figure 2-2: Electrical One-Line Diagram-Ship Service Electric Generator for 460 V Generation and 450 V Utilization

2.7.2   Electric propulsion system (ABS R2 redundancy)

Figure 2-3 below shows electric propulsion configuration with four prime movers, two high-voltage switchboards, two sets of propulsion motors with their drives. This configuration is in compliance with ABS R2-S notation for redundant propulsion system. In this configuration there is no requirement of physical separation by watertight and fireproof bulkhead in the machinery space, that means four engines and propulsion drives can be in the same machinery space.

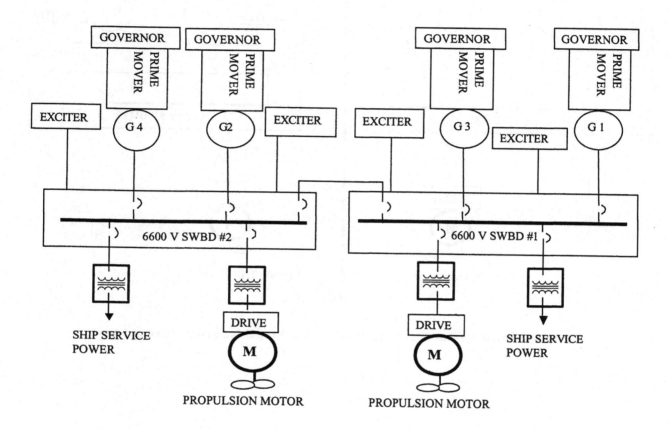

Figure 2-3: Electrical One-Line Diagram, Electrical Propulsion System (R2 Redundancy)

2.7.3   Electric propulsion system (Redundancy)

Figure 2-4 shows an electric propulsion configuration with four prime movers. Two high-voltage switchboards, two sets of propulsion motors with their drives. This configuration is an extension of Figure 2-3, with the additional requirement to meet ABS R2-S+ notation, which means there is a separation between two machinery spaces with a watertight centerline bulkhead.

Refer to Sections 14 and 21 for additional explanation for redundancy.

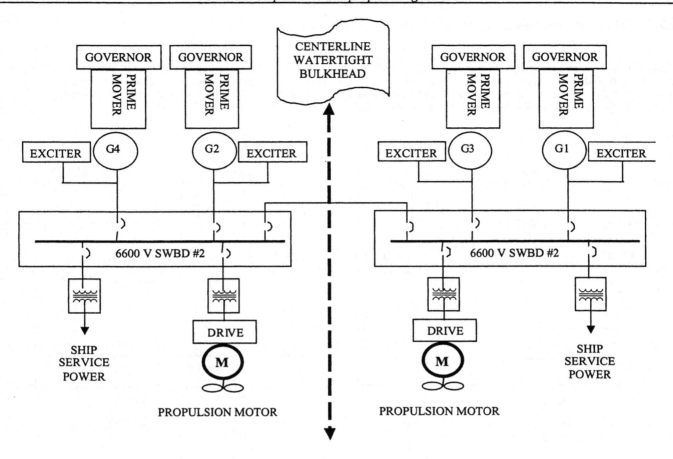

Figure 2-4: Electrical One-Line Diagram–Electrical Propulsion System (Redundancy)

3 Emergency power generation and distribution—shipboard (SWBS 310)

3.1 General

The provision of emergency power generation and distribution is a requirement by the regulatory bodies so that, in a shipboard electrical black-out situation, the emergency generator starts automatically after a predetermined interval and supplies power to dedicated emergency electric loads. The location of emergency generator and associated switchboard is also predetermined so that in case of damage of the engine room by flood or fire or electrical fault, the emergency generator room is not affected. The starting system of the emergency generator is also required to have multiple starting system, making sure that the emergency generator will start in black out situation.

3.2 IEEE Std 45-2002 Recommendations

Section 6 of IEEE Std 45-2002 refers to emergency power generation. Some of the most relevant comments are extracted below.

3.2.1 IEEE Std 45-2002, 6.1, General (extract)

… Every vessel should be provided with a self-contained emergency source of electric power, generally a diesel-engine generator, gas turbine-driven generator, or storage batteries. These emergency sources of power and an emergency switchboard should be located in a space separate and remote from the main switchboard, that is above the uppermost continuous deck, aft of the collision bulkhead, outside the main machinery compartment, and readily accessible from the open deck. The emergency switchboard should be located in the same space as the emergency power source, in an adjacent space, or as close as practical….

3.2.2 IEEE Std 45-2002, 6.2, Emergency Generators (extract)

Each emergency generator should be equipped with starting devices with an energy storage capability of at least six consecutive starts. A single source of stored energy, with the capacity for six starts, should be protected to preclude depletion by the automatic starting system, or a second source of energy should be provided for an additional three starts within 30 min. If, after three attempts, the generator set has failed to start, an audible and visual start failure alarm should be activated in the main machinery space control station and on the navigation bridge. The starting sequence should be automatically locked-out until an operator can initiate the final three starting attempts from the emergency generator space.

3.2.3 IEEE Std 45-2002, 6.4, Emergency Power Distribution System (extract)

The emergency switchboard should be supplied during normal operation from the main switchboard by an interconnecting feeder. This interconnecting feeder should be protected against short-circuit and overload at the main switchboard and, where arranged for feedback, protected for short-circuit at the emergency switchboard. The interconnecting feeder should be disconnected automatically at the emergency switchboard upon failure of the main source of electrical power. Means shall be provided to prevent auto closing of the emergency generator circuit breaker should a fault occur on the emergency switchboard….

Upon interruption of normal power, the prime mover driving the emergency power source should start automatically. When the voltage of the emergency source reaches 85% to 95% of nominal value, the emergency loads should transfer automatically to the emergency power source. The transfer to emergency power should be accomplished within 45 seconds after failure of the normal power source. If the system is arranged for automatic retransfer, the return to normal supply should be accomplished when the available voltage is 85% to 95% of the nominal value and the expiration of an appropriate time delay. The emergency generator should continue to run without load until shut down either manually or automatically by use of a timing device….

The emergency switchboard should be arranged to prevent parallel operation of an emergency power source with any other source of electrical power (i.e. main power), except where suitable means are taken for safeguarding independent emergency operation under all circumstances. This will allow for the emergency generator to be used to supply non-emergency loads. This is useful when exercising and testing the emergency generator(s).

3.2.4 Emergency shipboard electrical equipment roll and pitch withstanding requirement

IEEE Std 45-2002, 1.5.1, Table-1, Roll and pitch requirements extract to show emergency shipboard equipment only

Table -1: Roll and pitch requirement [emergency equipment] (Extract only)

Equipment	Roll		Pitch	
	Static (°)	Dynamic (°)	Static (°)	Dynamic (°)
Emergency equipment	22.5	22.5	10	10

3.3 Emergency Source of Electrical Power – ABS 2002 Requirement

ABS Section 4-8-2 refers to system design and also has several requirements for emergency power generation and distribution, from which we have quoted below.

3.3.1 ABS-2002, 4-8-2, 5.1.3 – Requirements by the Governmental Authority

Attention is directed to the requirements of governmental authority of the country, whose flag the vessel flies, for emergency services and accumulator batteries required in various types of vessels.

3.3.2 ABS-2002, 4-8-2, 5.3.3 – Separation from Other Spaces

Spaces containing the emergency sources of electrical power are to be separated from spaces other than machinery space of category A by fire rated bulkheads and decks in accordance with Part 3, Chapter 4 of ABS Rules or Chapter II-2 of SOLAS.

3.4 SOLAS (Safety of Life at Sea) Requirements

The SOLAS convention also sets requirements for emergency power generation and distribution, from which we have quoted.

3.4.1 Bulkhead and Deck Class "A" divisions (refer to SOLAS Chapter II-2)

a) They shall be constructed of steel or other equivalent material
b) They shall be suitably stiffened
c) They shall be so constructed as to be capable of preventing the passage of smoke and flame to the end of the one-hour standard fire test approved non-combustible materials such that the average temperature of the unexposed side will not rise more than 139 °C above the original temperature, nor will the temperature, at any point, including any joint, rise more than 180 °C above the original temperature, within the time listed below:
   — Class "A-60"   60 min
   — Class "A-30"   30 min
   — Class "A-15"   15 min
   — Class "A-0"   0 min

3.4.2 SOLAS requirements (Part D—Chapter II-1, Regulation 42) (extract only)

a) Emergency lighting for a period of 36 hours
b) Navigational lights and external Communication, for a period of 36 hours
c) Internal Communications for 36 hours (Refer to SOLAS for further details)
d) Fire fighting and emergency bilge for 36 hours
e) Steering gear as required by SOLAS Chapter II-1 Regulation 29
f) Watertight doors for 30 minutes

3.5 List of Emergency Loads (USCG CFR 46)

By regulation, the following emergency loads must be fed from the emergency switchboard. For emergency generator rating calculation the connecting loads must be calculated for 100 percent load factor: (Refer to USCG 46 CFR Chapter 1 Subpart 112.15 for details)

— Emergency fire pump (At least one fire pump with all auxiliaries must be fed from the emergency switchboard)
— Steering gear (For 1 steering gear system the power must be provided from ship service as well as emergency through automatic bus transfer switch)
— Watertight door closure (all)
— Emergency lighting
— Navigation and communication system—Emergency power supply with UPS back up
— Emergency power supply for the automation with UPS support
— Emergency bilge (1)

3.6 Use of Emergency Generator in port (ABS-2002, 4-8-2-5.17) (extract only)

In accordance with ABS rules the emergency generator can be used in port with the following provisions:

— Special fuel tank with level alarm for harbour operation with a change over feature
— Continuous rated prime mover with special features
— Special electrical system fault protection
— Safeguard from overload with automatic load sharing

(For details refer to ABS rules.)

3.7 Emergency generator and emergency transformer rating—load analysis (sample calculation)

Table 3-1 is a sample electrical load analysis in support of selecting the emergency generator kilowatt rating and emergency transformer kVA rating. Items 1 through 7 are required to be powered from the emergency generator. The load factor for these loads is required to be 100 percent.

1	2	3	4	5	6	7	8
Item	Service	Qty	kW – Each	kW- Connected	Load factor	Total -kW Generator	Total-kW transformer
1	Emergency Fire Pump System	1	125.0	125.0	1.0	125.0	
2	Steering Gear	1	98.0	98.0	1.0	98.0	
3	Watertight Dr Closure	5	3.0	15.0	1.0	15.0	

1	2	3	4	5	6	7	8
Item	Service	Qty	kW - Each	kW-Connected	Load factor	Total -kW Generator	Total-kW transformer
4	Emergency Lighting	all		50.0	1.0	50.0	50.0
5	Navigation & Communications System			15.0	1.0	15.0	15.0
6	UPS	2	5.0	10.0	1.0	10.0	10.0
7	Emergency Bilge	1	5.0	5.0	1.0	5.0	
	Grand total—generator					317.0	
	Grand total—transformer						75.0

NOTES

1—Emergency generator rating: (example only)
— Emergency generator load analysis load factor is 100 (1.0) percent for all loads as shown in column 7
— As shown in the calculation the generator output rating must be at least 317 kW.
— The engine output must be 317/0.95 = 333.68 kW [Considering (95 percent) 0.95 generator efficiency]
— The engine output rating should be equal to or greater than the calculated load 333.68 kW, or equivalent 447 Horsepower (HP)
— Additionally, large motor starting transient load on the generator must be considered. The engine must be able to start the largest load such as emergency fire pump, 125 kW.
— Sometimes the engine increase is necessary to support large motor starting transient.
2—Emergency transformer rating:
— Emergency transformer rating in kilovoltampere (kVA) = 75 kW/0.8 = 93.45 kVA (0.8 is the system power factor)
— The transformer nameplate rating 100 kVA

Table 3-1: Sample Load Analysis for Emergency Generator and Transformer Rating

3.8 Emergency power generation and distribution with ship service power and distribution system

Shipboard emergency generation and distribution consists of a dedicated emergency switchboard for a dedicated emergency generator. The emergency switchboard is tied to the ship service switchboard through a bus tie. The emergency switchboard feeds the 120 V section of the emergency switchboard through 450 V/120 V transformer. The 450 V section of the emergency switchboard provides power to all 450 V emergency loads. The 120 V section of the emergency switchboard provides power to all 120 V loads.

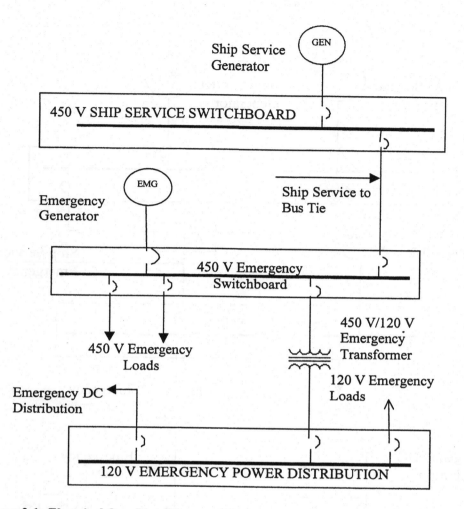

Figure 3-1: Electrical One-Line Diagram-Emergency Power Generation and Distribution with Ship Service Power

3.9 Emergency power generation and distribution with ABT bus tie

Shipboard power generation in a radial configuration and with an emergency generator generally configured such away that automatic bus transfer system in incorporated in the ship service and emergency generation control scheme. Figure 3-2 provides additional redundancy with a dedicated automatic bus transfer (ABT) function provided between two ship service generators and the emergency switchboard.

The emergency switchboard is tied to two ship service switchboards through a bus tie and automatic bus transfer switch (ABT), as shown in Figure 3-2.

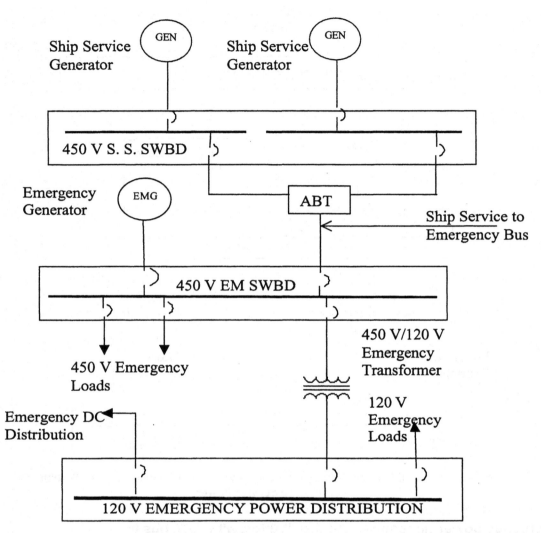

Figure 3-2: Electrical One-Line Diagram-Emergency Power Generation and Distribution with Ship Service Power Generation and Distribution with ABT Bus-Tie

3.10 Emergency Transformer 450 V/120 V (Per ABS 2002-4-8-5, Section 3.7.5)

In place of one three-winding transformer or three single-winding transformers in one enclosure with one primary breaker requirement, the ABS rules recommend the use of three separate single-phase transformers with dedicated circuit breakers, as shown in Figure 3-3.

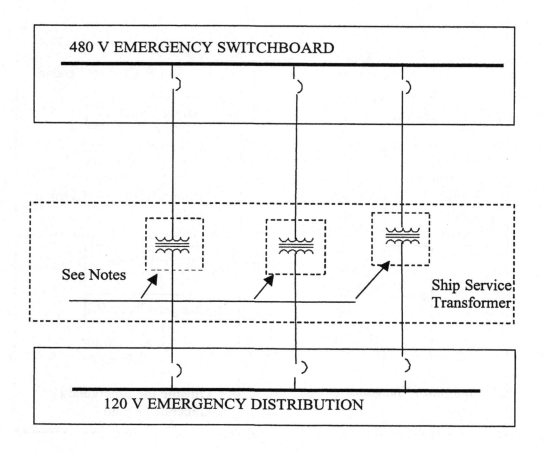

Figure 3-3: Emergency Transformer 480 V/120 V (Per ABS 2002- 4-8-5, Section 3.7.5)

NOTES:

1. Three single phase 450 V/120 V transformers, forming 120 V three-phase distribution. Each transformer is installed separately and each transformer is protected separately by dedicated circuit breaker. (ABS -2002, 4-8-5 Section 3.7.5)
2. One three phase transformer 450 V/120 V with appropriate protection

3.11 Emergency Generator Starting Block Diagram

The starting system of the emergency generator engine is required to have redundant engine starting capabilities, ensuring that the engine is ready, get started, and take over emergency loads within 45 seconds of the failure of the normal ship service power, by detecting zero voltage at the switchboard bus. The starting systems are usually combination of compressed air, hydraulic, and electrical UPS.

Figure 3-4 is a typical starting system with one hydraulic and two electrical systems. The second electrical starting system is additional capability to ensure redundancy if and when the criticality of the ship's performance is ship owner's requirement.

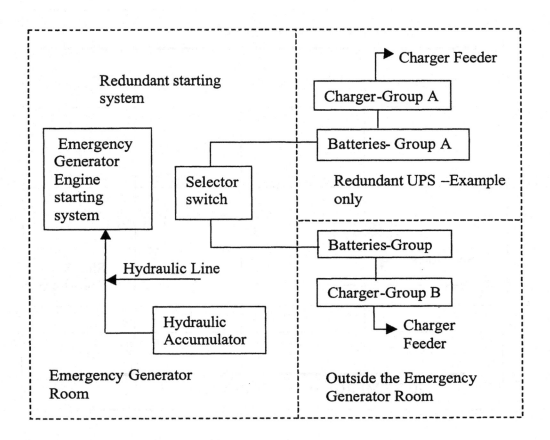

Figure 3-4: Emergency Generator Starting Block Diagram

4 Shipboard electrical system power quality

4.1 General

Shipboard AC power generation and distribution is theoretically perfect three-phase, 60 Hz or 50 Hz sinusoidal. In practice, the sinusoidal waveform gets distorted for reasons such as voltage transients, frequency transients, and harmonic distortion. The harmonic frequencies are multiples of fundamental frequencies such as 3rd, 5th, 7th, which are 180 Hz, 300 Hz, and 420 Hz for a 60 Hz system. The non-linear loads also contribute to the harmonic effects. Non-linear loads include:

- power electronics driven adjustable frequency drive AC motor for ship's propulsion system
- switching power supplies such as UPS
- microprocessor based controllers, computers, etc.

The non-linear loads produce electromagnetic interference (EMI) effects, electromagnetic susceptibility (EMS) effects, and radio frequency interference (RFI). These three effects together fall under the heading of electromagnetic compatibility (EMC). The EMI effect is exacerbated by reflected waves related to capacitive coupling, galvanic coupling, inductive coupling, and radiated coupling. EMS effects are increased by conducted susceptibility and radiated susceptibility. The non-linear loads also produce common-mode voltage and common mode current, which relates to EMI and EMC. These are the electrical noises produced and propagated in the system by high rate of dv/dt in the power electronic switching circuits.
For further detail on EMC, refer to Section 18.10.1 and Figure 18-4.

The effects of harmonic content in the distribution may include:

- Torque pulsation during motor application
- Overheating of transformer and rotating machines
- Nuisance tripping of protective devices
- Signal transmission interference
- Computer malfunction

The power supplies to the microprocessor-based equipment are non-linear, non-sinusoidal and they are also harmonic generators.

Variable frequency power electronic drives generate harmonics. The variable frequency drives, such as cycloconverter, load commutating inverter, and pulse width modulation (PWM) also produce harmonics. The harmonic distortion of these drives effects quality of power. Total harmonic distortion of each application should be calculated and then appropriate measures must be taken to minimize and manage harmonic contents within acceptable levels.

There are many different ways shipboard power system harmonics can be reduced, such as the use of transformer with shielding between primary and secondary winding, phase shift transformer for harmonic cancellation, 6-pulse, 12-pulse, 24-pulse configuration for harmonic cancellation, and harmonic filters. If these measures are not sufficient to reduce the harmonics to an acceptable level, motor generator installation is recommended. Sample electrical distribution system one-line diagrams are shown with different configurations in this section and other sections. These one-line diagrams are for guidance only. Electrical power distribution for each class of ship is to be set up to meet all operational requirements. It is the responsibility of the design engineer to design a system meeting such requirements and in compliance with applicable regulatory body requirements.

IEEE Std 519™-1992 provides required standards for harmonic measurements and acceptable harmonic levels in general. IEEE Std 45-2002, 4.5, Table 4 provided the same information for shipboard application, which is similar to Mil-Std-1399.

The EMI and EMS related noise propagation can be reduced by appropriate cable application with proper insulation, cable shielding, and cable grounding features. For additional details refer to the cable chapter 18.10 (Cable requirements for non-linear power equipment).

4.2   Total harmonic distortion calculation

Equations for calculating total harmonic distortion are given below.

$$\text{Total Harmonic Distortion (THD)} = THD_{\%Fundamentals} = (\frac{I_{rms(distortion)}}{I_{Fundamentals}})x100 \qquad \text{(a)}$$

$$I_{RMS} = \sqrt{I_1^2 + I_2^2 + I_3^2 + ...I_n^2} \qquad \text{(b)}$$

$$I_{RMS} = \sqrt{I_{1(Fundamentd-60Hz)}^2 + I_{2(5th)}^2 + I_{3(7th)}^2 + ...I_n^2} \qquad \text{(c)}$$

4.2.1   Current—harmonic distortion (sample calculation)

Figure 4-1 and Table 4-1 show a sample diagram and harmonic current calculation. Figure 4-2 is provided showing fundamental, 3rd harmonic, 5th harmonic wave forms, and resultant wave form of those wave forms.

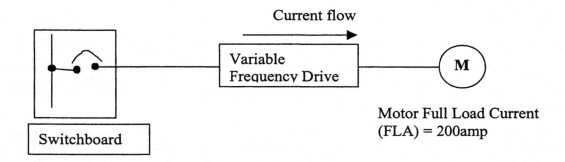

Figure 4-1: Harmonic Distortion Current – Simple Diagram

1st harmonic	200 A
5th harmonic	27 A
7th harmonic	18 A
11th harmonic	10 A
Distortion rms	33.9 A
Rms	202.86 A
Total harmonic distortion (THD)	16.98%

Table 4-1: Current Distortion Calculations (typical)

The harmonic current does not flow through the system; however it directly contributes to transformer winding heating, causing heating effects. The transformer should be oversized to prevent premature winding failure due to harmonic-related overheating

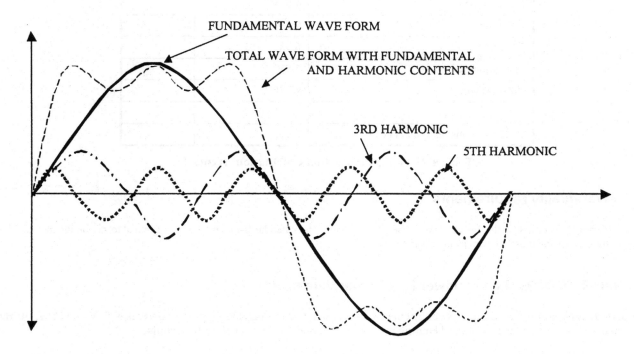

Figure 4-2: Sinusoidal Waves with Harmonic Contents

4.2.2 Voltage harmonic distortion (sample calculation)

Figure 4-3 and Table 4-2 show the result of voltage harmonic distortion calculations for the sample diagram shown.

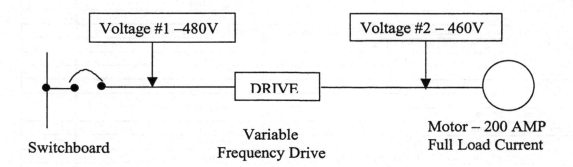

Figure 4-3: Harmonic Distortion Voltage – Diagram

1st harmonic	480 V	
5th harmonic	9.5V	
7th harmonic	6.5V	
11th harmonic	5.5V	
Distortion rms	12.76V	
rms	480.17V	
Total harmonic distortion (THD)	2.6%	

Table 4-2: Voltage Distortion Calculations (typical)

4.3 Power quality requirements

When considering power quality, the designer should be aware of what other standards-making bodies have had to say on the matter. The sections below discuss the most relevant standards in this area.

4.3.1 MIL-STD-1399, Type I power requirements (extract)

For shipboard application, power quality requirements are given in MIL-STD-1399. Type I power is 440 V or 115 V, 60 Hz ungrounded and is the standard shipboard electric power source. See Table 4-3 below for details.

Characteristics	Type I	
FREQUENCY		
a. Normal Frequency	60 Hz	
b. Frequency Tolerance	Plus or minus 3%	
c. Frequency Modulation	½ percent	
d. Frequency Transient		
1.Tolerance	Plus or minus 4%	
2.Recovery time	2 seconds	
e. Worst case, frequency excursion from nominal frequency	Plus or minus 5-1/2 %	
Resulting from b, c, and d(1) combined except under		
Emergency Cond		
VOLTAGE		
f. Normal user voltage	440, 115v	
g. User voltage tolerance		
1. Average of three line to line voltage	Plus or minus 5%	
2. Any one-line to line voltage including g(1) and	Plus or minus 7%	
Line voltage unbalance (h)		
h. Line voltage unbalance	3%	
i. Voltage modulation	2%	
j. Voltage transient		
(1) Voltage transient tolerance	Plus or minus 16%	
(2) Voltage transient recovery time	2 seconds	

k. Voltage spikes (peak value, includes fundamental)	Plus minus 2.500 V for 440 V sys and 1000 V for 115 V	
l. Maximum departure voltage resulting from g(1), g(2), (h), and (I) combined, except under transient or emergency conditions	Plus or minus 6 percent	
m. The worst case voltage excursion from nominal user voltage resulting from (g)(1), (g)(2), (h), (I), and j(1) combined except under emergency conditions.	Plus or minus 20 percent	
n. Insulation resistance test		
(a) surface ships	500 vdc	
WAVEFORM (VOLTAGE)		
(o) Maximum total harmonic distortion	5 percent	
(p) Maximum single harmonic	3 percent	
(q) Maximum deviation factor	5 percent	
EMERGENCY CONDITION		
(r) Frequency excursion	Minus 100 to plus 12 percent	
(s) Duration of frequency excursion	Up to 2 minutes	
(t) Voltage excursion	Minus 100 t0 plus 35percent	
(u) Duration of voltage excursion		
(1) lower limit (minus 100 percent)	Up to 2 minutes	
(2) upper limit (plus 35 percent)	2 minutes	

Table 4-3: Type I Power Requirement per MIL-STD-1399

4.3.2 IEEE Std 45-2002, 4.5, AC Power system characteristics

Power distribution systems should maintain the system characteristics described in Table 4 under all operating conditions. Power-consuming equipment should operate satisfactorily under the conditions described in Table 4, and it should be designed to withstand the power interruption, transient, electromagnetic interference (EMI), radio frequency interference (RFI), and insulation resistance test conditions inherent in the system. Power-consuming equipment requiring a nonstandard voltage or frequency for successful operation should have integral power conversion capability. Power-consuming equipment should not have inherent characteristics that degrade the power quality of the supply system described in Table 4-4.

Table 4-4, Alternating current (AC) power characteristics

Characteristics	Limits
Frequency	
a) Nominal frequency	50/60 Hz
b) Frequency tolerances	± 3%
c) Frequency modulation	½%
d) Frequency transient:	
1) Tolerance	± 4%
2) Recovery time	2 s
e) The worst-case frequency excursion from nominal frequency resulting from item b), item c), and item d) 1) combined, except under emergency conditions.	± 5½%
Voltage	
a) User voltage tolerance:	
1) Average of the three line-to-line voltages	± 5%
2) Any one line-to-line voltage, including item a) 1) and line voltage unbalances item b)	± 7%
b) Line voltage unbalance	3%
c) Voltage modulation	5%
d) Voltage transient:	
1) Voltage transient tolerances	± 16%
2) Voltage transient recovery time	2 s
e) Voltage spike (peak value includes fundamental)	± 2500 V (380–600 V) system; 1000 V (120–240 V) system
f) The maximum departure voltage resulting from item a) 1) and item d) combined, except under transient or emergency conditions.	± 6%
g) The worst case voltage excursion from nominal user voltage resulting from item a) 1), item a) 2), and item d) 1) combined, except under emergency conditions.	± 20%
Waveform voltage distortion[a]	
a) Maximum total harmonic distortion	5%
b) Maximum single harmonic	3%
c) Maximum deviation factor	5%

Emergency conditions	
a) Frequency excursion	−100 to +12%
b) Duration of frequency excursion	Up to 2 min
c) Voltage excursion	−100 to +35%
d) Duration of voltage excursion:	
1) Lower limits (−100%)	Up to 2 min
2) Upper limit (+35%)	2 min

Definitions:

a) Frequency

    1) Nominal frequency: The designated frequency in hertz.

    2) Frequency tolerance: The maximum permitted departure from nominal frequency during normal operation, excluding transient and cyclic frequency variations. It includes variations caused by load changes, environment (temperature, humidity, vibration, inclination), switchboard meter error, and drift. Tolerances are expressed in percentage of nominal frequency.

    3) Frequency modulation: The permitted periodic variation in frequency during normal operation that might be caused by regularly and randomly repeated loading. For purposes of definition, the periodicity of frequency modulation should be considered as not exceeding 10 s.

$$\text{Frequency Modulation (\%)} = \frac{\{f_{maximum} - f_{minimum}\}}{\{2 \times f_{nominal}\}} \times 100$$

    4) Frequency transient tolerance: A sudden change in frequency that goes outside the frequency tolerance limits, returns to, and remains inside these limits within a specified recovery time after initiation of the disturbance. Frequency transient tolerance is in addition to frequency tolerance limits.

    5) Frequency transient recovery time: The time period from the start of the disturbance until the frequency recovers and remains within frequency tolerance limits.

b) Voltage

    1) User voltage tolerance: The maximum permitted departure from nominal user voltage during normal operation, excluding transient and cyclic voltage variations. It includes variations such as those caused by load changes, environment (temperature, humidity, vibration, inclination), switchboard meter error, and drift.

    2) Line voltage unbalance tolerance (three-phase system): The difference between the highest and lowest line-to-line voltages.

$$\text{Line Voltage Unbalance Tolerance (\%)} = \frac{\{E_{maximum} - E_{minimum}\}}{\{E_{nominal}\}} \times 100$$

    3) Voltage modulation (amplitude): The periodic voltage variation (peak to valley) or the user voltage that might be caused by regularly or randomly repeated pulsed loading. The periodicity or voltage modulation is considered to be longer than 1 Hz and less than 10 s. Voltages used in the following equation shall be all-peak or all-rms:

$$\text{Voltage Modulation (\%)} = \frac{\{E_{maximum} - E_{minimum}\}}{\{2 \times E_{nominal}\}} \times 100$$

    4) Voltage transient

        i) Voltage transient tolerance: A sudden change (excluding spikes) in voltage that goes outside the user voltage tolerance limits and returns to and remains within these limits within a specified recovery time longer than 1 ms after the initiation of the disturbance. The voltage transient tolerance is in addition to the user voltage tolerance limits.

        ii) Voltage transient recovery time: The time elapsed from initiation of the disturbance until the voltage recovers and remains within the user voltage tolerance limits.

    5) Voltage spike: A voltage change of very short duration (less than 1 ms).

c) Waveform

    1) Total Harmonic Distortion (THD)(of a sine wave): The ratio in percentage of the rms value of the residue (after elimination of the fundamental) to the rms value of the fundamental.

    2) Single Harmonic (of a sine wave): The ratio in percentage of the rms value of that harmonic to the rms value of the fundamental.

    3) Deviation Factor (of a sine wave): The ratio of the maximum difference between corresponding ordinates of the wave and of the equivalent sine wave to the maximum ordinate of the equivalent sine wave when the waves are superimposed in such a way that they make the maximum difference as

$$\text{Deviation factor (\%)} = \frac{\{\text{maximum deviation}\}}{\{\text{maximum ordinate of the equivalent sine wave}\}} \times 100$$

small as possible.

d)   Emergency conditions

1)   A situation or occurrence of a serious nature that may result in electrical power system interruptions or deviations, such as the occurrence of ship service generator failure and the emergency generator coming on line.

[a]For ships with electric propulsion or other adjustable speed drive loads, higher voltage distortion can be accepted on a dedicated power bus if the equipment connected to the dedicated power bus is designed and tested for the actual conditions.

4.3.3   IEEE Std 45-2002, 4.6, Power quality and harmonics

Solid state devices such as motor controllers, computers, copiers, printers, and video display terminals produce harmonic currents. These harmonic currents may cause additional heating in motors, transformers, and cables. The sizing of protective devices should consider the harmonic current component. Harmonic currents in nonsensically current waveforms may also cause EMI and RFI. EMI and RFI may result in interference with sensitive electronics equipment throughout the vessel.

Isolation, both physical and electrical, should be provided between electronic systems and power systems that supply large numbers of solid state devices, or significantly sized solid state motor controllers. Active or passive filters and shielded input isolation transformers should be used to minimize interference. Special care should be given to the application of isolation transformers or filtering as the percentage of power consumed by solid state power devices compared with the system power available increases. Small units connected to large power systems exhibit less interference on the power source than do larger units connected to the same source. Solid state power devices of vastly different sizes should not share a common power circuit. Where kilowatt ratings differ by more than 5 to 1, the circuits should be isolated by a shielded distribution system transformer. Surge suppressers or filters should only be connected to power circuits on the secondary side of the equipment power input isolation transformers.

Notes:

1.   To preclude radiated EMI, main power switchboards rated in excess of 1 kV and propulsion motor drives should not be installed in the same shipboard compartment as ship service switchboards or control consoles.(This is per IEEE Std 45-1998, 4.6).

2.   To reduce the effect of radiated EMI, special considerations on filtering and shielding should be exercised when main power switchboards and propulsion motor drives are installed in the same shipboard compartment as ship service switchboards or control console.

3.   IEEE Std 519™-1992 provides additional recommendations regarding power quality.

4.4   IEEE Std 45-2002, 31.8, Propulsion power conversion equipment (power quality)

The following quote is an extract referring only to the power quality portion of this clause.

Whenever power converters for propulsion are applied to integrated electric plants, the drive system should be designed to maintain and operate with the power quality of the electric plant. The effects of disturbances, both to the integrated power system and to other motor drive converters, should be regarded in the design. Attention should be paid to the power quality impact of the following:

a)   Multiple drives connected to the same main power system.

b)   Commutation reactance, which, if insufficient, may result in voltage distortion adversely affecting other power consumers on the distribution system. Unsuitable matching of the relation between the power generation system's sub-transient reactance and the propulsion drive commutation impedance may result in production of harmonic values beyond the power quality limits.

c)   Harmonic distortion can cause overheating of other elements of the distribution system and improper operation of other ship service power consumers.

d)   Adverse effects of voltage and frequency variations in regenerating mode.

e)   Conducted and radiated electromagnetic interference and the introduction of high-frequency noise to adjacent sensitive circuits and control devices. Special consideration should be given for the installation, filtering, and cabling to prevent electromagnetic interference....

4.5   Typical system design with harmonic management

Figure 4-4 is a simple power generation and distribution diagram with non-linear loads. A harmonic filter is used to reduce the harmonic level within the required limits.

Figure 4-5 shows power system distribution with non-sensitive and sensitive loads. The sensitive loads are usually the loads that cannot tolerate harmonic effects. The sensitive bus is powered by a motor generator set to ensure quality power for the sensitive loads.

Figure 4-6 is a typical power distribution for a ship's central control system. The central control system requires redundant clean power with battery backup to ensure power at the console under all normal and emergency situations.

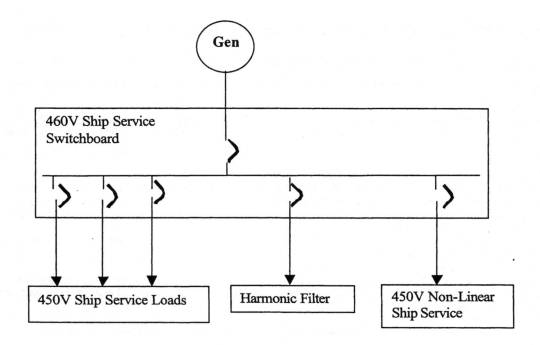

Figure 4-4: Electrical One-Line Diagram-Ship Service Power Generation and Distribution with Harmonic Filter

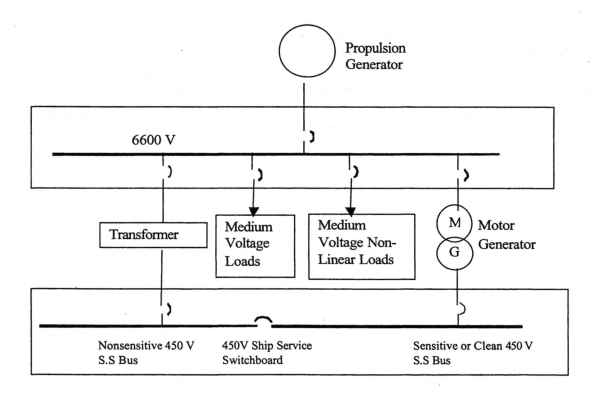

Figure 4-5: Electrical One-Line Diagram-6600 V Power Generation and 450 V Distribution with Clean (Sensitive/Clean Power) with Motor Generator (Sensitive Bus) and Transformer (Non-Sensitive Bus)

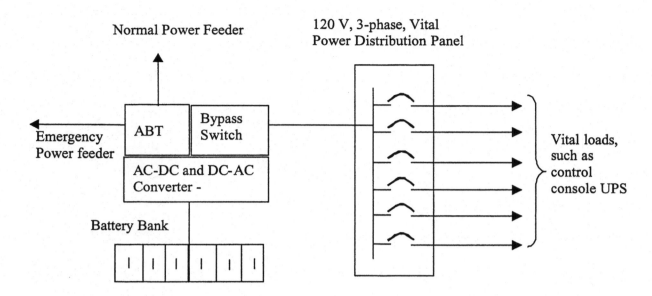

Figure 4-6: Clean Power—UPS with Redundant Power Feeders and Automatic Bus Transfer System

Note that the UPS has the following features:

1. Normally the vital load power supply is from the normal ship service power.
2. Normal power is backed up by emergency power source by automatic bus transfer switch (ABT)
3. The normal and emergency power is backed up by battery power.
4. The battery is usually trickled charged continuously to ensure the batteries are fully charged all the line
5. The batteries are also set for rapid recharge when ship's power is restored

5   Power system electrical one-line diagram—shipboard

5.1   General

An electrical one-line diagram is a simplified presentation of shipboard electrical power generation and distribution. The recommendation of IEEE Std 45 is to install at least two sources of power generation, one being stand-by to ensure availability of electrical power during all required operating profiles of the vessel: at sea cruising, at port loading, unloading, at anchor, special operations, other operating profiles as deemed necessary for the intended application of the ship, as well as during emergency situations. The shipboard emergency situation is an electrical blackout caused by failure of running generators, which in turn can be caused by flooding, fire, and electrical fault condition. The operating profile of ocean-going passenger ships is different from that of tankers; tankers differ from dry cargo carriers; and so on and so forth. The preparation of electrical power generation and distribution starts with electrical one-line diagrams at the concept design phase. The concept design leads power system generation and distribution equipment selection, then the contract design and finally the detailed design for procurement and installation.

A concept electrical one-line diagram is prepared from the operating profile and estimated load distribution requirement of the ship. The estimated load distribution is used to prepare a conceptual electrical load analysis for estimated ratings of the ship service generators, emergency generators, propulsion generators power, transformers etc.

At the detailed design phase, the power generation, power distribution requirements and load requirements are finalized by design development of systems such as mechanical, HVAC, electrical power, lighting, communications, and controls. Additionally, the electrical load analysis of the concept power system , and the electrical one-line diagram, are considered as final designs. The equipment selection process starts from this final electrical one-line diagram. The final design phase one-line diagram, and equipment selection process are undertaken in full compliance with the regulatory body requirements with respect to redundancy, criticality consideration, weight, location, volume, and cost.

The electrical one-line diagrams presented in this section are examples for better understanding of shipboard power generation and distribution outline configuration. However, these diagrams are in accordance with IEEE Std 45-2002 recommendations for power generations and distribution and the best design practice. As an aid to understanding the electrical power generation and distribution through these electrical one-line diagrams, IEEE Std 45-2002 recommendations are extracted below with respective clause numbers.

5.1.1   IEEE Std 45-2002, 7.1.1, Electric power generation—general (extract)

> Electric power generating systems discussed in Clause 7 consist of one or more generator sets. These recommendations include both sound engineering practices and special considerations for safe and reliable operation in marine applications. ...

5.1.2   IEEE Std 45-2002, 7.4.2, Selection and sizing of generators (extract)

> In determining the number and capacities of generating sets to be provided for a vessel, careful consideration should be given to the normal and maximum load demands (i.e., load analysis) as well as for the safe and efficient operation of the vessel when at sea and in port. The vessel must have at least two generating sources. For ships, the number and ratings of the main generating sets should be sufficient to provide one spare generating set (one set not in operation) at all times to service the essential and habitable loads. ...
>
> For vessels propelled by electric power and having two or more constant-voltage, constant-frequency, main power generators, the ship service electric power may be derived from this source and additional ship service generators need not be installed, provided that with one main power generator out of service, a speed of 7 knots or one-half of the design speed (whichever is the lesser) can be maintained. The combined normal capacity of the operating generating sets should be least equal to the maximum peak load at sea. If the peak load and its duration is within the limits of the specified overload capacity of the generating sets, it is not necessary to have the combined normal capacity equal to the maximum peak load. ...

5.1.3 IEEE Std 45-2002, 6.1 and 6.2, Emergency power systems (extracts)

… Every vessel should be provided with a self-contained emergency source of electric power, generally a diesel-engine generator, gas turbine-driven generator, or storage batteries. These emergency sources of power and an emergency switchboard should be located in a space separate and remote from the main switchboard, that is above the uppermost continuous deck, aft of the collision bulkhead, outside the main machinery compartment, and readily accessible from the open deck. The emergency switchboard should be located in the same space as the emergency power source, in an adjacent space, or as close as practical. … [6.1]

Emergency generator(s) should be sized to supply 100% of connected loads that are essential for safety in an emergency condition. Where redundant equipment is installed so that not all loads operate simultaneously, these redundant loads need not be considered in the calculation. … [6.2]

5.1.4 IEEE Std 45-2002, 31.3.1, Electrical propulsion—general (extract)

The design of an integrated electric power system should consider the power required to support ship service loads and propulsion loads under a variety of operating conditions with optimum usage of the installed and running generator sets. …

5.1.5 IEEE Std 45-2002, 31.4 and 31.5.1, Prime movers and generators for integrated power and propulsion plants

Prime movers, such as diesel engines, gas turbines, or steam turbines, for the generators in integrated electric power systems shall be capable of starting under dead ship conditions in accordance with requirements of the authority having jurisdiction. Where the speed control of the propeller requires speed variation of the prime mover, the governor should be provided with means for local manual control as well as for remote control. … [31.4])

The generators of an integrated propulsion and power system can be considered as emergency generators, provided that class rules and regulation and legislation requirements for emergency generators are fulfilled. … [31.5.1]

5.2 Electrical one-line diagram examples

Figure 5-1 through Figure 5-9 are the power system electrical one line diagrams showing various power system distribution set-ups. These drawings are examples only to explain fundamentals of power system design and development, with limited technical explanation. If these drawings resemble specific installation, it is only by coincidence.

5.2.1 Electrical one line diagram—Ship service power generation and distribution with one ship service generator and one emergency generator (Figure 5-1)

This power generation and distribution system design consists of one generator and one switchboard for ship service power and one emergency switchboard with one emergency generator.

- Ship service switchboard—460 V section
  - Generator breaker and feeder
  - Motor controllers
  - Ship service load section for propulsion auxiliaries, HVAC, deck machinery, and electrical services
  - Shore power (open option of feeding from ship service or emergency switchboard)
  - Emergency bus tie
- Ship service switchboard—115 V section
  - 115 V Lighting
  - Other 115 V feeders for 115 V services

— Emergency switchboard—480 V section
  — Emergency generator circuit breaker
  — 460 V emergency loads
  — Ship service bus tie breaker section
— Emergency switchboard—115 V section
  — Emergency lighting
  — Other 115 V emergency loads

NOTE—This configuration may not be in compliance with regulatory body requirements for certain classes of ships.

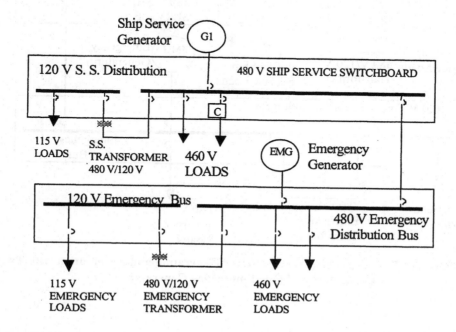

Figure 5-1: Electrical One-Line Diagram – Ship Service Power Generation and Distribution with One Ship Service Generator and One Emergency Generator

5.2.2 Electrical one line diagram—ship service power generation and distribution with two ship service generators and one emergency generator (Figure 5-2)

This design consists of one ship service switchboard and two ship service generators. The emergency switchboard is with one emergency generator.
— Ship service switchboard—480 V section
  — Generator breakers
  — Motor controllers
  — Ship service load section for machinery auxiliaries, HVAC, deck machinery, and other electrical services
  — Shore power (open option of feeding from ship service or emergency switchboard)
  — Emergency bus tie
— Ship service switchboard—120 V section
  — Lighting
  — Other 115 V services
— Emergency switchboard—480 V section
  — Emergency generator circuit breaker
  — 480 V emergency loads
  — Emergency bus tie to the ship service switchboard
— Emergency switchboard—120 V section
  — Emergency lighting and other 115 V emergency loads

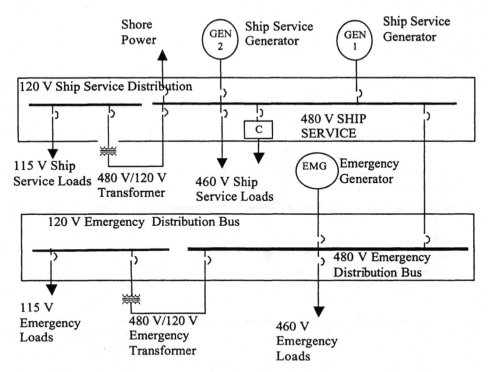

Figure 5-2: Electrical One-Line Diagram – Ship Service Power Generation and Distribution with Two Ship Service Generators and One Emergency Generator

5.2.3 Electrical one line diagram—ship service power generation and distribution with four ship service generators and one emergency generator (Figure 5-3)

In this example, there are four ship service generators, two ship service switchboards, one emergency generator, and one emergency switchboard. Each ship service switchboard is connected to two ship service generators. This set-up is for ships with large service load demand, such as cruise ships. Due to the number of generators, the load distribution and management of four ship service generators and one emergency generator can be complex. However, the design has proven to be very successful. The parallel operation and necessary protective control and their interlocking must however be clearly understood by the operator.

— Port—Ship service switchboard—480 V section
  — Generator breakers
  — Motor controllers
  — Ship service load section for machinery auxiliaries, HVAC, deck machinery, and electrical services
  — Shore power (open option of feeding from ship service or emergency switchboard)
  — Emergency bus tie
— Port—Ship service switchboard—120 V section
  — 115 V Lighting
  — Other 120 V feeders for 115 V services
— Starboard—Ship service switchboard—480 V section
  — Generator breakers
  — Motor controllers
  — Ship service load section for propulsion auxiliaries, HVAC, deck machinery, and electrical services
  — Shore power (open option of feeding from ship service or emergency switchboard)
  — Emergency bus tie
— Starboard—Ship service switchboard—120 V section

— 115 V lighting
— Other 115 V services
— Emergency switchboard—480 V section
    — Emergency generator circuit breaker
    — 480 V emergency loads
    — Ship service bus tie breaker section
— Emergency switchboard—120 V section
    — Emergency lighting
    — Other 115 V emergency loads

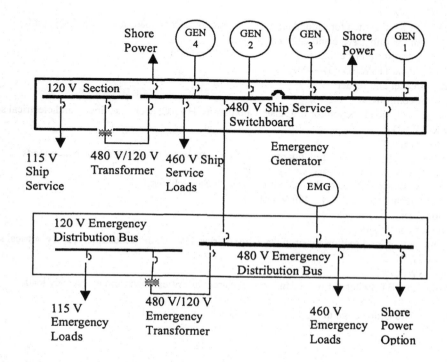

Figure 5-3: Electrical One-Line Diagram – Ship Service Power Generation and Distribution with Four Ship Service Generators and One Emergency Generator

5.2.4 Electrical one line diagram—ring bus configuration with four generators; each generator designated as ship service/emergency (Figure 5-4)

In this example design, there are four ship service generators and four ship service switchboards. Each ship service switchboard is connected to one dedicated generator. However, each switchboard is connected to another switchboard by a bus tie configuration. This configuration facilitates automatic load transfer in case of loss of voltage in one switchboard. In addition to that, an additional bus transfer device can easily be installed to comply with redundant power feeder requirements. This arrangement is well adopted in some classes of military ship design. The ship service power distribution profile meets all regulatory body requirements. The switchboard set-up is known as a *ring bus configuration*. The physical arrangement of the generators and switchboards are in different fire zones, watertight compartments, and different decks, providing the best possible redundant zonal power distribution. However, the arrangement requires complex power feeder assignment, and complex power management due to the fact that the generators and switchboards are not only called ship service; they are designated as ship service and emergency. The control scheme of the configuration is very complex which may not be the most cost-effective design.

— Ship service switchboard (SWBD #1)—480 V section
    — Generator breaker and feeder
    — Zonal motor controllers

- — Zonal ship service load section for propulsion auxiliaries, HVAC, deck machinery, and electrical services
- — Bus tie to SWBD #2
- — Bus tie to SWBD #3
- — Tie breaker to automatic bus transfer switch (ABT) for redundant power source for critical loads and emergency loads
- — 120 V section for ship service and emergency loads
- — 480 V/120 V transformer tie to the 120 V section of the switchboard
- — Ship service switchboard (SWBD #2)—480 V section
  - — Generator breaker and feeder
  - — Zonal motor controllers
  - — Zonal ship service load section for propulsion auxiliaries, HVAC, deck machinery, and electrical services
  - — Bus tie to SWBD #1
  - — Bus tie to SWBD #4
  - — Tie breaker to ABT switch for redundant power source for critical loads and emergency loads
  - — 120 V section for ship service and emergency loads
  - — 480 V/120 V transformer tie to the 120 V section of the switchboard
- — Ship service switchboard (SWBD #3)—480 V section
  - — Generator breaker and feeder
  - — Zonal motor controllers
  - — Zonal ship service load section for propulsion auxiliaries, HVAC, deck machinery, and electrical services
  - — Bus tie to SWBD #2
  - — Bus tie to SWBD #4
  - — Tie breaker to ABT switch for redundant power source for critical loads and emergency loads
  - — 120 V section for ship service and emergency loads
  - — 480 V/120 V transformer tie to the 120 V section of the switchboard
- — Ship service switchboard (SWBD #4)—480 V section
  - — Generator breaker and feeder
  - — Zonal motor controllers
  - — Zonal ship service load section for propulsion auxiliaries, HVAC, deck machinery, and electrical services
  - — Bus tie to SWBD #2
  - — Bus tie to SWBD #3
  - — Tie breaker to ABT switch for redundant power source for critical loads and emergency loads
  - — 120 V section for ship service and emergency loads
  - — 480 V/120 V transformer tie to the 120 V section of the switchboard

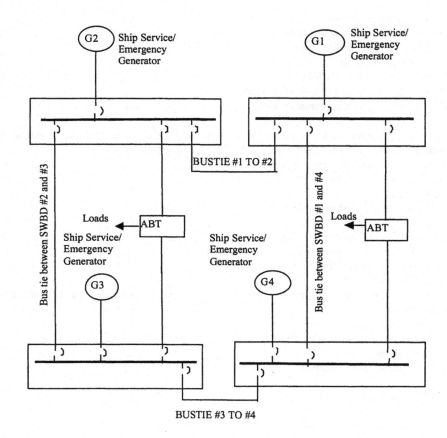

**Figure 5-4: Electrical One-Line Diagram-Ring Bus Configuration with Four Ship Service Generators –
Each Generator Designated as Ship Service/Emergency Generator**

5.2.5 Electrical one line diagram—electric propulsion system with four main generators and one emergency generator (Figure 5-5)

The electrical propulsion system power generation, distribution, management, and control features are different from typical ship service power generation and distribution. The ship service power distribution typically provides services to the propulsion system directly coupled to the prime movers, directly converting mechanically-generated power to the propulsion power. The electric propulsion system drives the propulsion power electrically, particularly by electric propulsion motor. The electric power generation requirement is the power required for the propulsion load plus the power required for the propulsion auxiliaries as well as miscellaneous ship service loads. In this case the propulsion load analysis and the ship service load analysis is to be performed separately. Total combined propulsion load and ship service load determines the power generation requirement for the ship.

The one line diagram shows four main generators, and one emergency generator. The main generator rating is in the range of many megawatts for supporting propulsion motor load. The ship service power demand is usually less than 10% of the total power generation and the emergency power requirement is much less than the ship service power requirement. The power distribution categories are as follows:

— Port—6.6 KV Propulsion switchboard (medium-voltage system)
  — Propulsion generators #2 and #4 breaker sections
  — Propulsion motor variable speed drive feeder section
  — Medium-voltage load starter section
  — Medium-voltage transformer feeder section
  — Bus tie to the starboard 6.6KV SWBD #2
— Starboard—6.6 KV Propulsion switchboard (medium-voltage system)

- — Propulsion generators #1 and #3 breaker sections
  - — Propulsion motor variable speed drive feeder section
  - — Medium-voltage load starter section
  - — Medium-voltage transformer feeder section
  - — Bus tie to the port 6.6KV SWBD #1
- — Port—480 V Ship service switchboard—480 V section
  - — 6.6 kV transformer secondary breaker section
  - — Motor controllers
  - — Ship service breaker section for propulsion auxiliaries, HVAC, deck machinery, and electrical services
  - — Shore power (open option of feeding from port or starboard ship service)
  - — Bus tie to the starboard ship service switchboard
  - — Ship service switchboard bus tie port to starboard
  - — Emergency bus tie
- — Port—480 V Ship service switchboard—120 V section
  - — Lighting
  - — Other 120 V feeders for 120 V services
- — Starboard—480 V Ship service switchboard—480 V section
  - — 6.6 kV transformer secondary breaker section
  - — Motor controllers
  - — Ship service breaker section for propulsion auxiliaries, HVAC, deck machinery, and electrical services
  - — Shore power
  - — Bus tie to the starboard ship service switchboard
  - — Ship service switchboard bus tie port to starboard
  - — Emergency bus tie
- — Starboard 480 V Ship service switchboard—120 V section
  - — Lighting
  - — Other 120 V feeders for 120 V services
- — Emergency switchboard—480 V section
  - — Emergency generator circuit breaker
  - — 480 V emergency loads
  - — Emergency bus-tie to the port ship service switchboard
  - — Emergency bus tie to the starboard ship service switchboard
- — Emergency switchboard—120 V section
  - — Emergency lighting and other 120 V emergency loads

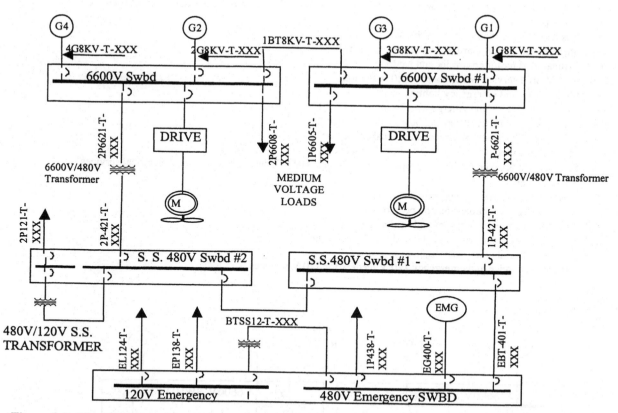

Figure 5-5: Electrical One-Line diagram-Electrical Propulsion System with Four Main Generators and One Emergency Generator

5.2.6 Electrical one line diagram—electric propulsion power (6600 V/480 V/120 V) generation and distribution one-line diagram (no emergency generator) (Figure 5-6)

This electric propulsion plant configuration is similar to the one described in Figure 5-5. However, in this design there is no emergency generator, nor an emergency switchboard. The propulsion generators and switchgears perform the functions of the ship service generator and the emergency generator. The propulsion generator power distribution and automatic stand-by starting capability of the main propulsion engine under black ship condition is required to fulfill the requirement of emergency generator.

The power distribution categories are as follows:

— Port—6.6 KV Propulsion switchboard (medium-voltage system)
  — Propulsion generators #2 and #4 breaker sections
  — Propulsion motor variable speed drive feeder section
  — Medium-voltage load starter section
  — Medium-voltage transformer feeder section
  — Bus tie to the starboard 6.6 kV SWBD #1
— Starboard—6.6 KV Propulsion switchboard (medium-voltage system)
  — Propulsion generators #1 and #3 breaker sections
  — Propulsion motor variable speed drive feeder section
  — Medium-voltage load starter section
  — Medium-voltage transformer feeder section
  — Bus tie to the port 6.6 kV SWBD #2
— Port—480 V Ship service switchboard—480 V section
  — 6.6 kV/480 V transformer secondary breaker section
  — Motor controllers

- — Ship service breaker section for propulsion auxiliaries, HVAC, deck machinery, and electrical services
- — Shore power
- — Bus tie to the starboard ship service switchboard
- — Ship service switchboard bus tie port to starboard
- — Critical load motor controller and feeder section
- — Port—480 V Ship service switchboard—120 V section
  - — Ship service and emergency lighting
  - — Other ship service and emergency 120 V feeders for 120 V services
- — Starboard—480 V Ship service switchboard—480 V section
  - — 6.6 kV/480 V transformer secondary breaker section
  - — Motor controllers
  - — Ship service breaker section for propulsion auxiliaries, HVAC, deck machinery, and electrical services
  - — Shore power
  - — Bus tie to the starboard ship service switchboard
  - — Ship service switchboard bus tie port to starboard
  - — Critical load motor controller and feeder section
- — Starboard—480 V Ship service switchboard—120 V section
  - — Ship service and emergency lighting
  - — Other ship service and emergency 120 V feeders for 120 V services

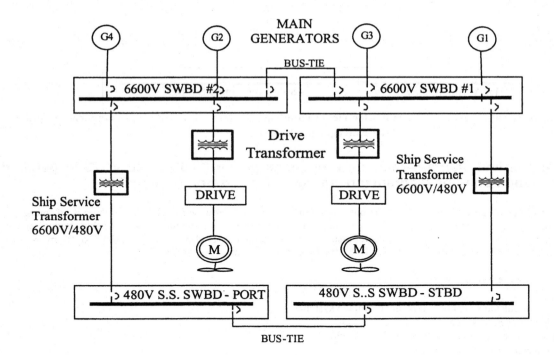

Figure 5-6: Electrical One-Line Diagram – Electrical Propulsion System with Four Generators and Ship Service Transformer

5.2.7 Electrical one line diagram—electrical propulsion (6600 V/480 V/120 V) generation and distribution one-line diagram in compliance with ABS R2-S notation (Figure 5-7)

This electric propulsion configuration is similar to Figure 5-6 with the following additional features to comply with the ABS requirements for ABS R2-S class notation. For full details of R2-S refer to the ABS Redundancy Document. The requirement is summarized as follows:

— Must have two completely independent engine rooms with watertight bulkhead separation at the centerline of the ship.
— Each engine room must be completely independence for operation so that in case of black out in one engine room, the other engine room remains fully functional.

The electrical distribution system is as follows:

— Port—6.6 KV Propulsion switchboard (medium-voltage system)
  — Breaker section for propulsion generators #2 and #4 in port engine room
  — Feeder section for port propulsion system variable speed drive feeder section
  — Medium-voltage load starter for all port engine room medium-voltage loads
  — Medium-voltage transformer feeder section for the port engine room
  — Medium-voltage bus tie breaker at the port engine room for tie to the starboard switchboard
— Starboard—6.6 KV Propulsion switchboard (medium-voltage system)
  — Breaker section for propulsion generators #1 and #3 in starboard port engine room
  — Feeder section for starboard propulsion system variable speed drive feeder section
  — Medium-voltage load starter for all starboard engine room medium-voltage loads
  — Medium-voltage transformer feeder section for the starboard engine room
  — Medium-voltage bus tie breaker at the starboard engine room for tie to the port switchboard
— Port—480 V Ship service switchboard—480 V section
  — Port engine room 6.6 kV transformer secondary breaker section
  — Motor controllers
  — Ship service breaker section for port engine room propulsion auxiliaries, HVAC, deck machinery, and electrical services
  — Shore power
  — Port ship service switchboard bus tie port to starboard ship service switchboard
  — Port ship service switchboard bus tie to the emergency switchboard
— Starboard—480 V Ship service switchboard—480 V section
  — Starboard engine room 6.6 kV transformer secondary breaker section
  — Starboard engine room motor controllers
  — Starboard engine room Ship service breaker section for all starboard engine room propulsion auxiliaries, HVAC, deck machinery, and electrical services
  — Shore power
  — Bus tie to the port engine room ship service switchboard
  — Starboard engine room ship service switchboard bus tie the emergency bus tie
— Emergency switchboard—480 V section
  — Emergency generator circuit breaker
  — 480 V emergency loads
  — Emergency bus tie to the port ship service switchboard
  — Emergency bus tie to the starboard ship service switchboard
— Emergency switchboard—120 V section
  — Emergency lighting and other emergency 120 V loads

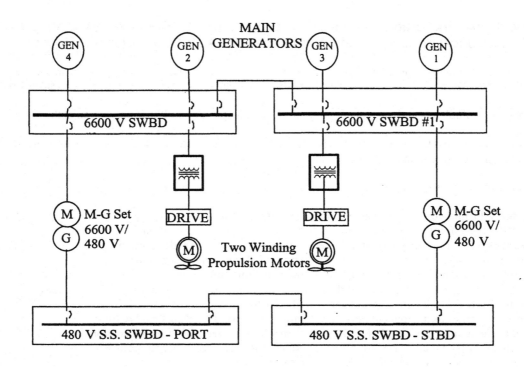

Figure 5-7: Electrical One-Line Diagram – Electrical Propulsion System with Redundant Propulsion, One Emergency Generator and Motor Generator (Emergency Generator not shown)

5.2.8 Electrical one line diagram—medium-voltage electric propulsion—ring bus configuration for Navy application (Figure 5-8)

In this design for Naval applications, there are four medium-voltage main generators and four medium-voltage main switchboards. Each ship service switchboard is fed by one dedicated generator. However, one switchboard is connected to another switchboard in a bus tie configuration. This configuration facilitates automatic load transfer in case of loss of voltage in one switchboard. In addition to that, additional bus transfer device can easily be installed to comply with redundant power feeder requirements. This arrangement is well adopted in some classes of Military ship design. This arrangement is known as a ring bus configuration. The physical arrangement of the generators and switchboards in different compartments, and different decks, provide the best possible redundant zonal power distribution.

Each generator is rated at 20 MW. In view of weight and space consideration, the most probable prime movers are gas turbines. The generators and switchboards are both rated at 11 000 V. The generator feeders and switchboard tie feeders and propulsion drive transformer feeders are rated at least 11 000 V.

Each electric propulsion motor is rated at 40 MW. Given today's technology, generators and motors at that power range are not suitable due to the weight and the volume. A good deal of research is underway to increase the power density of generators, motors, transformers and so forth. In this effort, permanent magnets and high temperature superconductor equipment are being considered.

Each propulsion motor is fed from two generators for redundancy and reliability reasons.

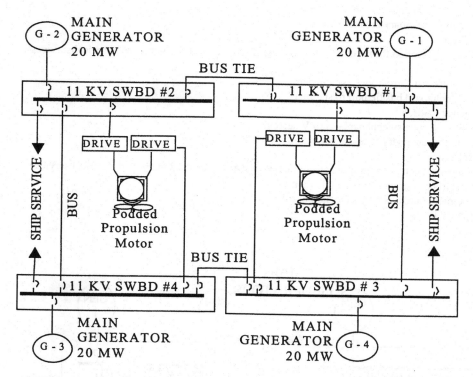

Figure 5-8: Electrical One-Line Diagram – Ring Bus Four Generators and Four Switchboards with Podded Propulsion

5.2.9 Electrical one line diagram—electric propulsion power (6600 V/4580 V/120 V) generation and distribution one-line diagram—ABS R2-S+ (without emergency generator) (Figure 5-9)

The electrical propulsion system power generation, distribution, management, and control features in this design are different from typical ship service power generation and distribution. The ship service power distribution typically provides services to the propulsion system directly couple to the prime movers, directly converting mechanically generated power to the propulsion power. The electric propulsion system drives the propulsion power electrically, particularly by electric propulsion motor. The electric power generation requirement is the power required for the propulsion load plus the power required for the propulsion auxiliaries as well as miscellaneous ship service loads. In this case the propulsion load analysis and the ship service load analysis are to be performed separately. Total combined propulsion load and ship service load determines the power generation requirement for the ship.

Figure 5-129 shows four main generators. The main generator rating is in the megawatts range for supporting propulsion motor load and ship service load and emergency load. The ship service power demand is usually less than 10% of the total power generation. The power distribution categories are as follows:

— Port—6.6 KV propulsion switchboard (medium-voltage system)
   — Propulsion generators #2 and #4 breaker sections
   — Propulsion motor variable speed drive feeder section
   — Medium-voltage load starter section
   — Medium-voltage transformer feeder section
   — Bus tie to the starboard 6.6 kV switchboard
— Starboard—6.6 KV Propulsion switchboard (medium-voltage system)
   — Propulsion generators #1 and #3 breaker sections
   — Propulsion motor variable speed drive feeder section
   — Medium-voltage load starter section
   — Medium-voltage transformer feeder section
   — Bus tie to the port 6.6 kV switchboard

- Port—480 V Ship service switchboard—480 V section
  - 6.6 kV transformer secondary breaker section
  - Motor controllers
  - Ship service breaker section for propulsion auxiliaries, HVAC, deck machinery, and electrical services
  - Shore power
  - Bus tie to the starboard ship service switchboard
  - Ship service switchboard bus tie port to starboard
  - 480 V emergency loads
- Starboard—480 V Ship service switchboard—480 V section
  - 6.6 kV transformer secondary breaker section
  - Motor controllers
  - Ship service breaker section for propulsion auxiliaries, HVAC, deck machinery, and electrical services
  - Bus tie to the starboard ship service switchboard
  - Ship service switchboard bus tie port to starboard
  - 480 V emergency loads

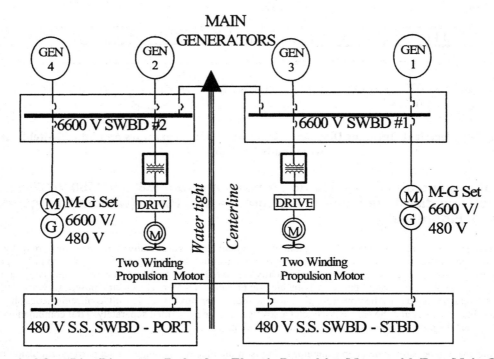

Figure 5-9: Electrical One-Line Diagram – Redundant Electric Propulsion Motors with Four Main Generators and Motor Generator for S. S. System (No Emergency Generator)

5.3   Electric bus transfer systems

In this section, Figure 5-10 – Figure 5-12 show an automatic bus transfer system (ABT), a manual bus transfer system (MBT), and a remote bus transfer system (RBT). These configurations are to ensure redundant power source for critical loads.

5.3.1 Electric power Automatic Bus Transfer system—ABT

The automatic bus transfer switch (ABT) is an automatic power transfer switching device to transfer the power feeder from the main source to an alternate power source when the main source detects low or zero line voltage. The automatic bus transfer switch has the capability of local manual transfer of power to the alternate source.

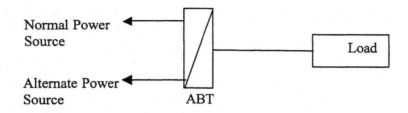

Figure 5-10: Electric Power Automatic Bus Transfer System

5.3.2 Electric power Manual Bus Transfer system—MBT

The manual bus transfer switch (MBT) is capable of supplying redundant power to the load. However, the power transfer scheme is not automatic as with the ABT. Normally, the power is supplied to the load by a normal power source. This can be manually transferred to the alternate source. The manual power source transfer capability is at the transfer switch only.

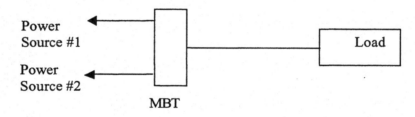

Figure 5-11: Electric Power Manual Bus Transfer Switch—MBT switch

5.3.3 Electric power Remote-Operated Bus Transfer system—RBT

An electric power remote-operated bus transfer switch (RBT) is capable of supplying redundant power to the load. However, the power transferring scheme is not automatic. Normally, the power is supplied to the load by normal power source, which can be manually transferred to the alternate source. The manual power source transfer capability is at the transfer switch as well as at a remote control station.

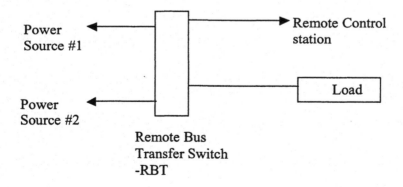

Figure 5-12: Electric Power Remote Operated Bus Transfer System—RBT switch

6 Electrical system load analysis—shipboard

6.1 General

The electrical load analysis is one of the documents required to be submitted to the regulatory bodies, to approve system design. See ABS-2002 Section 4-8-1-5.1.5 for ABS requirement details.

Electrical system load analysis is a detailed electrical load calculation for all ship's operating conditions, for instance:

- At Sea Cruising

- Maneuvering

- Docking/Undocking

- In Port

- Shore Power

As well, any other operating conditions must be accounted for. The load analysis is performed to establish total power requirement for those operations. Once the total power requirement is established, then the amount of power generating equipment and power distribution equipment such as the main generators, emergency generator, transformers, motor generators, shore power, can be selected.

For electrical load analysis, it is necessary to identify all electric power consuming devices, and understand the operating characteristics of those devices. The power consumption demand varies from one operating condition to another condition for the same equipment. The electric equipment service classification is very important for load analysis. The service classifications are *emergency, vital, non-vital, essential, non-essential, sensitive* and *non-sensitive.* The emergency loads are to be connected to the emergency distribution system. In some cases vital loads are also to be connected to the emergency switchboard, and in some cases these vital loads are to be provided with redundant power supplies. In some cases the redundant power supplies are to be provided with an automatic bus transfer (ABT) system, as described in the previous chapter.

Load analysis is performed for major load categories, such as propulsion, propulsion auxiliaries, HVAC system, normal lighting, emergency lighting, navigation, deck machinery, IC & Electronics, and hotel load. The categories are subject to different percentage loading requirements (called *load factor*) to support different loads. There are loads which are to be considered at a 100 percent load factor, and there are loads which are to be considered for less than 100 percent load factors; finally, there are loads with zero load factors for certain operating conditions. The number of generators and their ratings depend on the maximum demand load for the system set-up, such as full speed propulsion, anchor operation, shore side operation etc. The total demand load is the factored load calculated for the different operating requirements and the highest load demand calculated in kilowatts is considered total demand load for the power generation. If there is a requirement of additional load margin for certain growth category loads, that growth margin should be added to the calculated maximum to determine the power generation requirement.

6.2 Power flow and power efficiency

For an electric system consisting of an electric power generation plant, a distribution system, including distribution transformers and a variable speed drive, the power flow is shown in Figure 6-1.

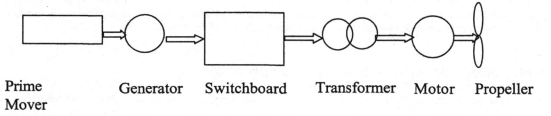

Prime Mover   Generator   Switchboard   Transformer   Motor   Propeller

Figure 6-1: Power Flow in a Simplified Electric Power System

The prime movers, e.g. diesel engines or gas turbines, supply power to the electric generator shaft. The electric motor, such as the propulsion motor, is loaded by power from its connected load. The power lost in the components between the shaft of the diesel engine and the shaft of the electric motor consists of mechanical and electrical losses, which in turn produces heat and temperature increase in the equipment as well as the ambient temperature.

Typical electrical equipment efficiency is

$$Electrical - Efficiency(\eta) = \frac{P_{output}}{P_{input}} = \frac{P_{output}}{P_{output} + P_{losses}}$$

For each of the components, the electrical efficiency can be calculated, and typical values at full (rated) power are

Generator:	$\eta = 0.95\text{-}0.97$
Switchboard:	$\eta = 0.999$
Transformer:	$\eta = 0.99\text{-}0.995$
Frequency converter:	$\eta = 0.98\text{-}0.99$
Electric motor:	$\eta = 0.95\text{-}0.97$

Hence, the efficiency of a diesel electric system, from diesel engine shaft, to electric propulsion motor shaft, is generally between 0.88 and 0.92 at full load. It must be noted that the electrical system efficiency depends on the loading of the system.

Typical electrical plant design starts with an electrical one–line diagram, as described in chapter 5, and electrical load analysis. In that chapter several example electrical one-line diagrams are presented in Figure 5-1 through Figure 5-9. These one-line diagrams are not representative of any specific application. However, they are representative configurations starting from a simple commercial ship with one 480 V ship service generator, to a complex electrical propulsion plant with four medium-voltage generators. These one-line diagrams are used in this chapter to prepare electrical load analysis. The load analysis takes into consideration all power consuming devices, and tabulated under load categories such as propulsion, propulsion auxiliaries, auxiliary machinery, deck machinery, lighting, Heating, Ventilation, and Air-Conditioning (HVAC), Interior Communications, Exterior Communications etc. For simplification, only the summary loads of these load categories are presented to provide better understanding of equipment size selection.

The load analysis is performed to calculate the rating and quantity of the following equipment:

a) Ship service generator, 480 V system (US) and 400 V (IEC),
b) Ship service transformer, 460 V/120 V (US) and 400/230 V (IEC)
c) Emergency generator, 480 V system (US) and 400 V (IEC),
d) Emergency transformer, 460 V system (US) and 400 V (IEC),
e) Propulsion generator for electric propulsion system (high-voltage system)
f) Propulsion motor converter equipment
g) Medium-voltage transformer
h) Shore power requirement

6.3 Kilowatt rating calculation for pump (example only)

Example: Calculate the pump kilowatts, generator kilowatts and engine kilowatts, given that four cargo pumps are to be supported by one diesel generator. This calculation is to show the pump rating calculation and generator rating calculation only. This arrangement may not meet regulatory body requirements. The equipment selection process should be in line with the service requirement of the pump, the pump manufacturer's agreement with the calculation and applicable regulatory approval of the system design.

6.3.1 Kilowatt power calculation—SI units (example only)

General: Calculate kilowatt power requirement of a crude oil pump motor for shipboard application given the following characteristics of the pump:

— Crude oil specific gravity is 1.025, crude oil discharge rating at the discharge manifold is 2150 m³/hr at a pressure head of 112 m.
— Manifold to the pump pressure head is 19 m.
— Pump to the manifold pressure loss is equivalent to 10.5 m.
— The crude oil pump efficiency is 84%.
— Pump manufacturer recommended calculation margin is 10%.

a) Total calculated pressure head is (112m + 19m + 10.5m) = 141.5 m.
b) Kilowatt calculation for crude oil pump

$$= \frac{0.00272 \times 141.5 \times 1.025 \times 100}{84}$$

= 1112.4 kW

c) Given the motor efficiency 97%

The motor rating is 1112.4 kW/0.97 = 1146.7 kW ~ 1200 kW.

d) If there are four crude oil pumps dedicated to a diesel generator, the generator rating:

Total pump load is 1200 kW x 4 = 4800 kW.

Total calculated generator output is 4800 kW

e) Engine rating calculation, for the generator efficiency of 96%

Generator input rating is 4800/0.96 = 5333 kW.

If the engine is to run at 90% maximum continuous rating, the engine kilowatt rating is 5333 kW/0.9 = 5925.9 kW.

6.4 Electrical load analysis (example only)

The examples given below are for guidance only, and not intended to be design restrictive or considered to be a proven design. These illustrations have limited details, mostly to clarify the load analysis issue related to equipment selection process, such as generator and transformer. Refer to Table 6-1: Summary Electrical Load Analysis— Commercial Ship for 480 V Power Generation with One Ship Service Generator and One Emergency Generator (see Figure 6-2 for Electrical One Line Diagram)

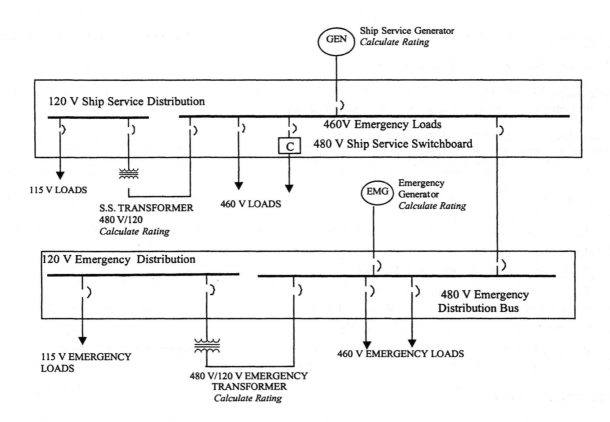

**Figure 6-2: One Line Diagram for Ship Service Power Generation and Distribution
with One Ship Service Generator and One Emergency Generator**

The ship's loads are grouped in seven categories in the table below:

— Row 4: Propulsion machinery
— Row 5: Auxiliary machinery
— Row 6: Cargo equipment
— Row 7: Deck machinery
— Row 8: Hotel loads
— Row 9: Electronics
— Row 10: HVAC equipment

The load groups for different operating profiles are given in columns C through F, as follows:

— Column C: At sea cruising operation
— Column D: Maneuvering operation
— Column E: Anchor and standby service
— Column F: Emergency generator loading

1:A	B	C	D	E	F
2				Anchor and standby	Emer-Connected Load total
		At sea	Maneuvering		
3	Load group	(kW)	(kW)	(kW)	(kW)
4	Propulsion machinery	300	225	124.5	120
5	Auxiliary machinery	80.2	65	43	78
6	Cargo equipment				
7	Deck machinery			18.4	0
8	Hotel loads	122.5	122.5	112	0
9	Electronics	24.9	24.9	15.3	48
10	HVAC equipment	240.8	240.8	145.00	0
11					
12	Total calculated – generator	768.4	678.20	458.2	246.00
13	Total for generator selection	768.4			
14	Gen required margin 15%	883.66			
15					
16	S.S. generator—nameplate (see Note 1)	1000.00			246.00
17					
18	Total – calculated transformer –kVA	250.00			
19					
20	S.S. transformer design margin 10% - kVA	275.00			
21	S.S. transformer nameplate rating (see Note 2) –kVA	300.00			
22					
23	Total calculated emergency generator				246.00
24	Emergency generator nameplate rating (see Note 3)				250

NOTES

1—By calculation, the ship service generator rating is 1000 kW.
2—By calculation, the ship service transformer rating is 250kVA and nameplate rating is 300kVA .
3—The emergency generator calculated rating is 250 kW.

Table 6-1: Summary Electrical Load Analysis— Commercial Ship for 480 V Power Generation with One Ship Service Generator and One Emergency Generator (see Figure 6-2 for Electrical One Line Diagram)

One ship service generator is for the 450 V ship service power service, and one emergency generator is for emergency service. Total generator load, as shown in row 12, for "at sea" operation, is 768.4 kW, for "maneuvering" is 678.2 kW, and for "anchor and standby" operation is 458.2 kW. The maximum load of 768.4 kW is taken under consideration for the selection of generator rating. Generator required margin of 15% (arbitrary) has been set in row 14, for a total load of 883.66 kW. In row 16, the nameplate rating of the generator is selected as 1000 kW in support of the load demand.

The ship service transformer (450 V/120 V) load at the "at sea" condition has been selected as a calculated value of 250kVA (arbitrary) in row 18 and the transformer nameplate rating of 300 kVA is selected to support the 250kVA load demand and it is shown in row 21. The ship service transformer load is the total 120 V ship service load.

The emergency generator total load of 246 kW, is shown in row 23. The emergency generator load is the total connected load calculated kilowatt load with 100% load factor. There is no load margin added for calculating the emergency generator load. The nameplate rating of the generator is 250 kW, which is shown in row 24.

The ship service generator and emergency generator ratings are direct interpolation of calculated loads. However, these generators and their prime movers must be able to sustain transient motor starting loads. During the load calculation, different load factors are taken under consideration. However, the starting transient, which can be six to 12 times the full load current, can be significant and the prime mover may need to supply required power of the connected largest motor connected across the line. If the 1000 kW is not enough for sustaining the largest motor transient starting load the next size generator should be selected.

The emergency generator rating is 250 kW, which is usually much smaller than the ship service generator. The large loads such as emergency fire pump and steering gear motors are connected to the emergency generator. The motor starting transient must be taken under consideration for the selection of emergency generator and engine. Additionally, the emergency generator must be able to pick up all load connected to the emergency bus under emergency situation, that means if there is a black out in the ship service bus, emergency generator must automatically get started and provide power to all emergency loads. Therefore, the emergency generator must be capable of sustaining large block load as required by regulatory bodies.

6.4.1 Electrical load analysis example (mechanical propulsion – direct prime mover driven) (Figure 6-3 and Table 6-2)

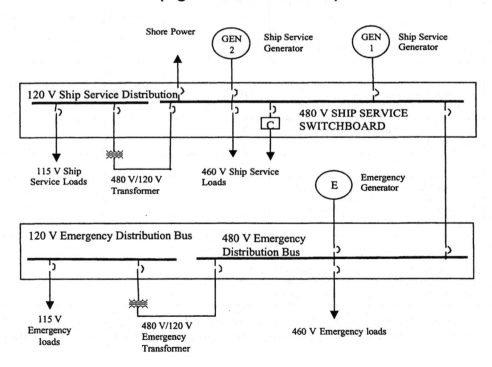

Figure 6-3: One Line Diagram for Ship Service Power Generation and Distribution with Two Ship Service Generators and One Emergency Generator

The ship's loads are grouped in seven categories:

— Row 4: Propulsion machinery
— Row 5: Auxiliary machinery
— Row 6: Cargo equipment
— Row 7: Deck machinery
— Row 8: Hotel loads
— Row 9: Electronics
— Row 10: HVAC equipment

These load groups are assigned to different operating profiles in columns C through G as follows:

— Column C: At sea cruising operation
— Column D: Maneuvering operation
— Column E: Docking and undocking operation
— Column F: Anchor and standby service
— Column G: Emergency generator loading

1:A	B	C	D	E	F		G
2				Docking and undocking	Anchor and standby		
		At sea	Maneuvering				Emer
3	Load group	(kW)	(kW)	(kW)	(kW)		(kW)
4	Propulsion machinery	862.4	907.3	687.2	300.4		194.2
5	Auxiliary machinery	100.1	87.5	84.2	77.6		146.4
6	Cargo equipment						
7	Deck machinery			18.4	18.4		33.2
8	Hotel loads	160.2	165	166.5	165.2		50
9	Electronics	24.9	24.9	24.9	15.3		51
10	HVAC equipment	760.5	554	554	374.7		2.6
11							
12	Total calculated - generators (2)	1908.1	1738.7	1684	951.5		477.4
13	Total for generator selection	1908.1					
14	Generator required margin 15%	2194.32					
15	Generator design margin 6%	2325.97					
16	S.S. generator	2500.00					
17	For each generator—nameplate kW	1250.00			951.5		477.4
18							
19							
20	Total – calculated transformer each -kVA	800.00	738.7	684			
21	S.S. transformer required margin 15% -kVA	920.00					

22	S.S. transformer design margin 6%	975.20						
23	S.S. transformer nameplate rating - kVA	1000.00						
24								
25	Total calculated emergency generator -kW							477.4
26	Emergency generator nameplate rating -kW							500

**Table 6-2: Summary Electric Load Analysis - Commercial Ship with Two Ship Service Generators
and One Emergency Generator (see Figure 6-3 for Electrical One Line Diagram)**

Two ship service generators are for the 450 V ship service power service, and one emergency generator is for emergency service. Total generator load, as shown in row 12, for at sea operation is 1908.1 kW, for maneuvering is 1738.7 kW, for docking and undocking load is 1684 kW, and for anchor and standby operation is 951.5 kW. The maximum "at sea" load of 1908.1 kW is selected for generator rating selection. Generator required margin of 15% (arbitrary) is used in row 14, and a design margin of 6% (arbitrary) is taken under consideration for a total load of 2325.97 kW for the two generators. In row 17, the nameplate rating of each ship service generator is shown as 1250 kW.

The ship service transformer load in the "at sea" condition has been selected arbitrarily as 800 kVA in row 20, and the transformer nameplate rating of 1000 kVA is in row 23 with a 15% required margin and 6% design margin. The ship service transformer load is the total 120 V ship service load.

The emergency generator total load of 477.4 kW is shown in row 25. The emergency generator load is the total connected load calculated with 100% load factor. There is no load margin added for emergency generator. The nameplate rating of the generator is 500 kW, which is shown in row 26.

6.4.2 Electrical load analysis example (mechanical propulsion – direct prime mover driven) (Figure 6-4 and Table 6-3)

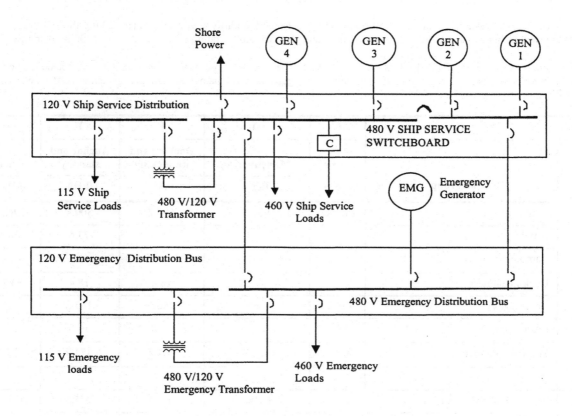

Figure 6-4: One Line Diagram for Ship Service Power Generation and Distribution with Four Ship Service Generators and One Emergency Generator

The ship's loads are grouped in seven categories:

— Row 4: Propulsion machinery
— Row 5: Auxiliary machinery
— Row 6: Cargo equipment
— Row 7: Deck machinery
— Row 8: Hotel loads
— Row 9: Electronics
— Row 10: HVAC equipment

These load groups are assigned to different operating profiles in columns C through G as follows:

— Column C: At sea cruising operation
— Column D: Maneuvering operation
— Column E: Docking and undocking operation
— Column F: Anchor and standby service
— Column G: Emergency generator loading

Four ship service generators are for ship service distribution, and one emergency generator is for emergency power distribution. In this example, one generator is considered a spare generator. The total load is calculated for three generators. Total generator load, as shown in row 12, for "at sea" operation is 2285.2 kW, for "maneuvering" is 1738.7 kW, for "docking and undocking" load is 1684 kW, and for "anchor and standby" operation is 951.5 kW. The maximum "at sea" load of 2285.2 kW is selected for generator rating selection. Generator required margin of 10% (arbitrary) is used in row 14, and a design margin 6% (arbitrary) is used in row 15 for a total load of 2664.54 kW for the three generators. In row 18, the nameplate rating of each ship service generator is shown as 900 kW.

The ship service transformer load at the "at sea" condition has been selected arbitrarily at 900 kVA in row 20 for the transformer load with a 15% required margin (arbitrary) and 6% design margin (arbitrary) gives nameplate rating of the transformer 1200 kVA in row 23.

The emergency generator total load of 427.4 kW is shown in row 25. The emergency generator load is the total connected load calculated with 100% load factor. There is no load margin added for emergency generator. The nameplate rating of generator is 450 kW, which is shown in row 26.

1:A	B	C	D	E	F	G
2		At sea	Maneuvering	Docking and undocking	Anchor and standby	Emer
3	**Load group**	(kW)	(kW)	(kW)	(kW)	(kW)
4	Propulsion machinery	1450.00	907.3	687.2	300.4	194.2
5	Auxiliary machinery	100.10	87.5	84.2	77.6	146.4
6	Cargo equipment					
7	Deck machinery			18.4	18.4	33.2
8	Hotel loads	160.20	165	166.5	165.2	
9	Electronics	24.90	24.9	24.9	15.3	51
10	HVAC equipment	550.00	554	702.7	374.7	2.6
11						
12	Total calculated – generators (3)	2285.20	1738.7	1684	951.5	427.4
13	Total for generator selection	2285.20				
14	Generator required margin 10%	2513.72				
15	Generator design margin 6%	2664.54				
16	S.S. generator - load (3 generators)	2664.54				
17	For each generator—average	888.18				427.4
18	For each generator—nameplate	900.00				
19						
20	Total – calculated transformer each (kVA)	900.00				
21	S.S. transformer required margin 15% (kVA)	1035.00				
22	S.S. transformer design margin 6%	1097.10				
23	S.S. transformer nameplate rating (kVA)	1200.00				
24						

25	Total calculated emergency generator					427.4
26	Emergency generator nameplate rating					450.00

Table 6-3: Summary Electrical Load Analysis with Four Ship Service Generators and One Emergency Generator (see Figure 6-4 for Electrical One Line Diagram)

6.4.3 Electrical load analysis summary (ring bus configuration with no dedicated emergency generator) (Figure 6-5 and Table 6-4)

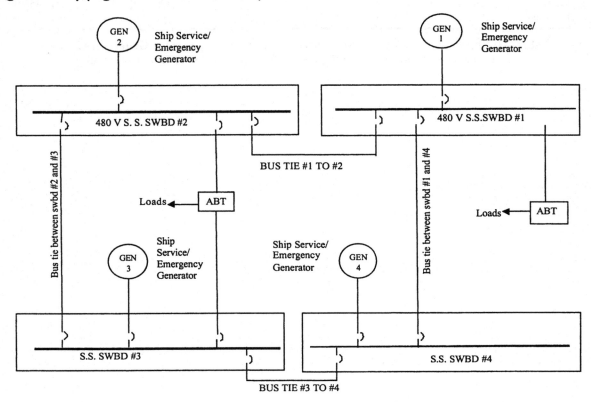

Figure 6-5: One Line Diagram for Ring Bus Configuration with Four Generators in the System and Each Generator Designated as Ship Service/Emergency

The ship's loads are grouped in seven categories:

— Row 4: Propulsion machinery
— Row 5: Auxiliary machinery
— Row 6: Cargo equipment
— Row 7: Deck machinery
— Row 8: Hotel loads
— Row 9: Electronics
— Row 10: HVAC equipment

These load groups are assigned to different operating profiles in columns C through F as follows:

— Column C: At sea cruising operation
— Column D: Maneuvering operation
— Column E: Docking and undocking operation
— Column F: Anchor/ Standby service

There are four ship service generators in a ring bus configuration for ship service and no emergency generator. In this configuration one generator is considered to be spare. Therefore, the total load is for three generators, which is shown in row 12. The "at sea" load is 2217.90 kW, "maneuvering" is 1738.7 kW, "docking and undocking" load is 1684 kW, and "anchor and standby" operation is 951.5 kW. The maximum "at sea" load of 2217.90 kW is selected for generator rating selection. Generator design margin of 10% (arbitrary) is used in row

13, for a total load of 2439.69 kW for the three generators, in row 14. The nameplate rating of each ship service generator is shown as 850 kW in row 17.

There is no dedicated emergency generator in the ring bus configuration. Each generator must have the dead ship starting capability similar to the emergency generator starting. The generators should be identified as ship service/emergency generator. Each generator is assigned to a dedicated switchboard (SWBD). SWBD #1 is connected to SWBD #2 , SWBD #2 is connected to SWBD#3, SWBD # 3 is connected to SWBD # 4 , and SWBD # 4 is connected to SWBD#1 by dedicated bus ties and bus tie breakers. Additionally, the vital loads are fed by two dedicated feeders from two switchboards with automatic bus transfer (ABT) switches. This ring bus arrangement is especially for ship's electrical power generation and distribution redundant configuration requirement. Each generator, with its dedicated switchboard, can be installed in different fire zones and different decks to provide zonal distribution as well as alternate power supply.

1:A	B	C	D	E	F
2		At sea	Maneuvering	Docking and undocking	Anchor and standby
3	Load group	(kW)	(kW)	(kW)	(kW)
4	Propulsion machinery	962.4	907.3	687.2	300.4
5	Auxiliary machinery	200.1	87.5	84.2	77.6
6	Cargo equipment				
7	Deck machinery			18.4	18.4
8	Hotel loads	240.00	165	166.5	165.2
9	Electronics	54.9	24.9	24.9	15.3
10	HVAC equipment	760.5	554	702.7	374.7
11					
12	Totals	2217.90	1738.7	1684	951.5
13	Design margin 10%	2439.69			
14	Generators—load (3)	2439.69	1738.7	1684	951.5
15	Generator—load (1)	813.23			
16					
17	Generator nameplate rating	850.00			

NOTES

1—Total four generators in the system. One generator is considered spare.
2—Each generator has a dedicated switchboard. Each switchboard is connected to two other switchboards by bus tie.
3—The load centers and vital loads will have two power sources by ABT.

Table 6-4: Summary Electrical Load Analysis—Ring Bus Configuration with No Dedicated Emergency Generator (see Figure 6-5 or Electrical Load Analysis)

6.4.4 Electrical load analysis example (commercial ship with four main generators and one emergency generator) (Figure 6-6 and Table 6-5)

Refer to the following figure, 6-6 (electric propulsion—direct electric motor driven).

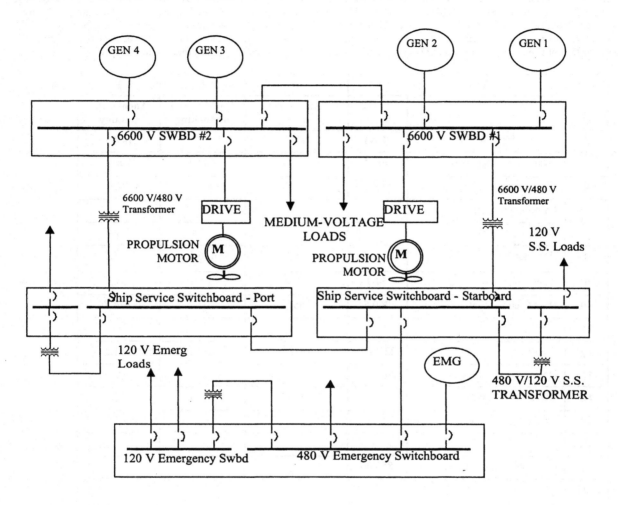

**Figure 6-6: One Line Diagram for Commercial Electrical Propulsion Drive with
Four Main Generators and One Emergency Generator**

The load analysis is performed to determine the calculated kilowatt ratings and nameplate ratings of four 6600 V main propulsion generators, KVA rating for two 6600 V/480 V ship service transformers, the kilowatt rating for 480 V emergency generator and the KVA rating of 480 V/120 V emergency transformer.

The ship's loads are grouped in seven categories:

— Row 6: Propulsion machinery
— Row 7: Auxiliary machinery
— Row 8: Cargo equipment
— Row 9: Deck machinery
— Row 10: Hotel loads
— Row 11: Electronics
— Row 12: HVAC equipment

These load groups are assigned to different operating profiles in columns C through K as follows:

— Columns C and D: At sea cruising operation
— Columns E and F: Maneuvering operation
— Columns G and H: Docking and undocking operation
— Columns I and J: Anchor and standby service
— Column K: Emergency generator loading

The load analysis is performed for summer operation as well as winter operation to establish the loading requirement for the worst possible operating scenario. Each load group and its corresponding service total kilowatt value is the summary value extrapolated from its relationship database where each and every load is evaluated with its load factor. The propulsion machinery group total "at sea" connected 6600 V load is 32 761.6 kW (see row 6). The propulsion machinery group total "at sea" connected 460 V load is 862.4 kW (See row 6).

Total medium-voltage load for "at sea" operation (as shown in row 14) is 32 761.6 kW. This medium-voltage load (32 761.7 kW) plus the 460 V ship service load (1908.1 kW) for "at sea" operation for a total of 34 669.7 kW. This value is the factored service load accumulated value for the four main propulsion medium-voltage generators. Therefore, the output rating of each generator is 8667.43 kW as shown in row 19 .

For the 6600 V/480 V ship service transformer, the "at sea" calculated maximum load requirement is 1908.10 kW, which is shown in row 21, which is equivalent to 2650.1kVA. For transformer calculated rating of 2650.1kVA, the nameplate rating 3000kVA is selected as shown in row 27.

The emergency generator loads are given in column K. The total emergency load is 677.3kW, given in row 14. The nameplate rating of the emergency generator is selected for 700 kW as shown is row 23.

1:A	B	C	D	E	F	G	H	I	J	K
2		At sea		Maneuvering		Docking and undocking		Anchor and standby		Emer
3		Bus		Bus		Bus				
4		MV	LV	MV	LV	MV	LV	MV	LV	Bus
5	Load group	(kW)	(kW)	(kW)	(kW)	(kW)	(kW)	(kW)	(kW)	(kW)
6	Propulsion machinery	32761.6	862.4	14819.7	907.3	8359.5	687.2	19.6	300.4	394.2
7	Auxiliary machinery		100.1		87.5		84.2		77.6	146.4
8	Cargo equipment							459		
9	Deck machinery						18.4		18.4	33.2
10	Hotel loads		160.2		165		166.5		165.2	50
11	Electronics		24.9		24.9		24.9		15.3	51
12	HVAC equipment		760.5		554		702.7		374.7	2.6
13										
14	Totals	32761.6	1908.1	15278.7	1738.7	8818.5	1684	478.6	951.5	677.3
15	Total (LV)		1908.1		1738.7		1684		951.5	
16	Total (MV + LV)	34669.7		17017.4		10502		1430.1		
17										
18	Main generators (4)	34669.7		17017.4		10502				

1:A	B	C	D	E	F	G	H	I	J	K
2		At sea		Maneuvering		Docking and undocking		Anchor and standby		Emer
3		Bus		Bus		Bus				
4		MV	LV	MV	LV	MV	LV	MV	LV	Bus
5	Load group	(kW)	(kW)	(kW)	(kW)	(kW)	(kW)	(kW)	(kW)	(kW)
19	Main generator (1)	8667.4		8508.7		5251.2				
20										
21	S.S. transformers (2)		1908.1		1738.7		1684			
22										
23	Emergency generator (1)									700
24										
25	Maximum S.S. transformer load	1908.1		1738.7		1684				
26	Maximum S.S. transformer load (kVA)	2650.1		2414.9		2338.8				
27	Max. S.S. transformer load (kVA)	3000								

Table 6-5: Summary Electrical Load Analysis—Commercial Ship Electric Propulsion System with Four Main Generators and One Emergency Generator (see Figure 6-6 for Electric Load Analysis)

6.4.5 Electrical load analysis (electric propulsion with no emergency generator) (Figure 6-7 and Table 6-6)

Figure 6-7 (electric propulsion—direct electric motor driven) shows an example of 6600 V, 480 V, and 120 V systems for electric propulsion with four main generators (6600 V), propulsion power distribution, stepping down to 460 V distribution and stepping down to 120 V distribution with no emergency generator.

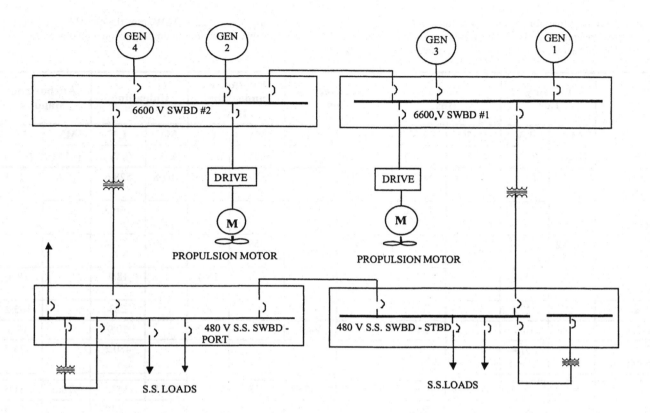

Figure 6-7: One Line Diagram for Commercial Ship Electrical Propulsion System with No Emergency Generator

The load analysis is performed to determine the calculated ratings and nameplate ratings of four 6600 V main propulsion generators, two 6600 V/480 V ship service transformers.

The ship's loads are grouped in seven categories:

— Propulsion machinery
— Auxiliary machinery
— Cargo equipment
— Deck machinery
— Hotel loads
— Electronics
— HVAC equipment

These load groups are assigned to different operating profiles as follows:

— Loading
— At sea cruising operation
— Maneuvering operation
— Docking and undocking operation
— Unloading
— Anchor and standby service

The load analysis is performed for summer operation and winter operation to establish the loading for the worst possible operating scenario; however, this analysis is an arbitrary example only. Each load group and its corresponding service total kilowatt value are the summary value extrapolated from its relationship database where each and every load is evaluated with its load factor. The propulsion machinery group total "at sea" connected 6600 V load is 32 761.6 kW. The propulsion machinery group total "at sea" connected 460 V load is 862.4 kW.

Total medium-voltage load for "at sea" operation is 33220.60 kW. This medium-voltage load (33220.60 kW) plus the 480 V load (1908.1 kW) for "at sea" operation totals 35128.70 kW and is the factored load accumulated value for the four main propulsion medium-voltage generators. Therefore, the output rating of each generator is 8782.20 kW.

Load group	Loading Bus MV (kW)	Loading Bus LV (kW)	At sea Bus MV (kW)	At sea Bus LV (kW)	Maneuvering Bus MV (kW)	Maneuvering Bus LV (kW)	Docking and undocking Bus MV (kW)	Docking and undocking Bus LV (kW)	Unloading MV (kW)	Unloading LV (kW)	Anchor and standby MV (kW)	Anchor and standby LV (kW)
Propulsion machinery		191.6	32 761.6	862.4	14 819.7	907.3	8359.5	687.2	107.8	528.6	19.6	300.4
Auxiliary machinery		79		100.1		87.5		84.2		166.5		77.6
Cargo equipment	2355.2		459		459		459		459		459	
Deck machinery		18.4						18.4		18.4		18.4
Hotel loads		227.1		160.2		165		166.5		223		165.2
Electronics		16.9		24.9		24.9		24.9		20.9		15.3
HVAC equipment		347.4		760.5		554		702.7		480.2		374.7
Totals	2355.2	880.4	33 220.6	1908.1	15 278.7	1739	8818.5	1684	566.8	1438	478.6	951.5
Total (LV)		880.4		1908.1		1739		1684		1438		951.5
Total (MV)	3235.6		35 128.7		17 017.4		10 502.4		2004.3		1430.1	
Main generators (4)			35 128.7		17 017.4		10 502.4		2004.3			
Main generator (1)	3235.6		8782.2		8508.7		5251.2					
S.S. transformers (2)		880.4		1908.1		1739		1684		1438		
Maximum main generator load			8782.2		8508.7		5251.2					
Maximum S.S. transformer load			1908.1		1738.7		1684		1437.5			
Maximum S.S. transformer load (kVA)			2650.1		2414.9		2338.8		1996.5			

Table 6-6: Summary Electrical Load Analysis - Commercial Electric Propulsion System with No Emergency Generator (see Figure 6-7 for Electrical One Line Diagram)

6.4.6 Electrical load analysis (electrical propulsion system with four main generators and one emergency generator) (Figure 6-8 and Table 6-7)

The following figure, 6-8 (electrical propulsion—direct electric motor driven) shows an example of 6600 V, 480 V, and 120 V systems for electric propulsion with four main generator (6600 V) propulsion power distribution, stepping down to 480 V distribution and stepping down to 120 V distribution, and with one emergency generator for 480 V and 120 V emergency distribution for ABS R2-S+ redundancy.

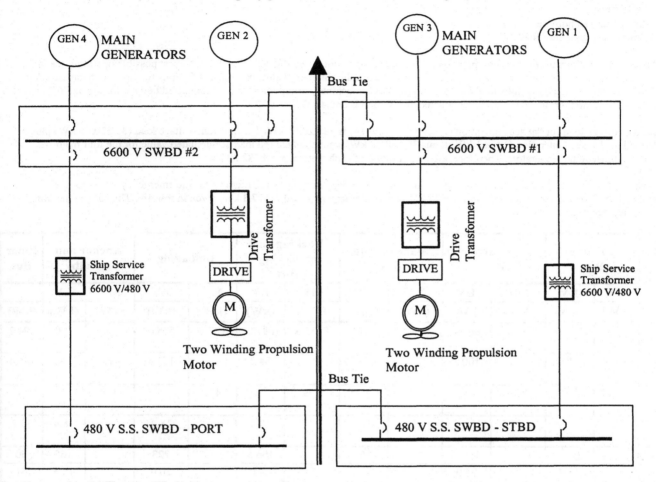

Figure 6-8: One-Line Diagram for Electric Propulsion System with Four Main Generators and One Emergency Generator—ABS Class R2-S+ Redundancy

The load analysis is performed to determine the calculated ratings and nameplate ratings of four 6600 V main propulsion generators, two 6600 V/480 V ship service transformers, the 480 V emergency generator and the 480 V/120 V emergency transformer. The 480 V emergency generator and the 480 V/120 V emergency transformer configuration are not shown in Figure 6-8.

The ship's loads are grouped in seven categories:

— Propulsion machinery
— Auxiliary machinery
— Cargo equipment
— Deck machinery
— Hotel loads
— Electronics
— HVAC equipment

These load groups are assigned to different operating profiles as follows:

— Loading
— At sea cruising operation
— Maneuvering operation
— Docking and undocking operation
— Unloading
— Anchor and standby service
— Emergency generator loading

The load analysis is performed for summer operation and/or winter operation to establish the loading for worst possible operating scenario. Each load group and its corresponding service total kilowatt value are the summary value extrapolated from its relationship database where each and every load is evaluated with its load factor. The propulsion machinery group total "at sea" connected 6600 V load is 32 761.6 kW. The propulsion machinery group total "at sea" connected 480 V load is 862.4 kW.

Total medium-voltage load for "at sea" operation (as shown in row 14) is 33 220.60 kW. This medium-voltage load (33 220.60 kW) plus the 480 V load for "at sea" operation (1908.1 kW) totals 35 128.70 kW and is the factored load accumulated value for the four main propulsion medium-voltage generators. Therefore, the output rating of each generator is 8782.20 kW.

The emergency generator load calculation in shown in Column O. The regulatory body requirement for the emergency generator load calculation is to use 1.0 load factor for all connected loads. The total emergency load is 677.3 kW, given in row 14 . The nameplate rating of the emergency generator is 750 kW.

Load group	Loading Bus		At sea Bus		Maneuvering Bus		Docking and undocking Bus		Unloading		Anchor and standby		Emer Bus
	MV	LV	MV	LV	MV	LV	MV	LV	MV	LV	MV	LV	
	(kW)	(kW)	(kW)	(kW)	(kW)	(kW)	(kW)	(kW)	(kW)	(kW)	(kW)	(kW)	(kW)
Propulsion machinery		191.6	32 761.6	862.4	14 820	907.3	8359.5	687.2	107.8	528.6	19.6	300	394.2
Auxiliary machinery		79		100.1		87.5		84.2		166.5		77.6	146.4
Cargo equipment	2355.2			459		459		459		459		459	
Deck machinery		18.4						18.4		18.4		18.4	33.2
Hotel loads		227.1		160.2		165		166.5		223		165	50
Electronics		16.9		24.9		24.9		24.9		20.9		15.3	51
HVAC equipment		347.4		760.5		554		702.7		480.2		375	2.6
Totals	2355.2	880.4	33 220.6	1908.1	15 279	1739	8818.5	1684	566.8	1437.5	478.6	952	677.3
Total (LV)		880.4		1908.1		1739		1684		1437.5		952	
Total (MV)	3235.6		35 128.7		17 017		10 502.4		2004.3		1430.1		
Main generators (4)			35 128.7		17 017		10 502.4		2004.3				
Main generator (1)	3235.6		8782.2		8509		5251.2						
S.S. transformers (2)		880.4		1908.1		1739		1684		1437.5			
Emergency													750

Load group	Loading Bus		At sea Bus		Maneuvering Bus		Docking and undocking Bus		Unloading		Anchor and standby		Emer Bus
	MV (kW)	LV (kW)	MV (kW)	LV (kW)	MV (kW)	LV (kW)	MV (kW)	LV (kW)	MV (kW)	LV (kW)	MV (kW)	LV (kW)	(kW)
generator – Rating in KW (1)													
Maximum main generator load			8782.2	-4	8509	-2	5251.2						
Maximum S.S. transformer load			1908.1		1739		1684		1437.5				
Maximum S.S. transformer load (kVA)			2650.1		2415		2338.8		1996.5				

Table 6-7: Electrical Load Analysis for Electrical Propulsion System with Four Main Generators and One Emergency Generator—ABS Class R2-S+ (see Figure 6-8 for Electrical Load Analysis)

6.5 Example of typical electrical loads

The following examples of representative electric loads should be taken into consideration when performing electrical system load analysis.

6.5.1 Propulsion plant options

The following are examples of propulsion plant options:

— Fixed pitch propeller (FPP) directly driven by prime mover
— Controllable pitch propeller (CPP) directly driven by prime mover
— Fixed pitch propeller (FPP) directly driven by electric motor and variable frequency drive
— Podded propulsion with fixed pitch propeller (FPP) directly driven by electric motor and variable frequency drive

For the total fuel efficiency, one must consider the hydrodynamic efficiency of the propeller and mechanical transmission, and the fuel consumption of the prime mover . The hydrodynamic losses vary significantly, depending on the operational condition for a controllable pitch propeller (CPP) used in direct driven diesel solutions compared to fixed pitch propellers (FPP). In low load condition, it is a rule of thumb that the zero-load hydrodynamic losses for a CPP is about 15%, while it is close to 0 for a speed controlled FPP. In most CPP configurations, the propeller speed has to be kept constant on quite high rotations per minute (RPM) even though the thrust demand is zero. For FPP, the variable speed drive will allow zero RPM at zero thrust demand. The advantage with CPP is that the propeller pitch ratio will be hydro-dynamically optimized for a wider speed range. A propeller designed for high transit speed, will have reduced efficiency at low speed and vice versa.

6.5.2 Main engine lube oil service pumps are considered vital auxiliaries

There are three pumps, one pump for each engine, with the third pump as standby. During at-sea full power cruising conditions, the lube oil pressure must be maintained. The load factor for these pumps should be close to 1.0 depending on the selection of the pump for the application. The requirement of the third pump is stand-by for the min pumps, which should take the load when one of the running pumps goes down. Therefore the third pump load factor for the load analysis can be considered zero. For both auxiliary equipment -- normal as well as emergency -- operating load profile understanding is very important to perform electrical system load analysis.

6.5.3 Steering gear system

In accordance with regulatory body requirements, two redundant steering gear pumps must be installed. One steering gear pump must be connected to the emergency switchboard. For emergency generator power rating calculation the load factor for the steering gear pump must be considered 1.0. The other steering gear pump can be considered stand-by and connected to the ship service generator.

6.5.4 Emergency generator

For calculating the emergency generator rating, all connected load to the emergency switchboard to be calculated at 1.0 load factor per USCG 46CFR subpart 112.15. Additionally, the emergency generator and engine should be sized for 100% connected load, and the engine should be able to start and pick up 100% connected load in an event of ship service power blackout.

6.5.5 Emergency fire pump

The emergency fire pump load factor can be considered around 0.5 for ship service switchboard, but for emergency generator rating the same emergency fire pump load factor must be 1.0.

6.5.6 120 V ship service loads

The 120 V ship service loads with appropriate load factor will contribute to the selection of ship service transformers as well as ship service generators. All emergency 120 V loads are to be considered with 1.0 load factor for emergency transformer and the emergency generator.

6.5.7 Electric shore power

Electrical shore power is the power demand to support ship service load when the ship is at the pier side. Grand total of those loads with appropriate load factor is the shore power requirement for this ship. The shore power facility must be selected to support those loads.

6.6 Generator rating calculations-block diagram approach (refer to Figure 6-9 and Table 6-8)

Table 6-8 is a simplified format to calculate generator rating by using the efficiencies and losses in the system.

Figure 6-9 represents a single propeller high-speed electric propulsion motor and two main generators. It is assumed that the propulsion power demand is 10 000 kW, 15% at sea load margin, and the system losses as shown in Table 6-8. The calculated rating of each generator is 6957.69 kW and the engine is 7279.65 kW. The theoretical kilowatt ratings are to show the calculation process only. The engine calculated kilowatt rating leads to manufacturer's next matching higher rating and then the generator size is selected to match the engine size.

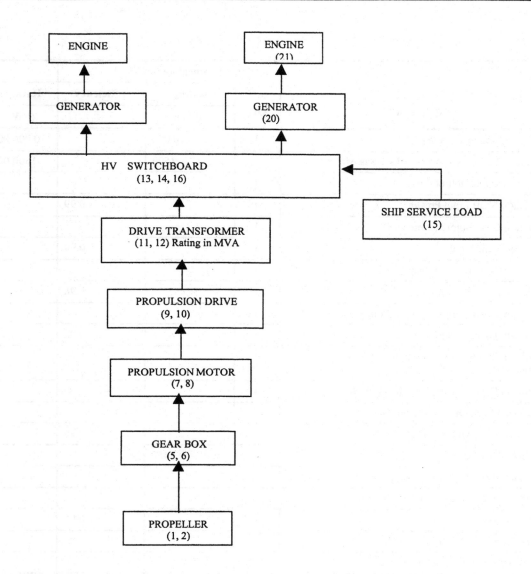

Figure 6-9: Typical Block Diagram for Electric Propulsion System—Electrical Load Calculation

Item #	Equipment	Qty	Nameplate Rating (kW)	Efficiency	Rating (kW	Total Rating (kW)
1	Propeller Calculated Rating	1	10 000		10 000.00	10 000.00
2	Additional Propeller Power for 15% Sea Margin	1	1500		1500.00	1500.00
3	Propulsion Shaft Output Power	1				11 500.00
4	Propulsion Shaft Input Power	1		0.99		11 616.16
5	Reduction Gear Output (same as 4)	1				11 616.16
6	Reduction Gear Input	1		0.99		11 733.50
7	Propulsion Motor Output (same as 6)	1				11 733.50
8	Propulsion Motor Input Power	1		0.98		11 972.96
9	Propulsion Drive Output Power (same as 8)	1				11 972.96
10	Propulsion Drive Input Power	1		0.98		12 217.31
11	Propulsion Drive Transformer Output Power (same as 10)	1				12 217.31
12	Propulsion Drive Transformer Input Power	1		0.99		12 340.72
13	Switchgear Transmission Output (same as 12)					12 340.72
14	Switchgear Transmission Input			0.99		12 465.37
15	Ship Service Load		1450			1450.00
16	Switchgear Load Input					13 915.37
17	Generator Output for Two (same as 16)					13 915.37
18	Generator Input for Two			0.99		14 055.93
19	Engine Output for Two (same as 18)					14 055.93
20	Generator Output Rating Each					6957.69
21	Engine Output Rating Each					7279.65

NOTES

1—Propulsion motor rating is 11 733.50 kW (see item 7).
2—Main generator rating 6957.69 kW (see item 20).
3—Engine output rating 7279.65 kW (see item 21).

Table 6-8: Generator Rating Calculations—Electric Propulsion (Single Shaft) Load Calculation (see Figure 6-9 for the block diagram, showing the item numbers)

7 Voltage drop calculations – shipboard electrical system

7.1 General

The line voltage drop calculation for an electrical distribution system is required to ensure that the voltage drop is within the limits set by regulatory bodies. The AC power generation of 120 V, 480 V, 4160 V, 6600 V, and 13 800 V corresponds to the utilization voltage of 115 V, 440 V, 4000 V, 6000 V, and 13 200 V respectively. The difference is usually the transmission line drop of the generating system. Additionally, sudden load application-related transient voltage drop, and the associated recovery time, are also important factors to consider in ensuring continuity of power at the load.

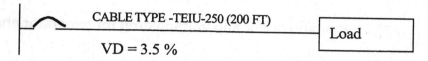

CABLE TYPE -TEIU-250 (200 FT)

VD = 3.5 %

Load

Figure 7-1: Feeder One-Line Diagram for Voltage Drop (example only)

7.2 Voltage drop calculations

The line voltage drop calculation for an electrical distribution system is required to make sure that the line voltage drop is within the limits set by regulatory body requirements.

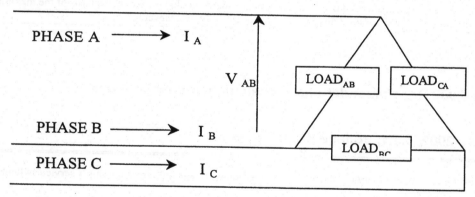

Figure 7-2: Schematic for Three Phase Load Voltage Drop Calculation

The voltage drop calculations are as follows:

a) For three-phase AC distribution system:

$$VD\%(3Phase) = \frac{100x1.732\{I\max \, xLength(cablelength(ft)xZ(Cable\,\mathrm{Im}\,pedence)xCos\theta}{E(Volt_{L-L})}$$

b) For single-phase AC distribution system:

$$VD\%(SinglePhase) = \frac{100x2\{I\max \, xLength(cablelength(ft)xZ(Cable\,\mathrm{Im}\,pedence)xCos\theta}{E(Volt_{L-L})}$$

c) For DC distribution system

$$VD\%(DC) = \frac{100x2\{I\max \, xLength(cablelength(ft)xR(Cable\,\mathrm{Re}\,sis\,\tan ce)}{E(Volt_{L-L})}$$

A typical voltage drop calculation is shown in Table 7-1 for information: (not a calculated value)

Circuit #	Cable type	Cable length (ft)	Total voltage drop (%)	Reference
4P- 423	TEIU-250	200	3.5	Figure 7-1

Table 7-1: Typical Voltage Drop Calculations

7.2.1 IEEE Std 45-1977, 21.11, Demand factor and voltage drop for generator and bus tie circuits (extract)

The conductors from each generator to the generator switchboard should be calculated for not less than 115 percent of the generator rating for continuous rated machines and 15 percent above the overload rating for 125 percent 2hour overload or special rated machines.

Conductors between separate ship service switchboards having connected generators (ship service generator switchboards) should be sized on the basis of 75 percent of the switchboard having the greatest generating capacity. The drop in voltage from each generator to its switchboard should not exceed 1%, and the drop in voltage between switchboards should not exceed 2%.

Conductors between ships service switchboards and the emergency switchboard should be sized on the basis of the maximum operating load of the emergency switchboard or 125% of the emergency generator capacity, whichever is larger.

(The same requirement is in IEEE-1983, 21.11)

7.3 IEEE Std 45-2002 Recommendations for calculating voltage drops

7.3.1 IEEE Std 45-2002, 5.5.1, Voltage drop—general

For all distribution circuits, unless stated otherwise, the combined maximum voltage drop from the ship's service switchboard to any point in the system should not exceed 5%.

7.3.2 IEEE Std 45-2002, 5.5.2, Feeder and branch circuit continuity

Except as permitted in 5.5.3 [see 8.2.3 of this Handbook] and in 5.6.1 [not quoted in this Handbook], each feeder and branch circuit supplying a single energy -consuming appliance should be continuous and uniform in size throughout its length. In instances of feeders of large size and exceptional length, junction boxes or splices may be used for ease of installation. When authorized by the appropriate regulatory agency, splices or junction boxes may be permitted to repair damaged cable (see 25.11 [not quoted in this Handbook]).

7.3.3 IEEE Std 45-2002, 5.5.3, Feeder connections

Where a feeder supplies more than one distribution panel, it may be continuous from the switchboard to the farthest panel or it may be interrupted at any intermediate panel. If the bus bars of any distribution panel carry "through" load, the size of the buses should be suitable for the total current. The size of feeder conductors will ordinarily be uniform for the total length but may be reduced at any intermediate distribution panel, provided that the smallest section of the feeder is protected by the overload device at the distribution switchboard.

IEEE Std 45-2002		ABS-2002		USCG CFR-46	
Reference	**Voltage Drop**	**Reference**	**Voltage Drop**	**Reference**	**Voltage Drop**
Table 4 (in 4.5)	Frequency ± 3% Voltage ± 5%	4/8/2–7.7.1(d)	6%	110.01–17	Frequency ± 5% Voltage +5% to –10%
5.5.1	Voltage drop not to exceed 5%	4/8/3–1.9	Frequency: ± 5% Permanent and ± 10% Transient (5 sec) Voltage: +6% to –10% Permanent and ± 20% (1.5 sec)		
25.12	Propulsion cable voltage drop not to exceed 5%	4/8/4–21.4	6%		

Table 7-2: Voltage Drop Calculations

8   Short-circuit analysis—shipboard electrical power system

8.1   General

One of the most important criteria in electric power system design is to ensure continuous operation of the electrical system under normal operating condition, as well as ensuring the survival of the electrical system under stresses such as electrical fault in the system. The basic current flow relation is system voltage divided by line resistance. Under line fault condition the resistance level diminishes to a very low level, making allowance for very high level current to flow. The short circuit conditions and related current flow must be calculated so that appropriate measures can be taken to disable the downstream line from the service, eliminating collateral damage and protecting the upstream distribution from fault propagation. Appropriate protective measures are to use power equipment with appropriate fault-withstanding capability, and to use appropriate protective devices and settings to provide fault current protection.

The system level fault current (over current) protection is provided by using circuit breakers, contactors and fuses. The circuit breaker is a switching mechanism capable of making, carrying and breaking normal current for a specific time. The circuit breaker is also capable of breaking current under abnormal condition without causing damage to the breaker within its operating range. The circuit breaker must be able to switch repeatedly under various conditions. The following definitions are provided to give you a better understanding of the fault current behavior:

a)   **Symmetric current**: A periodic alternating current in which point ½ a period apart are equal and have opposite sign

b)   **Asymmetric current**: Combination of the symmetric current plus the direct current component of the current

c) $I_{Average}$ is the average of the maximum asymmetrical rms current of the three-phase system fault at ½ cycle measured in amp.

d) I_{max} is the maximum value of short circuit current at ½ cycle in the phase having the max asymmetrical current.

e)   **Symmetrical fault current**: The symmetric fault current is rms current, which is identified as $I_{symmetrical}$. $I_{symmetrical}$ is the initial available symmetrical short circuit current, determined at ½ cycle after inception of the fault. This value is the peak-to-peak value divided by $2 * \sqrt{2}$.

f) I_{min} is the minimum available rms asymmetrical current at the point of application of each circuit breaker.

The short-circuit current is found by numerical simulation or by analytical methods, as required by IEC Publication -600363 and similar standards and publications. In the United States, commercial standards, commercial software, and Navy Department of Defense (DOD) design data sheets are available for calculating the fault current.

The USCG CFR46-111.52-5 fault current calculation requirement is as follows:

Short-circuit calculations must be submitted for systems with an aggregate generating capacity of 1500 kW or more by utilizing the following methods:

a)   Exact calculations using actual impedance and reactance values of system components

b)   Estimated calculations using the Naval Sea Systems Command Design Data Sheet DDS 300-2

c)   Estimated calculation using IEC Publication 600363

d)   Estimated calculations using a commercially established analysis procedure for utility or industrial applications.

8.2   Short-circuit calculation—modeling of components in electric power generation and distribution

For short circuit calculation of electric components in the electric power generation and distribution system, simplified models consisting of voltage sources and impedances are used. For most calculations, linear models are used, which means that the discrete voltage sources and impedances are regarded to be constant, independent on voltage, current, and frequency. For more accurate analysis, one might need to model non-linear phenomena, such as magnetic saturation of inductors and frequency dependent resistor.

8.2.1   Short circuit generator model

The generator is constructed of a rotating magnetization winding on the rotor, which also includes damper windings for dynamic stability. The stator consists of three-phase windings connected to a common neutral point. The induced voltage in the windings is proportional to the rotational speed of the magnetic flux (in stationary, equal to the rotational speed of the rotor) and the magnitude of the magnetic flux. The total magnetic flux in the stator windings consists of magnetic flux from the magnetizing current in the rotor, and armature reaction, i.e. magnetic flux set up by the currents in the stator windings.

For accurate modeling of the generator, a time domain model must be applied. However, for most analysis, one represents the generator as a voltage source behind an inductor and a resistor. The inductance represents the inductive behavior of the windings and the resistance represents the ohmic losses in the windings. Usually, the resistance is small and can be neglected. One can distinguish between stationary, transient, and sub-transient models of the generator, because this simplified method of modeling neglects dynamic behavior of the generator, and tries to represent it with stationary models. For fast transients, in the order of 10 ms or shorter, the sub-transient model is applied.

Typically, sub-transient models are used to find peak values of short circuit currents (in first period after short circuit), while transient models are used for calculating voltage variations during start-up of motors and transformers, and stationary models are used to calculate stationary load flow. The motivation is to give an engineering tool which is representative and sufficiently simple for calculation by analytic means. Figure 8-1 below shows such a three-phase equivalent diagram for a generator. Physically, the induced voltages represents magnetic flux that is characteristic and can be assumed to be quasi-constant for the transient period of interest, and the stator inductance is representative for the dynamically induced voltages in this time period. It must be noted that the diagram is simplified by disregarding the effects of saliency in the rotor due to the non-cylindrical construction. Also, the capacitive coupling between the windings and the ground is neglected--for load flow calculations this is acceptable.

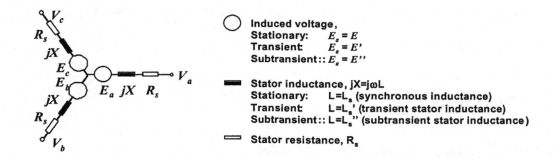

Figure 8-1: Model of Generators (resistances are normally neglecte d – small, assumed zero)

The voltage and currents are phase shifted. In the three phases, it is normal to draw the diagrams for the components in one of the phases, a so-called per phase diagram, as for the generator is shown in Figure 8-2 below.

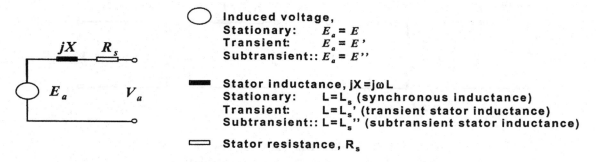

Figure 8-2: Per Phase Diagram of Generator

The phasor diagram will then be as shown in Figure 8-3 below.

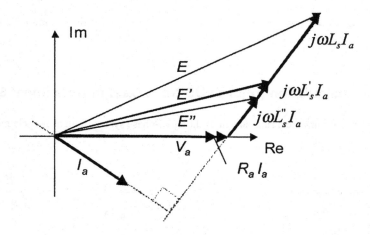

Figure 8-3 Phasor Diagram for Generator, Phase A Only.
(The diagram is drawn for an arbitrary inductive current with power factor 0.8 [phase angle j=36.8°])

The stator voltage equals to the induced voltage minus voltage drops over stator inductance and resistance, i.e.,

$$E = V_a + R_s I_a + j\omega L_s I_a$$

$$E' = V_a + R_s I_a + j\omega L_s' I_a$$

$$E'' = V_a + R_s I_a + j\omega L_s'' I_a$$

8.2.2   Short circuit calculation—transformer model

The purpose of the transformer is normally to convert between different voltage levels. It also provides possibilities for isolating different parts of the system in order to reduce the spreading of electromagnetic and harmonic noise, and to operate with different earthing philosophy in different parts of the power system.

The equivalent per phase diagram for the transformer is shown in Figure 8-4 below.

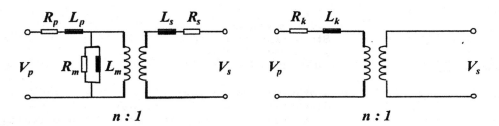

Figure 8-4: Equivalent Diagrams for Transformer.
(a) Left: Complete diagram for loss calculations
(b) Right : Simplified diagram for load flow and transient calculations.

For an accurate calculation of losses, the complete diagram should be used, which includes winding losses in stator primary and secondary (R_p and R_s), and magnetizing losses (R_m). L_p and L_s represents the so-called leakage inductance in primary and secondary windings, while L_m represents the magnetizing inductance.

From a load flow calculation, the magnetizing currents and losses are negligible and one can use the simplified diagram. Here, the primary and secondary resistances and inductances are represented by common equivalents, with subscript k.

The voltage transformation ratio is described by the *n:1* ratio, e.g. a 6600 V to 460 volt transformer has *n= 14.35*.

8.3   Short circuit current calculation (refer to US Navy Design Data Sheet 300-2 for details)

I_{Average} is the average of the maximum asymmetrical rms current of the three-phase system fault at ½ cycle measured in amp. The equation is

$$I_{AVE} = \frac{0.71\, Eg}{\sqrt{\left(\dfrac{0.7\, Eg}{Im} + R_C\right)^2 + \left(\dfrac{0.19\, Eg}{Im} + X_C\right)^2}} + \frac{K_1\, Eg}{\sqrt{R^2 + X^2}}$$

I_{max} is the maximum value of short circuit current at ½ cycle in the phase having the max asymmetrical current

$$I_{MAX} = \frac{0.81\,Eg}{\sqrt{\left(\dfrac{0.7\,Eg}{\mathrm{Im}} + R_C\right)^2 + \left(\dfrac{0.19\,Eg}{\mathrm{Im}} + X_C\right)^2}} + \frac{K2\,Eg}{\sqrt{R^2 + X^2}}$$

$I_{symmetrical}$ is the initial available symmetrical short circuit current, determined at ½ cycle after inception of the fault. (The rms value is equal to the peak-to-peak value divided by 2)

$$I_{SYM} = \frac{0.63\,Eg}{\sqrt{\left(\dfrac{0.7\,Eg}{\mathrm{Im}} + R_C\right)^2 + \left(\dfrac{0.19\,Eg}{\mathrm{Im}} + X_C\right)^2}} + \frac{Eg}{\sqrt{R^2 + X^2}}$$

I_{min} is the minimum available rms asymmetrical current at the point of application of each circuit breaker. The I min current should be determined on the basis of one generator of the lowest rating connected to the system. The motor contribution should be negligible. For simplification K2 is considered to be 1.

$$I_{MIN} = \frac{K2_1\,Eg}{\sqrt{R^2 + X^2}} \qquad\qquad I_{MIN} = \frac{Eg}{\sqrt{R^2 + X^2}}$$

8.3.1 Generator per unit base calculations

Select Base Voltage (Line voltage), for 460 V, three-phase system, where the line voltage is 460/1.732 = 256.7 V

Select Base Power (Per Phase kVA).

$$Z_{Base(\Omega)} = \frac{(E_g)^2}{KVA/3}, \quad X(\Omega) = (Xpu) \times (Z\,base), \quad R(\Omega) = (R\,pu) \times (Zbase)$$

Example 8-1: Ship service generator rated 2000 kW, 460 V, three-phase, 60 Hz, 0.8 pf. Synchronous reactance 0.88 pu and resistance 0.042 pu

Calculate (1) Z_{Base} in ohms, (2) Synchronous Reactance in ohms, (3) Resistance per Phase in ohms

$$Z_{Base(\Omega)} = \frac{(E_g)^2}{KVA/3} = \frac{(E_g)^2}{\dfrac{2000}{0.8x3}} = \frac{(256.7)^2}{833000\,VA} = 0.079\,\Omega \tag{1}$$

$X(\Omega) = 0.88 \times 0.079 = 0.0695\ \Omega$ (2)

$R(\Omega) = 0.042 \times 0.079 = 0.003318\ \Omega$ (3)

8.3.2 Per unit conversion

$$Z_B = \frac{BaseVoltage}{\sqrt{3}\,BaseVoltage} \qquad\qquad Z_\Omega = (Zpu)(Z_{Base})$$

$$Z_B = \frac{(V_{L-L})^2}{KVA(SinglePhase)}$$

8.3.3 Explanation of the terms

— **K1** is the ratio of the average asymmetrical rms current in the three phases at one-half cycle to the rms value of the symmetrical current

— **K2** is the ratio of the maximum rms asymmetrical current in one phase at one-half cycle to the rms value of the symmetrical current

— **E_g** is the line to neutral voltage of the system. For 480 V, three-phase system Eg is 480 divided by square root of 3

— **I_m** is the motor current contribution. For preliminary calculation Im is estimates 2/3 of the total connected generator ampere capacity

— **R** is the resistance of per phase circuit in ohms, looking from the fault to the generator

— **X** is the reactance per phase in ohms looking from the fault to the generator

— **Rc** is the per phase circuit resistance in ohms from the switchboard to the point of fault

— **Xc** is the per phase circuit reactance in ohms from the switchboard to the point of fault

8.4 IEEE Std 45-2002, 5.9.2, AC system (fault current)

The maximum available short-circuit current should be determined from the aggregate contribution of all generators that can be simultaneously operated in parallel, and the maximum number of motors that will be in operation. The maximum short-circuit current is calculated assuming a three-phase fault on the load terminals of the protective device. If the system is grounded and the zero sequence impedance is lower than the positive sequence impedance, a line-to-ground fault should be calculated in place of the three-phase fault. The protective device selected should have withstanding and interrupting capabilities, including peak current capability, that exceeds these calculated. Circuit breakers rated on a symmetrical basis should be applied on the basis of the symmetrical rms fault current. The system power factor at point of application should be greater than the power factor used in establishing the circuit breaker symmetrical rating. If it is not, consideration needs to be given to the circuit breaker's capability to withstand the asymmetrical value. The symmetrical RMS values of current can be obtained by applying the values of K_1 and K_2, as given in the decrement curve in Section 5, Figure 4 of IEEE Std 45-2002. The X/R ratio is determined from the reactance (X) and resistance (R) of the circuit under consideration. The X/R ratio is given in tabular form in Table 8-1 in Section 8.8 of this Handbook.

The short-circuit currents calculations guide lines are as follows:

Maximum asymmetrical rms current: Generator contribution based on circuit impedance including direct-axis sub-transient reactance of generators. Motor contribution based on four times the rated current of induction motors.

Average asymmetrical rms current: Generator contribution based on circuit impedances including direct-axis sub-transient reactance of generators. Motor contribution based on 3.5 times the rated current of motors.

Estimated short-circuit currents: For a preliminary estimate of short-circuit currents, pending the availability of generator reactances, the following may be used for estimating the generator contribution:

Maximum asymmetrical rms current: 10 times generator full-load current.

Average asymmetrical rms current: 8.5 times generator full-load current.

NOTE—These values for estimating generator contribution should not be used where unusually stringent transient voltage dip limitations have been specified for the generator. Minimum short-circuit current: The minimum available short-circuit current should also be determined to ensure that selectivity and fault clearing will be obtained under these conditions. The minimum short-circuit current is based on the least number of generators in operation and no motor load for a phase-to-phase fault at the load end of the cable connected to the protective device on an ungrounded system.

8.5   Typical short-circuit calculation examples

Example 8-2 provides basic steps for shipboard electrical system short circuit calculation or fault current calculation. The example should not be taken as complete calculation of any proven system. The electrical system fault calculation and system co-ordination requires special expertise in this field. However this example provides generator characteristics, cable characteristics, and impedance and x/r ratios and provides short circuit currents.

8.5.1   Example 8-2 (Figure 8-5)

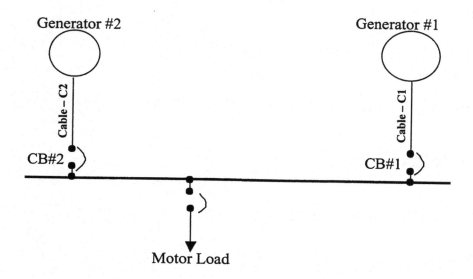

Figure 8-5: Typical One-Line for Short Circuit Calculation

a.   Generator Rating (#1 & #2)

1500 kVA, three-phase, 60 Hz, 0.8 PF, FLA = 1883A

   $Xd'' = 14\%$, $Xd = 18\%$, $Xd''/R = 29$

Im =2/3 (2 X 1883) = 2510 A

b. Technical characteristics of cable - C1 and C2

   C1 - T-400(qty-4) (120 ft)

   C2 - T-400(qty-6), (200 ft)

Resistivity of cable T-400 is 0.027 ohm per 1000 ft and per conductor

$$Z_B = \frac{Voltage(L-L)}{KVA(PerPhase)} \qquad X_{d'}.pu = 14\%$$

$$Z_B = \frac{\frac{450}{\sqrt{3}}}{\frac{(1500)}{3}} = 0.141\ \Omega$$

$$Xd''\ \Omega = 0.14 x 0.141 \Omega = 0.0197 \Omega$$

$$\frac{Xd''}{R} = 29$$

c.   Generator Impedance :

$$R + jX = (0.00068 + j\ 0.0197)\Omega$$

$$R = \frac{0.0197}{29} = 0.00068 \Omega$$

8.5.2   Example 8-3

Consider two generators running parallel (neglect line reactance for simplification):

$R_T = 0.00034;$ $X_T = 0.00985$

$X/R = 28.9$ $K1 = 1.331$ and $K2 = 1.625$

I Ave = 38,308 A

I Max = 43,772 A

8.5.3   Example 8-4

Consider two generators on line and reactances of the feeder cables are included in the short circuit calculation.

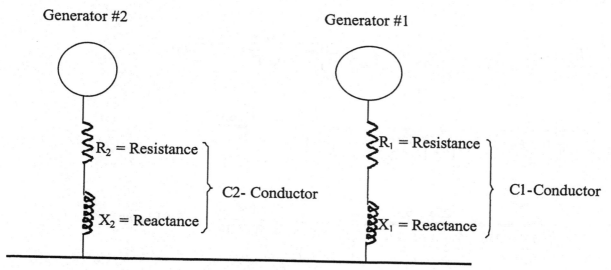

Figure 8-6: Typical One-Line for Impedance Calculation

Fault Current Calculation – Details

a)   Conductor C1 is T400, 120 ft Long, Qty 6
b)   Conductor C2 is 200 ft Long T400 Qty 6
c)   T-400   - Resistance R = 0.027 X 10^{-3}Ω/1000 ft/Conductor
              - Reactance X = 0.025 X 10^{-3}Ω/1000 ft/Conductor

R_{C1} = (0.027 X 10^{-3} X120) / 6 = 540 X 10^{-6}

X_{C1} = (0.025 X 10^{-3} X 120) / 6 = 500x 10^{-6} Ω

d)   Generator #1 and Cable total resistance and reactance:

GR1 + CR1 = (0.00068 + 0.00054) =0.00122

GX1 + CX1 =0.0197 + 0.00050 = 0.0202

For Generator #1 X/R =16.6 $K1_{(G1)}$ = 1.28 $K2_{(G1)}$ = 1.54

e)   Generator #2 is the Same as Generator #1

RC2 = 900 X 10-6

XC2 = 833 X 10-6

Generator #2 and cable =0.00158 + j0.0205

f) G1 impedance is in parallel with G2 impedance.

Total impedance value for Generators and cables = (G1+G2) R_T = 0.00316

For G1 and G2 total reactance X_T = 0.041
X_T/R_T =12.98   K1 = 1.26   K2 = 1.49

$$Isym = \frac{Eg}{\sqrt{R^2 + X^2}} = \frac{256.6}{\sqrt{(0.00316)^2 + (0.041)^2}} = \frac{256.6}{0.041} = 6285amp$$

$$I_{AVE} = \frac{0.71\,Eg}{\sqrt{\left(\dfrac{0.7\,Eg}{Im} + R_c\right)^2 + \left(\dfrac{0.19\,Eg}{Im} + X_c\right)^2}} + \frac{K_1\,Eg}{\sqrt{R^2 + X^2}}$$

$$Iave = \frac{256.6}{\sqrt{\left(\dfrac{0.7\,x\,256.6}{2511}\right)^2 + \left(\dfrac{0.19\,x\,256.6}{2511}\right)^2}} = \frac{256.6}{0.0735} = 3491 \quad amp$$

$$Im\,ax = \frac{Eg}{\sqrt{\left(\dfrac{0.7\,Eg}{Im} + R_c\right)^2 + \left(\dfrac{0.19\,Eg}{Im} + X_c\right)^2}}$$

8.6 Simplified wave forms (Figure 8-7 through Figure 8-12)

The waveforms in Figures 8-7 through 8-12 are provided to show simple sinusoidal symmetric wave, asymmetric wave, instantaneous peak-decay, prospective short circuit and melting/arcing current and asymmetric, symmetric and DC components of short.

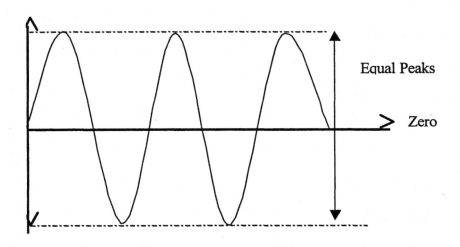

Figure 8-7: Symmetric Wave Form

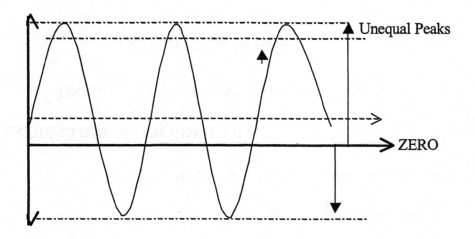

Figure 8-8: Asymmetric Wave Form

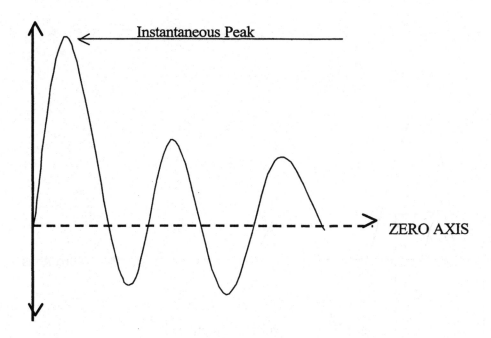

Figure 8-9: Instantaneous Arc Voltage Peak

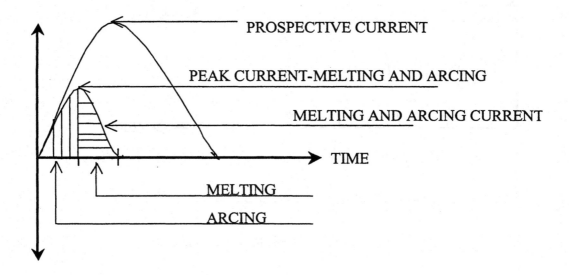

Figure 8-10: Instantaneous Peak Current First Half (½) Cycle (180 Electrical Degrees)

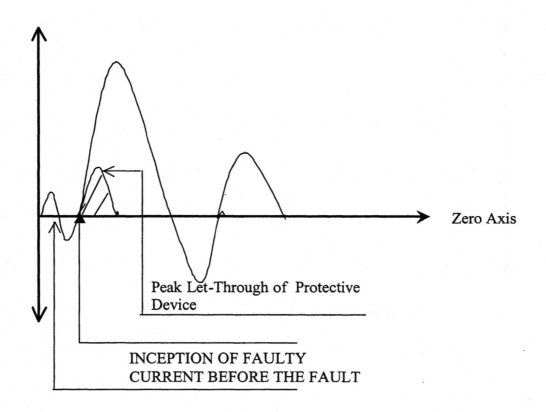

Figure 8-11: Protective Device Function on Short Circuit Condition

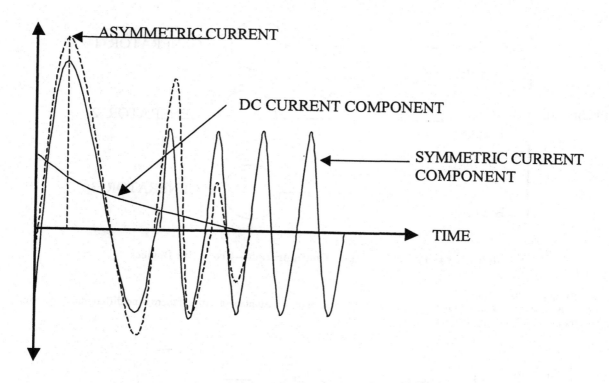

Figure 8-12: Asymmetric, Symmetric and DC Component

8.7   Generator/breaker coordination

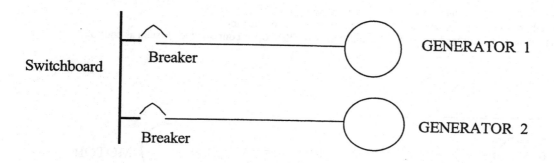

Figure 8-13: Two Generator Configuration—Protective Devices

NOTES ON FIGURE:

1—Generator circuit breaker will have long time delay (LTD) characteristics only.
2—Circuit breaker rating LTD rating: 115% FLA of the generator.
2—Breaker short circuit rating is based on average asymmetric rms current of the system.
3—If fuse is used with the breaker, the fuse short circuit rating should be on the basis of maximum asymmetric rms rating of the system.
4—Fuse and Breaker should be coordinated so that fuse interrupts the fault within 3 ms.
5—Breaker interruption is at the range of 15 ms.

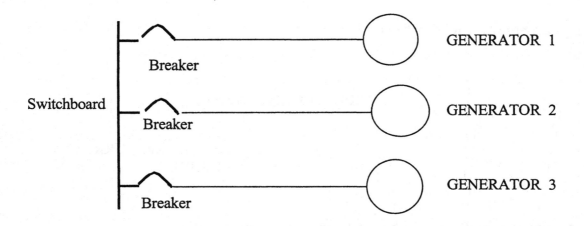

Figure 8-14: Three Generator Configuration—Protective Devices

NOTES (See IEEE Std 45-2002, 7.5.1)
1—Generator circuit breaker will have long time delay (LTD) characteristics instantaneous trip, if there is no differential protection
2—Circuit breaker rating LTD rating 115% FLA of the generator.

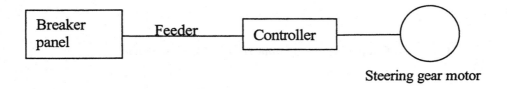

Figure 8-15: Steering Gear Motor Overload Protection

NOTES
1—The protective device with instantaneous trip is only for fault current protection.
2—The instantaneous setting is not less than 200% of the locked rotor current of the motor.
3—The feeder rating is 125% of the full load current of the motor plus 100% of other connected loads such as controls.

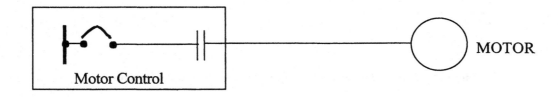

Figure 8-16: Motor Controller Center—Motor Protective Device for the Load

NOTES
1—Protective device with instantaneous trip only for short circuit protection.
2—Instantaneous trip set at 10 to 13 times the full load current of the load.

8.8 Short circuit calculation related decremental factors K1 and K2 values (Table 8-1)

— K1 = ratio of the average asymmetrical rms current in the three phases at one-half cycle to the rms value of the symmetrical current.
— K2 = ratio of the maximum rms asymmetrical current in one phase at one-half cycle to the rms value of the symmetrical current.
— For value of x/r less than 1.0, K1=1.00 and K2=1.00.
— X/R ratio values between those listed are to be obtained by linear interpolation.

X/R	K1	K2		X/R	K1	K2
1.0	1.001	1.002		11.0	1.241	1.459
1.2	1.003	1.005		11.2	1.244	1.463
1.4	1.006	1.011		11.4	1.246	1.467
1.6	1.010	1.020		11.6	1.248	1.471
1.8	1.015	1.030		11.8	1.250	1.475
2.0	1.021	1.042		12.0	1.252	1.478
2.2	1.028	1.056		12.2	1.254	1.482
2.4	1.036	1.070		12.4	1.255	1.485
2.6	1.043	1.086		12.6	1.257	1.488
2.8	1.051	1.101		12.8	1.259	1.491
3.0	1.059	1.116		13.0	1.261	1.494
3.2	1.067	1.132		13.2	1.262	1.498
3.4	1.075	1.147		13.4	1.264	1.500
3.6	1.082	1.162		13.6	1.266	1.503
3.8	1.090	1.176		13.8	1.267	1.506
4.0	1.097	1.190		14.0	1.269	1.509
4.2	1.104	1.203		14.2	1.270	1.512
4.4	1.111	1.216		14.4	1.272	1.514
4.6	1.118	1.229		14.6	1.273	1.517
4.8	1.124	1.241		14.8	1.274	1.519
5.0	1.130	1.253		15.0	1.276	1.522
5.2	1.136	1.264		15.2	1.277	1.524
5.4	1.142	1.275		15.4	1.278	1.526
5.6	1.147	1.285		15.6	1.280	1.529
5.8	1.153	1.295		15.8	1.281	1.531
6.0	1.158	1.305		16.0	1.282	1.533
6.2	1.163	1.314		16.2	1.283	1.535
6.4	1.167	1.323		16.4	1.284	1.537
6.6	1.172	1.331		16.6	1.286	1.539
6.8	1.176	1.330		16.8	1.287	1.541
7.0	1.181	1.347		17.0	1.288	1.543

X/R	K1	K2		X/R	K1	K2
7.2	1.185	1.335		17.2	1.289	1.545
7.4	1.189	1.362		17.4	1.290	1.547
7.6	1.192	1.369		17.6	1.291	1.549
7.8	1.196	1.376		17.8	1.292	1.551
8.0	1.200	1.383		18.0	1.293	1.553
8.2	1.203	1.389		18.2	1.294	1.555
8.4	1.206	1.395		18.4	1.295	1.556
8.6	1.210	1.401		18.6	1.296	1.558
8.8	1.213	1.407		18.8	1.296	1.559
9.0	1.216	1.412		19.0	1.297	1.560
9.2	1.219	1.418		19.2	1.298	1.562
9.4	1.222	1.423		19.4	1.298	1.564
9.6	1.224	1.428		19.6	1.299	1.566
9.8	1.227	1.433		19.8	1.300	1.567
10.0	1.230	1.438		20.0	1.301	1.568
10.2	1.232	1.442		25.0	1.318	1.598
10.4	1.235	1.447		30.0	1.336	1.630
10.6	1.237	1.447		35.0	1.338	1.634
10.8	1.239	1.455		40.0	1.341	1.642

Table 8-1: Fault Current Decremental Conversion Factors (K1 & K2)

9 Switchboards—shipboard electrical power system (SWBS 324)

9.1 General

The shipboard switchboards are usually distributed or split in two, three, or four sections, in order to meet the redundancy requirements of the vessel. According to rules and regulations, shipboard electric generation and must be able to withstand the consequences of one section failing due to a short circuit or other related electrical failures.

In propulsion mode of an electric propulsion ship, the switchboards are normally connected together, which gives the best flexibility in configuration of the power generation plant. The load transients are distributed on a large number of diesel-generators, and the most optimal number of units can be connected to the network.

As the installed power increases, the normal load increases, as does the short circuit current. With the physical limitations on handling the thermal and mechanical stresses in bus-bars and the switching capacity of the switchgear, it is necessary to increase the system voltage and hence reduce the current levels. Medium voltage has become a necessity to handle the increasing power demand in many applications.

In US, or where the ANSI standard applies, several additional voltage levels are recognized, such as 120 V, 208 V, 230 V, 240 V, 380 V, 450 V, 480 V, 600 V, 690 V, 2400 V, 3300 V, 4160 V, 6600 V, 11 000 V, and 13 800 V. Additionally, 33 000 V is a commonly used system voltage.

Recommended (IEC) voltage levels corresponding to the power level may be considered as follows:

- **11 kV:** Medium-voltage generation and distribution. Total installed generator capacity exceeds 30 MW, and motors from 400 kW and above.
- **6.6 kV:** Medium-voltage generation and distribution. Total installed generator capacity is between 4 MW and 30 MW, motors from 300 kW and above.
- **690 V:** Low-voltage generation and distribution. Total installed generator capacity is below 4 MW, and consumers below 400 kW

For ship service distribution lower voltage is used, e.g. 460 V/120 V (US) and 400 V/230 V (IEC).

Safety is an issue of concern when changing from low to higher voltages. In the context of safety, the high voltage switchboard is designed to protect personnel from contact with conductors, even in maintenance of the switchgears. NEMA arc-resistant switchboards and IEC 60298-3 arc-proof switchboards are available to prevent personal injury and limit equipment damage.

The following switchboards for shipboard application are addressed in this section:

- **Ship service and propulsion switchboards** (600 V AC or less for ANSI and 1000 V AC or less for IEC)
- **Emergency switchboard** (600 V AC or less for ANSI and 1000 V AC or less for IEC)
- **Medium-voltage switchboard** (601 V AC to 38 kV for ANSI and 1001 V AC to 35 kV IEC)

9.2 Switchboard major changes in IEEE Std 45-2002

The switchboard recommendations are found in IEEE Std 45-2002, 10, which has gone through considerable revisions, including the addition of IEC switchboard requirements along with ANSI requirements. Some of those additions and deletions from IEEE Std 45-1998 to IEEE Std 45-2002 are given in Table 9-1 through Table 9-3:

Item	IEEE Std 45-1998 - Clause	IEEE Std 45-1998 Requirements	IEEE Std 45-2002 - Clause	IEEE Std 45-2002 Requirements	Remarks
1	7.2	Construction for low-voltage (1000 V or less) switchboards	8.3	Low-voltage switchboards (600 V AC and less for ANSI; 1000 V AC and less for IEC) – Description and requirements (The title only)	Note that ANSI switchboard above 600 Vac is medium voltage and IEC switchboard above 1000 V AC. ANSI switchboard manufacturing requirements such as busbar ratings, busbar spacing, and assembly procedures are not the same as IEC requirements. ANSI switchboards must have UL recognized devices with UL labels to get UL label for the switchboard.
2	7.3	Construction for medium voltage (1001 V to 15 kV rms) switchboards	8.4	Medium-voltage switchboards (0.601-38.0 kV AC and less for ANSI and 1.01-35.0 kV AC and less for IEC) – Description and requirements (The title only)	Note that ANSI medium voltage switchboard is 0. 601-38.0 kVac and IEC is medium voltage switchboard is 1.01 – 35 kV
3	Table 7-1	Ampacity rating of rectangular bus bar. Bus bars are given in inches	None		This table is not in the IEEE Std 45-2002 as it is switchboard bus bar requirement for switchboard construction
4	Table 7-2,	Recommended Torque Value of bus bar joints	None		See Table 9-2
5	Table 7-3	Recommended Torque Value of breaker termination connection	None		See Table 9-3

Table 9-1: Switchboard Voltage Classifications per IEEE Std 45-1998 and IEEE Std 45-2002

Bolt Size (in)	Steel		Silicon Bronze	
	Minimum (ft-lbf)	Maximum (ft-lbf)	Minimum (ft-lbf)	Maximum (ft-lbf)
3/8	14	16	10	11
½	30	33	15	17
5/8	50	35	35	39

**Table 9-2: Recommended Torque Value for Bus Bar Joints
per IEEE Std 45-1998, Table 7-2 (removed from IEEE Std 45-2002)**

Copper Stud Size (in)	Steel Cap Screw Size	Torque (ft-lbs)	
		Minimum	Maximum
3/8		7	8
1/2		15	17
3/4		25	28
1-1/8		40	44
	5/8	50	55
	1	130	145

Table 9-3: Recommended Torque Value for Breaker Terminal Connections per IEEE Std 45-1998, Table 7-3 (removed from IEEE Std 45-2002)

9.2.1   Ship service switchboard and propulsion switchboard

The ship service switchboards provide for control, operation, and protection of the ship service generators, control and operation of remote generators, parallel operation of ship service generators, parallel operation of ship service generators with shore power, transfer of load and control and protection through appropriate circuit breakers of the electric power distribution system.

The Ship service switchboards are comprised of a combination of the following units (as required) :

- Generator control unit
- Bus tie unit
- Shore power unit
- 480 V and 120 V ship service loads

The voltage classifications of ANSI and IEC are as follows:

— Ship service switchboard (600 V AC or less for ANSI and 1000 V AC or less for IEC)
— Emergency switchboard (600 V AC or less for ANSI and 1000 V AC or less for IEC)
— Medium-voltage switchboard (601 V AC to 38 kV for ANSI and 1001 V AC to 35 kV IEC)

In propulsion mode, the switchboards are normally connected together, which gives the best flexibility in configuration of the power generation plant. The loads are distributed by multiple generators, and the most optimal number of units can be connected to the network.

Another possibility is to sail with independent switchboards supplying two or more independent networks. In this case the ship is often assumed to be virtually blackout proof, which could be attractive in congested waters. In this operating mode one network including its connected propulsion units is lost if one switchboard section fails, the other, however, remaining operable. In practice, there are also other considerations to be made in order to obtain such independence, especially all auxiliaries, such as lubrication, cooling, and ventilation must be made independent.

9.2.2   Emergency switchboard

The emergency switchboard provides control, operation and protection of the emergency generator and the emergency distribution system.

The emergency switchboards are arranged to supply power from the ship service switchboards in normal operating condition. Upon failure of ship service source, the emergency generator will start and the emergency loads will be transferred to the emergency source.

The emergency switchboard is located near the centerline of the ship, above the waterline if practicable, with maximum separation from the ship service switchboards.

The emergency switchboards are comprised of a combination of the following (as required) :

- Emergency generator control and bus transfer section.
- Bus tie to ship service switchboard
- Emergency 480 V distribution loads
- Emergency 120 V distribution loads

9.3   IEEE Std 45-2002, 8.2, Switchboard arrangement criteria (extract)

... The switchboard should be accessible from the front and rear, except for switchboards that are enclosed at the rear and can be fully serviced from the front. Access entrances, and operating clearances in front and rear of the switchboards for personnel safety, should be in accordance with Table 10 of IEEE Std 45-2002. Additional space or clearance may be required for maintenance of withdrawable elements.

Switchboards should be installed on a foundation that rises above the adjacent deck plating. To avoid excessively high foundations, consideration may be given to protecting the lower portion of the switchboard by watertight barriers and suitable drains. Switchboards may be self-supporting or braced to the bulkhead or deck above. If a switchboard is braced, the means of bracing must be flexible to allow deflection of the ship's structure without buckling the switchboard assembly. Nonconductive electrical floor matting meeting ASTM D-178-2001 J6-7 Type 2, Class 2, shall be installed in front of switchboards. The matting shall extend the entire length of the switchboard and be of sufficient width to suit the operating clearance specified.

When the space in the rear of the switchboard could be accessible to unauthorized personnel, the rear spaces should be protected within an enclosed lockable area. The area enclosure may be constructed of expanded metal sheets, or solid metal sheets with suitable ventilating louvers at the top and bottom. Where this arrangement is not feasible, rear covers that are bolted on and easily removable may be mounted on the switchboard framework. Alternatively, hinged covers may be used such that the cover may be fully opened. The hinged panels shall have provisions for locking

9.4   IEEE Std 45-1998 and IEEE Std 45-2002 recommendations

In this section, IEEE Std 45 recommendations for switchboard protective devices are provided from 1998 and 2002 editions. This is to support better understanding of existing as well as new switchboard design, both low voltage and medium voltage type.

9.4.1   IEEE Std 45-2002, 8.6, Circuit breakers—application

The continuous current rating of circuit breakers should be the current value that the circuit breakers will carry continuously without exceeding the specified temperature rise. Molded case breakers are normally rated to continuously carry 80% of their nominal rating; however, 100% rated molded case circuit breakers are available.

Low-voltage circuit breakers (open-frame or insulated or molded case) may be equipped with overcurrent trip devices of the electromechanical, electronic, or microprocessor type. The circuit breaker should be equipped with overcurrent trip devices that have long-time and instantaneous; or long-time and short-time; or long-time, short-time, and instantaneous characteristics. The time overcurrent devices should be adjustable on power circuit breakers but need not be adjustable on molded-case circuit breakers.

Medium-voltage class circuit breakers should use separate electromechanical- or electronic-type protective relays. In some cases, low-voltage power circuit breakers may use separate relays for overcurrent sensing. Separate overcurrent relays should have a current sensor in each phase.

Control power for tripping medium-voltage circuit breakers and, in some cases, low-voltage power circuit breakers should be from a battery source with automatic battery chargers. Where the battery can be disconnected for inspection and maintenance, the battery charger should be equipped with a battery eliminator type feature. Alternatively, capacitor trip devices may be used as a source of tripping power when DC power is not available. The capacitor trip device size should be sufficient for a minimum of two trip operation cycles.

For low-voltage circuit breakers, arcing contacts, except those used in molded case circuit breakers, should be easily renewable.

Note that, in IEEE Std 45-2002, circuit breaker application was revised from IEEE Std 45-1998 to have the following features:

— Circuit breaker continuous rating is 80% of the nameplate rating.
— UPS power such as battery and battery charger is required for circuit breaker trip devices.

9.4.2 IEEE Std 45-1998, 7.5, Circuit breakers

The current rating of circuit breakers should be the current value that the circuit breakers will carry continuously without exceeding the specified temperature rise. Low-voltage power (open-frame or insulated case) circuit breakers may be equipped with over-current trip devices of the electromechanical or electronic type. The over-current devices may be equipped with time over-current trips (long time, short time, or both), instantaneous trips, and any combination thereof. The time over-current trips should be adjustable on power circuit breakers but need not be adjustable on molded-case circuit breakers.

Low-voltage circuit breakers should have an integral trip device with tripping power being derived internally from the over-current.

Medium-voltage class circuit breakers normally utilize separate electromechanical or electronic type relays for overcurrent sensing, but direct acting overcurrent units may be used. Separate overcurrent relays should have a current sensor in each phase. Control power for tripping medium-voltage circuit breakers should be from an internal battery source with an automatic charger, or capacitor type storage may be used on AC circuit breaker trip units.

9.4.3 IEEE Std 45-1998, 7.8.2.2, Alternating current (AC) generators

Generators should be protected by means of trip-free, low-voltage power circuit breakers having long-time overcurrent trip units, or by medium-voltage circuit breakers and over-current relays with long-time pickup characteristics. The pickup setting of the long-time over-current trip should not exceed 115% of the generator rating for continuous rated machines and should not exceed 15% above the overload rating for specially rated machines.

Generator low-voltage circuit breakers should also have short time-delay trips. For medium-voltage circuit breakers, the over-current relay short-time feature may not be required if the long-time characteristic provides the needed protection. Where three or more generators are operated in parallel, the generator breakers should also have instantaneous trips that are set at a value in excess of the maximum short-circuit contribution of the individual generator.

In order to provide the maximum obtainable degree of protection for the generator, the over-current trips should be set at the lowest values of current and time that will comply with the recommendations of 11.28.

When two or more generators are to operate in parallel, each generator should be protected against reverse power flow. The reverse power relays should operate on reverse power values of 15% or greater of the generator rating for diesel driven generators, and of 5% or greater for turbine-driven generators. If reverse power relays are used that may operate at very low values of power reversal, a time-delay feature should be incorporated to prevent the tripping of generator circuit breakers during switching operation.

ABS reverse power protection requirements for prime movers are: Diesel engine range 8% to 15% and Turbine 2% to 6%. It is recommended to set the reverse power towards the high end of these given ranges. However, it is also very important to provide appropriate protection as required by the applicable prime mover configuration for specific application. The system configuration and prime mover reverse power sustaining capability must be coordinated with the prime mover manufacturer. When the reverse power is set for a specific application, the appropriate authority having jurisdiction should approve it.

9.4.4 IEEE Std 45-1998, 7.8.2.3.3, Generator reverse current

Where generators are operated in parallel, the generator circuit breakers should be provided with a reverse-current relay that will operate t o trip the breaker prior to current reversals of sufficient magnitude to cause injurious effects on the machine or prime mover. The reverse-current relays should operate on reverse current values of 15% or greater of the generator rating for diesel-driven generators, and of 5% or greater for turbine-driven generators.

When reverse-current relays are used that may operate at very low values of power reversal, a time-delay feature should be incorporated to prevent the tripping of generator circuit breakers during switching operations.

9.4.5 IEEE Std 45-2002, 8.9.4, Shore power

The shore power feeder should have a circuit breaker with a pole for each ungrounded conductor installed in the switchboard for connecting power from the shore connection panel to the ship service distribution bus.

An indicating light should be illuminated when power is available from shore, and one of the switchboard voltmeters should have selector switch capability to read shore power voltage.

Mechanical or electrical interlocking of the shore power circuit breaker with the generator circuit breakers should be installed unless load transfer paralleling capability is provided.

9.4.6 IEEE Std 45-2002, 8.9.5, Grounding

The switchboard ground bus and framework shall be solidly grounded by positive means to the hull or common ground reference. Metal cases of instruments, relays, or other devices and secondary winding of current and voltage transformers shall be grounded to the switchboard's equipment ground bus in accordance with the ANSI and IEC standards for switchboards.

9.4.7   IEEE Std 45-2002, 8.10, Switchboard phase and ground bus

Bus bars and connections should be copper and braced to withstand vibration and the maximum mechanical forces imposed by inrush current and all available system short-circuit currents. All bus supports and bus sleeves shall be high dielectric, non-hygroscopic, track resistant, and high strength.

All bus-bar connection points should be silver- or tin-plated. Plating should not peel off under normal operating conditions.

Bus-bar current-carrying capacity should be determined on the basis of generator and feeder full-load currents. For a single generator, the generator bus should have a current capacity equal to the full load rating of the generator plus any overload rating in excess of 30 min duration.

For multiple generator installations, the bus shall be sized to carry the maximum current that can be imposed on it as determined by load flow analysis, including any overload rating in excess of 30 min duration.

When the main source of electrical power is necessary for propulsion of the ship, and the aggregate generating capacity connected to a generator switchboard exceeds 3000 kW, the switchboard bus should be divided into at least two sections. The connections of the generators and any other duplicated distribution services should be divided equally between the bus sections as far as is practicable. The bus sectioning device may be an automatic or non-automatic circuit breaker, disconnect switch, or other suitable device.

For distribution sections of generator switchboards and distribution switchboards (without generation), the buses shall be sized to carry the maximum continuous load as determined by a load flow analysis plus the overload rating of the largest load and all identified future loads.

All bus bars should be accurately formed, and all holes should be made in a manner that will permit bus bars and connections to be fitted into place without being forced.

Bus bars, connection bars, and wiring on the back of the switchboard should be arranged so that maximum accessibility is provided for cable connections. Consideration should also be given to the arrangement of cables so they may be connected to the switchboard in an orderly manner.

9.4.8   IEEE Std 45-1998, 7.10, Bus bar and connections

The bus size should be selected on the basis of limiting the bus-bar temperature rise to 50 °C (122 °F) at rated current. Table 7-1 (IEEE Std 45-2002) may be used in determining the required bus-bar size.

Bus-bar current carrying-capacity should be determined on the basis of generator and feeder full-load -currents. For a single generator, the generator bus should have a current capacity equal to the full-load rating of the generator plus any overload rating in excess of 30 min duration. For more than one generator with all generating capacity feeding through one section of the bus, the capacity of the bus for the first generator should be the same as for a single-generator installation. For each subsequent generator, the bus capacity should be increased by 80% of the continuous rating of the added generator. The capacity of connection buses for each generator unit should be equal to the continuous rating of the generator plus any overload rating in excess of 30 min duration. In order to limit the size of bus for generator switchboards, it is recommended that consideration be given to locating generator sections in the center or at each end of the switchboard.

If the aggregate generating capacity connected to a generator switchboard exceeds 3000 kW, the switchboard bus should be divided into at least two sections. The connections of the generators and any other duplicated distribution services should be divided equally between the bus sections as far as is practicable. The bus sectioning device may be a bolted splice connection, an automatic or non-automatic circuit breaker, bus-disconnect links, or other suitable device.

9.4.9 IEEE Std 45-2002, 8.3, Low-voltage switchboards (600 V AC and less for ANSI; 1000 V AC and less for IEC)—description and requirements (extract)

Switchboards operating at a root-mean-square (RMS) voltage less than 1000 V should meet the requirements of UL Std 891 or IEC 60947 for dead-front switchboards or IEEE Std C37.20.1-1993, UL 1558-1999, or IEC 60947 for low-voltage, metal-enclosed power circuit breaker switchgear.

Circuit breakers installed in low-voltage switchboards should meet the following requirements for the class of service intended:

— Power circuit breakers installed in low-voltage switchboards should meet the requirements of IEEE Std C37.13-1990 or IEC 60947-2. When installed in low-voltage, metal-enclosed switchgear in accordance with IEEE Std C37.20.1-1993 or IEC 60947-2, these breakers shall be draw-out type.

— Power circuit breakers with proper insulation barriers may also be installed in dead-front switchboards per UL 891-1998 or IEC 60947-2. These breakers shall be draw-out type.

— Low-voltage molded or insulated case circuit breakers installed in switchboards shall meet the requirements of UL 489-1996 including all marine supplements, or shall meet the requirements of IEC 60947-2 including the additional performance requirements as defined in the marine supplements of UL 489-1996. The insulated case circuit breakers shall be draw-out type, and the molded case breakers shall be mounted on marine dead-front removable (plug-in) connectors (both line and load) to facilitate maintenance and replacement without a complete switchboard outage. ...

9.5 Switchboard circuit breaker requirements by IEEE Std 45, ABS, and USCG

The switchboard circuit breaker selection process requires a good understanding of ampacity ratings as well as requirements for ambient thermal characteristics. These requirements are quoted here in Table 9-4 from IEEE Std 45, ABS and USCG CFR.

IEEE Std 45-2002		ABS-2002		USCG CFR-46	
Reference	**Requirement**	**Reference**	**Requirement**	**Reference**	**Requirement**
5.8.1	The maximum connected load should neither exceed conductor ampacity nor 80% over the protective device setting or rating.	9.13	Over-current protective device to correspond to the feeder rating, if not, no more than 150%	111.54.1	(b) Each molded case breaker must meet UL 489 and its Marine supplement, or IEC 947-2 Part 2—except-Breakers in the Engine room, machinery space calibrated for 50 °C
5.9.5.8	Over-current device not to exceed 150%	9.17.2	Motor protection device 100%, but not more than 125%		
7.5.1	Generator power breaker not to exceed 115% and molded case breaker 100%				
8.3	Molded case breaker shall meet requirements of UL 489-1996, including its marine supplements, or IEC 60947-2 including marine supplements of UL 489-1996				

IEEE Std 45-2002		ABS-2002		USCG CFR-46	
Reference	Requirement	Reference	Requirement	Reference	Requirement
8.6	Molded case circuit breakers — Normally available at 80% rating — Also 100% rated breakers available				

Table 9-4: Switchgear—Molded-Case Breaker Requirements (IEEE, ABS, and USCG)

9.6 Note for UL 489-1996 Marine Supplements SA, SB, SC, D

9.6.1 Supplement SA

Supplement SA is for molded-case circuit breakers and circuit-breaker enclosures (marine use) for use under USCG Electrical Systems Regulations Subchapter–S (33 CFR, part 183) and USCG Electrical Engineering Regulations Subchapter J (46 CFR, Part 110–113).

The following requirements are extracts from marine supplement SA:

— SA3.1: Circuit breakers to be used in marine service shall be rated for an ambient temperature 40 °C (104 °F)

— SA6 - Vibration test, SA7-Shock test, SA8-Short circuit test etc, for marine circuit beaker to be used on vessels 65 ft or less

9.6.2 Additional requirements:

Supplement SB – Molded case circuit breakers and circuit breaker enclosures for Naval Use

Supplement SC – Molded case circuit breakers and circuit breaker enclosures for use with uninterruptible power supply

Supplement D – Classified breakers for use with specified panel boards as an alternate for specified circuit breakers

9.7 Medium-voltage circuit breaker selection information (for detail refer to ANSI 37.06)

Nominal rms voltage	Nominal three-phase MVA class	Impulse withstand (kV)	Continuous rms current at 60 Hz (A)	Short circuit capability (3 s) (kA)
4160	250, 350	60	1200, 2000, 3000, 3500, 4000	36, 49
7200	500	95	1200, 2000, 3000, 3500, 4000	41
13800	500, 750, 1000	95	1200, 2000, 3000, 3500, 4000	23, 36, 48

Table 9-5: Medium-Voltage Circuit Breaker Selection

9.8 IEC rating for 50 Hz system—medium-voltage circuit breaker selection information

Nominal rms voltage	Nominal three-phase MVA class	Impulse withstand (kV)	Continuous rms current at 60 Hz (A)	Short circuit making current (kA rms) (symm)	Short circuit breaking current—peak (kA)	Short circuit capability (3 sec) (kA)
3600		60	1250, 2000	25, 40	63, 100	25, 40
7200		60	1250, 2000	25, 40	63, 100	25, 40
12000		75	1250, 1600, 2000	25, 40	63, 100	25, 40
13800		95	1250, 2000	25, 40	63, 100	25, 40
15000		95	1250, 2000	63	170	63

Table 9-6: IEC Rating for 50 Hz System—Medium-Voltage Circuit Breaker

9.9 Switchboard device identification

Device #	Device function/identification
1	Master Element
2	Time-Delay Rely (Starting or Closing)
25	Synchronizing or Synchronism-Check Device
27	Under voltage Relay
37	Undercurrent or Under power Relay
46	Reverse-phase or Phase Balance Current Relay
47	Phase Sequence Relay
50	Instantaneous Over current or Rate of Rise Relay
51	AC Time Over current Relay
53	Exciter or DC Generator Relay
56	Field Application Relay
58	Rectification Failure Relay
59	Over voltage Relay
60	Voltage or Current Balance Relay
62	Time-Delay Stopping or Opening Relay
64	Ground Protective Relay
65	Governor
67	AC Directional Over current Relay
68	Blocking Relay
74	Alarm Relay
79	AC Reclosing Relay
81	Frequency Relay

Device #	Device function/identification
83	Automatic Selective Control or Transfer Relay
86	Locking-Out Relay
87	Differential Protective Relay

Table 9-7: IEEE/ANSI Standard Device Function Numbers

(The numbers above are based on a system adopted as standard for automatic switchgear by IEEE and ANSI. Refer to ANSI 37.2 for details. This list is not complete.)

9.10 Switchboard layouts (Figure 9-1 through Figure 9-6)

The figures 9-1 through 9-6 are to show various switchboard configurations as examples. Each configuration has its unique characteristics with dedicated service requirement. It is very important to understand the system requirements prior to preparing switchboard configurations.

Figure 9-1: Ship service 450 V and 120 V distribution switchboard and 450 V and 120 V emergency switchboard in radial configuration

Figure 9-2: Ship service 450 V distribution switchboard in ring bus configuration

Figure 9-3: 6600 V main propulsion switchboards and 450 V and 120 V distribution Ship service distribution switchboards

Figure 9-4: 6600 V main propulsion switchboards and 450 V and 120 V distribution Ship service distribution switchboards and 450 V and 120 V emergency switchboard in radial configuration

Figure 9-5: 6600 V main propulsion switchboards showing the propulsion motor loads

Figure 9-6: 6600 V main propulsion switchboards and 450 V and 120 V distribution Ship service distribution switchboards

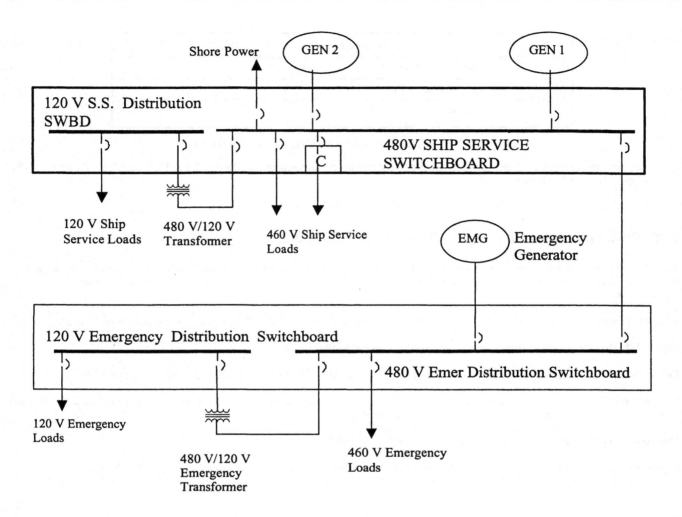

Figure 9-1: Electrical One–Line Diagram-Switchboards with Ship Service and Emergency Distribution

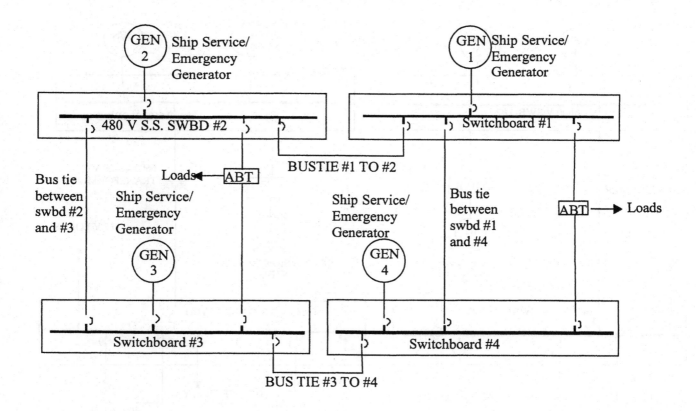

Figure 9-2: Electrical One-Line Diagram-Switchboard in Ring Bus Configuration

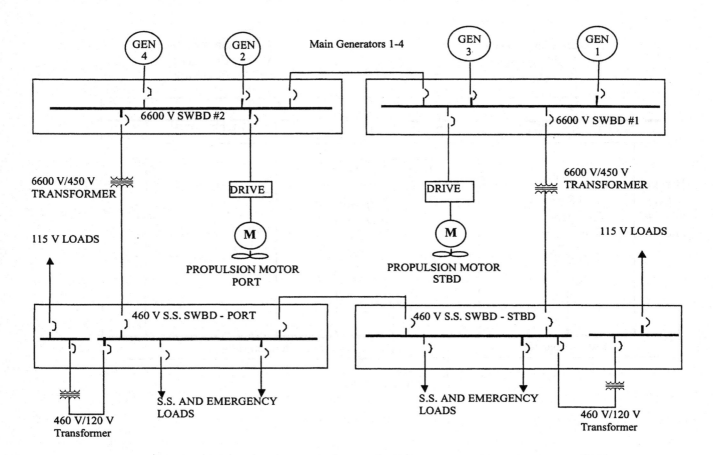

Figure 9-3: Electrical One-Line Diagram-Main Swbd and S.S. Swbd for Electrical Propulsion System (No Emergency Generator)

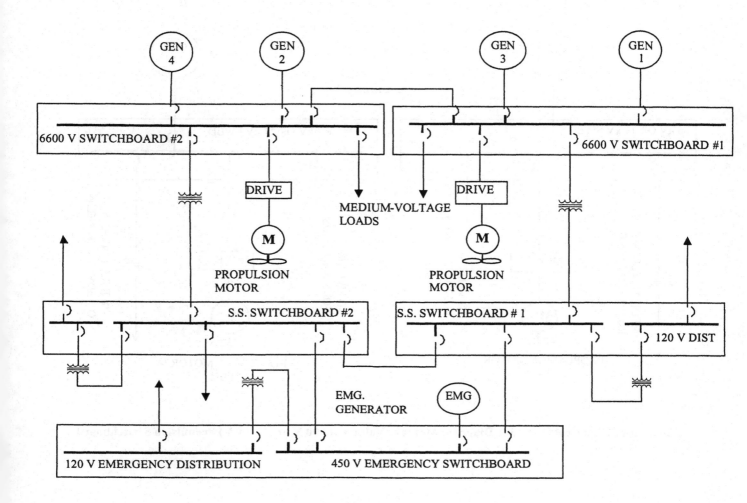

Figure 9-4: Electrical One-Line Diagram-Main Switchboards, Ship Service Switchboards, and Emergency Switchboard for Electrical Propulsion System

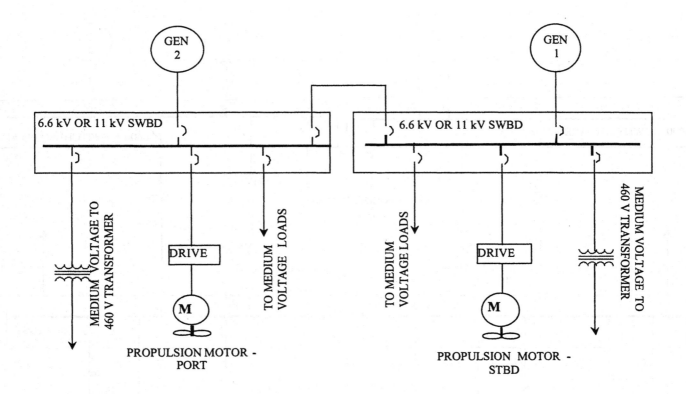

Figure 9-5: Electrical One-Line Diagram-Medium-Voltage (6600 V or 11 000 V) Propulsion Switchboard

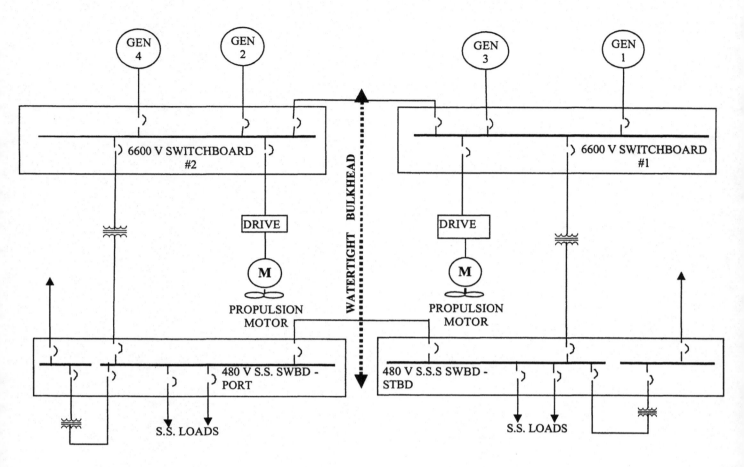

Figure 9-6: Electrical One-Line Diagram–6600 V Swbd and 480 V S.S. Swbd in Compliance with Class R2-S+

Engine rooms are fully segregated for fire integrity and watertight integrity. Only electrical cross-connections are the bus tie cross connections with watertight penetrations.

10 Motor and motor controller—shipboard (SWBS 302)

10.1 General

Shipboard auxiliary services are mostly electric motors such as squirrel cage or three phase air-cooled induction motors. Motor enclosures are NEMA classified as watertight, drip-proof etc, or fall into equivalent IEC IP classifications. The motor speed ranges are two-pole or four-pole nominal speeds of 3600 RPM or 1800RPM at 60 Hz, or 3000RPM or 1500RPM at 50 Hz. Typical shipboard motor applications are: ventilation fan motors, fire pump motors, lube oil pump motors, cooling pump motors, steering gear motors, compressor motors, etc. Motor speed–torque must match the duty requirement so that the motor minimum torque, accelerating torque, maximum torque are all in accordance with the service requirements.

Shipboard propulsion motors are considered large machines, compared to other shipboard motors. Propulsion motors are mostly two winding synchronous motors with variable speed drive, and forced cooling with a heat exchanger.

10.2 Induction motor

The induction motor consists of a stator and a rotor and mounting bearings. The motor stators and the rotors are separated by an air gap. The stator consists of a core made up of laminations for carrying current-carrying conductor embedded in the slot. The current-carrying conductors are connected, forming the armature winding. The AC current is supplied to the armature windings and current in the rotor winding is induced. This is known as an induction motor. The induction motor rotor does not require any external source of power. The induction motor operates on the basis of interaction of induced rotor current and the air gap field.

The induction motors are suitable for across the line starting, although the breakaway starting torque-related starting current can be 6 to 10 times the full load current. Due to limitations on shipboard generator size, the large motor requires further consideration due to high inrush current requirements. Additionally, motors with their inherently large starting current may result in an excessive voltage drop across the motor terminals. The reduced voltage starters are used to reduce starting current, but will also produce reduced starting torque. Therefore the starting system must be able to sustain inrush current and must provide the required breakaway torque.

Example 10.1: A variable frequency driven induction motor for shipboard auxiliary pump service, rated at 100 kW, 4 Pole, 460 V, three-phase, 60 Hz. The motor full load RPM is 1780. Calculate the frequency requirement for the same motor running at 1450RPM for delivering 60 kW, assuming the motor operation at the rated flux.

$$RPM = \frac{120 \, x f(Hz)}{P}$$

Where f is the power supply frequency and P the motor Pole number, and RPM is the motor synchronous speed.

Motor Synchronous Speed (RPM) = $\frac{120 x 60}{4} = 1800$

The slip speed at full load power = 1800-1780 = 20

The motor torque is proportional to power divided by the speed.

$$\frac{Torque(1450)}{Torque(1780)} = \frac{60kW}{100kW} \, x \, \frac{1780 \, RPM}{1450 \, RPM} = 0.74$$

When the motor is operating at constant flux, the slip speed is proportional to the torque. There, the slip speed at 0.74 x rated torque

= 0.74 x 20 = 14.80 (Slip Speed)

For the motor to run at 1450RPM, the applied frequency must correspond to a synchronous speed of 1450+14.80 = 1464.80

Therefore, the corresponding frequency for RPM $1464.80 = \frac{120 \, x Frequency}{4}$

$$Frequency = \frac{1464.80 \, x 4}{120} = 48.83 \, Hz$$

10.3 Synchronous motor

The synchronous motor stator winding is the same as the induction motor stator winding. In addition, the rotor has a DC winding which produces the air gap flux directly. The synchronous motor torque generation occurs when the rotor field winding is supplied with power and the rotor is synchronized with the rotating field caused by the stator magnetomotive force (MMF). The synchronous motor is supplied with three-phase AC power to the stator and DC power to the rotor field. The AC power source to the stator and DC power the rotor can be adjusted independently so that the synchronous motor speed can be controlled properly. The synchronous motor runs in absolute synchronism with the stator rotating MMF. Where variable drive motor is required, appropriate frequency and voltage controllers such as cycloconverters, load commutating inverters, or pulse width modulating converters can be used.

Example 10.2: A variable frequency synchronous motor for shipboard auxiliary drive, rated at 200 kW, 6 Pole, operates at 460 V, 60 Hz, supply under rated condition. At what speed the motor would have to operate to ensure approximate rated flux conditions when 230 V line is applied to the terminal.

Constant flux is indicated if the voltage per hertz is constant, therefore

(Operating Frequency) $\quad f = \dfrac{230 x f_r}{460} = \dfrac{230 x 60}{460} = 30$ hertz (f_r = normal frequency 60)

The speed (S) for the 30 Hz frequency = $\dfrac{120 x 30}{6} = 500 \, RPM$

10.4 Motor locked rotor information (adapted from NEC®)

NEMA Motor locked rotor indicating "Code Letter" per NEC 430-7(b) – this is a code letter marked on the motor nameplate to show motor kVA/HP.

Code letter	kVA/hp with locked rotor	Code letter	kVA/hp with locked rotor
A	0-3.14	L	9.0–9.99
B	3.15–3.54	M	10.0–11.19
C	3.55–3.99	N	11.2–12.49
D	4.0–4.49	P	12.5–13.99
E	4.5–4.99	R	14.0–15.99
F	5.0–5.59	S	16.0–17.00
G	5.6–6.29	T	18.0–19.99
H	6.3–7.09	U	20.0–22.39
J	7.1–7.99	V	22.4 and UP
K	8.0–8.99		

Table 10-1: NEC Table 430-7(b)—Locked rotor indicating code letters

The locked rotor current calculation can be performed based on this code letter as follows:

Locked rotor current for three-phase motor = (hp × kVA/hp × 1000 X 1.732Volts)/(Voltage)

Locked rotor current for single-phase motor = (hp × kVA/hp × 1000)/(Voltage)

10.5 IEEE Std 45-2002 recommendations for motors

10.5.1 IEEE Std 45-2002, 13.4, Installation and location of motors

Motors exposed to the weather or located where they would be exposed to seas, splashing, or other severe moisture conditions should be watertight or protected by watertight enclosures. Such enclosures should be designed to prevent internal temperatures in excess of motor ratings. Motors should be located so they cannot be damaged by bilge water. Where such a location is unavoidable, they should be either submersible or provided with a watertight coaming to form a well around the base of the equipment and a means for removing water. A suitable NEMA or IEC enclosure should be used (reference Annex C of IEEE Std 45-2002]) as required by the authority having jurisdiction. Fan-cooled motors should not be installed in locations subject to ice formation.

For motors in hazardous locations, see Clause 33 of IEEE Std 45-2002.

Horizontal motors should be installed, as far as practicable, with the rotor or armature shafts in the fore and aft direction of the vessel. If motors must be mounted in a vertical or an athwartship position, the motors should have bearings appropriate for the intended installation.

Motors (except for submersible motors, sealed motors, or motors for hazardous locations) should be designed to permit ready removal of the rotor, stator components, and bearings, and to facilitate bearing maintenance. All wound-rotor AC motors, synchronous motors, and commutating motors of the enclosed type, except fractional horsepower motors, should have viewing ports or readily removable covers at the slip ring or commutator end for inspection of the rings, commutator, and brushes while in operation.

Eyebolts or equivalent should be provided for lifting motors.

10.5.2 IEEE Std 45-2002, 13.5, Insulation of windings

Insulating materials and insulated windings should be resistant to moisture, sea air, and oil vapor. All form-wound coils for AC motors and assembled armatures, and the armature coil for open-slot DC construction, should utilize vacuum pressure impregnated solventless epoxy insulating systems.

For AC motors, all random-wound stators and rotors having insulated windings should utilize vacuum pressure impregnated solventless epoxy Class F or H insulation systems, with Class B rise, or be of the encapsulated type.

Most standard NEMA frame motors are fabricated using non-hygroscopic NEMA Class F or H insulation. In totally enclosed motors, the normal insulation can be expected to provide satisfactory service. If open drip-proof or weather-protected motors are selected, it is recommended that the insulation be a sealed system. Motors with NEMA Class F insulation and with a NEMA Class B rise at rated motor horsepower are available in most motor sizes and types and are recommended to provide an increased service factor and longer insulation life.

All field coils for DC motors should be treated with varnish or other insulating compound while being wound, or should be impregnated. The finished winding should be water- and oil-resistant.

Abnormal brush wear and commutator maintenance may occur in motors containing silicon materials. Silicon materials should not be used in motors with brushes.

10.5.3 IEEE Std 45-2002, 13.6, Locked rotor kilovoltampere (kVA)

Three-phase induction motors normally are designed for a starting kVA of five to six times the horsepower rating. This starting kVA corresponds to NEMA locked rotor Codes F and G and is suitable for most applications. IEC motors exhibit related characteristics, but not 100% equivalent. It may be desirable that large motors be specified with lower inrush currents to minimize the effects of starting on the power source. Consult the motor manufacturer for specific details.

10.5.4 IEEE Std 45-2002, 13.7, Efficiency of motors

It is recommended that designers of new installations consider the use [of] energy-efficient motors. For a given horsepower rating and speed, the efficiency of a motor is primarily a function of load. The full load efficiency generally increases as the rated horsepower or speed increase.

Efficiency increases with a decrease in slip (difference between synchronous speed and full load speed of an induction motor, divided by the synchronous speed). High slip motors usually yield higher overall efficiency for applications involving pulsating, high inertia loads.

Reference ANSI/NEMA MG 1-1998 or IEC 60034 for additional guidance. Generally, the inrush current on energy-efficient motors is higher than that for standard motors.

10.5.5 IEEE Std 45-2002, 13.12, Ambient temperature of motors

Motors for main and auxiliary machinery spaces containing significant heat sources such as prime movers and boilers, except machine tools, should be selected on the basis of a 50 °C ambient temperature or the actual expected maximum. Motors for machine tools may be selected on the basis of 40 °C ambient temperature. Motors in locations where the ambient temperature will not exceed 40 °C may be selected on the basis of 40 °C ambient temperature.

Motors that must be installed where the temperature normally will exceed 50 °C should be considered as special and should be designed for 65 °C or the actual expected ambient temperature. Consideration should be given to ensuring satisfactory lubrication at high temperatures.

If a machine is to be utilized in a space in which the machine's rated ambient temperature is below the assumed ambient temperature of the space, it should be used at a derated load. The assumed ambient temperature of the space plus the machine's actual temperature rise at its derated load should not exceed the machine's total rated temperature (machine's rated ambient temperature plus its rated temperature rise).

10.5.6 IEEE Std 45-2002, 13.13, Limits of temperature rise (extract)

It is recommended that AC and DC motors have a design temperature rise of 80 °C, by resistance, in a 40 °C ambient temperature (NEMA Class B), but be constructed with a minimum of NEMA Class F insulation to provide optimum balance between initial cost and long-life operations. Reference 1.5.1 [IEEE Std 45-2002)] for the applicable ambient temperature requirements.

AC and DC motors normally are designed for 40 °C ambient temperatures, and thus, they should be derated in accordance with manufacturer's recommendations if operated in higher ambient temperatures. Insulation of motor windings with quality insulation materials that are designed to be resistant to the salt laden moist atmosphere locations is recommended.

Deck-winch and direct-acting capstan motors should be rated on a full load run of at least 1/2 h; direct-acting windlass motors should be rated on a full load run of at least 1/2 h; and those operating through hydraulic transmission should be rated for 30-min idle pump operation, followed by full load for 1/4 h. Steering-gear control motors should be rated on a full load run of 1 h.

The temperature rise of each of the various parts of AC motors and DC motors when tested in accordance with the full load rating should not exceed values given in Table 17 [IEEE Std 45-2002]. ...

10.6 Motor controls—general

Motor starters are classified as direct across the line starter (DOL), reduced voltage starter, soft starter and so forth. The starters are further classified as to whether they are low voltage protection (LVP) or low voltage release (LVR) type. The use of LVP or LVR starter depends on the criticality of the system and equipment.

Refer to Figure 10-13 for LVP starter details. In an LVP starter, the *start* pushbutton is of the spring-return type and is normally open. The pushbutton must be pressed to close, to initiate the starting sequence. When the start pushbutton is closed, the appropriate contactor is energized and the open contact across the start pushbutton closes to seal in the electrical circuit. However, in case of voltage failure, the seal-in contact opens up, stopping the motor. If the motor stops, the starting sequence must be manually initiated again by pressing the spring return start pushbutton. If LVP starting is not appropriate, then an LVR type starter should be considered.

In the LVR starter, the *start* switch is not a spring-return switch which is normally open, but rather a normal ON-OFF type switch. When the LVR starter is in the "ON" position, the motor will start whenever voltage is present. If the system loses voltage (blackout situation), the motor stops. When the voltage comes back, the motor starts again automatically.

For selecting a motor starting system, the driven equipment, such as pump motor, ventilation fan motor, hydraulic pump motor, starting torque and inertia requirement must be considered. *A direct across the line starter* is a full voltage starter, where the motor can draw six to ten times its full load current during starting (starting kilo-volt-ampere requirement). A *reduced voltage starter* is limited by the kilo-volt-ampere inrush.

The LVR controllers can be further classified as low-voltage release effect (LVRE). LVRE controllers function as LVR, but starting event is not only the low-voltage release, but is also affected by other external sensing devices such pressure, level switches, and temperature switch. These controllers may or may not be starting time sequenced for low-voltage starting effect as well as for system operational reliability.

The shipboard electric motor starting methods are as follows:
— Direct On Line (DOL) starting (also referred to as across-the-line starting)
— Wye-Delta Starting, Open Transition
— Wye-Delta Starting Close Transition
— Auto Transformer starting
— Solid state—Soft Starting
— Variable Frequency Drive

10.6.1 DOL starting

A direct on-line (DOL) starting system is simple. It consists of a contactor and an overload protection relay. The disadvantages of a DOL starter are high starting current of a range 6 to 12 times the full load current of the motor, and high starting torque depending on the motor locked rotor code rating (See NEC table 430-7(b) in Table 11-1). The high starting current puts high stress on the contactor and requires protection against short circuit, such as upstream circuit breaker protection of the electric system. The high torque puts high stress on the mechanical system. Due to very large starting current demand, the ship service power generating system must be robust enough so that the voltage transient during the motor stating remains within the allowable voltage dip of the system. However, the starting system should also be for multi-speed starting, forward and reverse starting etc. For typical DOL starter schematics and speed/torque characteristics refer to Figure 10-1 through Figure 10-6 and Figure 10-8 through Figure 10-11.

10.6.2 Wye-Delta starting – reduced voltage

The Wye-Delta starting is considered mainly to manage the starting current of the motor. The starting current is only one-third of the DOL starting current. However this also produces 25% torque reduction. Wye-Delta starter are further classified as Closed Transition and Open transition. Refer to Figure 10-15 and Figure 10-16 for typical Wye-Delta starter schematics, and torque, current and speed characteristics in Figure 10-5 through Figure 10-8.

10.6.3 Auto-transformer starting

This is based on the principle of reducing the voltage across the motor during starting. This means that a substantial reduction is the available starting torque and pull-out torque. Refer to Figure 10-14 for typical auto-transformer schematics.

10.6.4 Solid state – soft starter

The soft starter is supplied with power electronic devices, which can control and increase voltage to the motor during starting and the starting time can be adjusted to achieve a soft starting process. The voltage can be controlled to match the voltage requirement during starting so that the power loss is minimized contributing to smooth acceleration.

10.6.5 Variable frequency drive

A by-pass contactor is an integral part of the starter, which shorts out the starting circuit when the motor attains the normal speed. When the stop function is selected the bypass contactor opens up to bring the stopping circuit in line. The start time and stop time range is adjustable which can be set to a specific application requirement. An inherent characteristic of a power electronics driven solid-state starter is the electric line loss-related heat generation. The power electronics also generate harmonics affecting the power quality. Refer to chapter 4 for harmonics details. For further details of the variable frequency drive, refer to chapter 15.

10.7 IEEE Std 45-2002 recommendations for control apparatus

10.7.1 IEEE Std 45-2002, 10.6, Manual starters and controllers

Manually operated controllers require manual operation of the main contact-switching device to start or stop a motor. The device should contain a trip-free overload relay. The controller should be of rugged construction and arranged for operation without opening the enclosure. Low-voltage protection should be included.

10.7.2 -IEEE Std 45-2002, 10.7, Magnetic starters and controllers

Magnetic starters and motor controllers utilize electromagnetic main contactors to control the motor circuits that remotely start and stop the motors. The magnetic controllers should be operated by control circuits that include overload protection devices and are actuated by local or remote pushbuttons or similar master switches to suit the intended motor application. Either low-voltage release or low-voltage protection should be used, depending on the motor application and regulatory requirements.

10.7.3 IEEE Std 45-2002, 10.8, Solid-state starters and controllers

Solid-state motor controllers utilizing static power converters have a wide range of applications in controlling shipboard machinery. Variable speed or soft starting motor controls can be used for pumps, fans, cranes, deck machinery, or any other application where variable speed, reversing, or precise speed control is desired. Converters can be DC/DC, DC/AC, AC/DC, or AC/AC and be either regenerative (four quadrant) or non-regenerative (two quadrant).

The location of solid-state motor controllers should be given careful consideration, especially for units requiring forced air-cooling. Weather deck locations or locations subject to spray or washdown should be avoided. Engine room locations are acceptable as far as moisture is concerned, but elevated temperatures should be considered when sizing power components in high ambient temperatures. Units in machinery spaces should be rated for operation in a 45 °C environment[1]. Solid-state motor controllers should be capable of operating in moist, dusty, and oil-laden atmospheres. Equipment should be able to withstand humidity levels of up to 95% and the vibration levels in 1.5.1 [IEEE Std 45-2002]. The units should be capable of continuous operation when the steady state power supply voltage and frequency variations are within the limits specified in Table 4 [see Section 5.4.1 of this Handbook].

Enclosures should be drip-proof or watertight. Enclosure openings for cooling air should have filters. Devices requiring forced air cooling or controllers utilizing multiple load sharing power bridges should be equipped with over-temperature detectors for alarm (local and remote) and with remote shutdown. Backup equipment should be considered for units controlling vital machinery or other critical functions. Power semiconductors should have a peak inverse voltage (PIV) of 1200 V when connected to a power source of 480 V AC or less. Series connection of lower voltage devices should not be used to achieve the 1200 V rating. Semiconductors may be connected in parallel to increase current capacity provided circuitry is included to ensure equal load sharing.

In applications such as cranes, load lowering will cause power to be generated in the motors. This power must be absorbed by other power-consuming devices or the generator sets, or it must be dissipated as heat into dynamic braking resistors. Failure to absorb regenerated power may cause reverse current relays to operate or cause an over-speed condition on the generator prime mover. Multiple step braking resistor assemblies are recommended over single-step units. Multiple step control connects the minimum amount of resistance required to absorb a given level of regenerative power without causing undue stress on the generators or distribution system. Dynamic braking resistors should be protected against overheating.

Power may be supplied to a motor through a solid-state switching device that is controlled locally or remotely by suitable master switches. Soft-start or variable speed control may also be applied by the solid-state switching device. An electromagnetic device, such as a contactor, or a circuit breaker with shunt trip, or equivalent, should be provided for disconnecting the motor under fault conditions.

[1] The regulatory body may require equipment rated at 50 degrees C ambient for engine room installation and lower ambient for other locations.

10.7.4 IEEE Std 45-2002, 10.9, Medium-voltage controllers

Medium-voltage motor controllers (1000 V rms or greater) should be electromagnetic controllers utilizing current-limiting power fuses, vacuum, sulfur hexaflouride (SF_6), or air break contactors and a medium-voltage isolating means to disconnect the medium-voltage supply. Generally, the features should be similar to those previously described for low-voltage controllers, but with the following additional recommendations:

  a)   The isolating means should be externally operable with position indication.
  b)   The isolating means should be capable of interrupting the no-load current of the control-circuit transformer supplied with the controller.
  c)   The isolating means may be a three-pole switch or a drawout-type contactor.

The medium- and low-voltage sections of a controller should be mechanically segregated by means of permanent barriers or movable shutters, where possible.

Mechanical means, or a combination of mechanical and electrical means, should provide the following interlocking features:

  —   Prevent the isolating means from being opened or closed unless all line contactors are open
  —   Prevent the opening of a medium-voltage compartment door when the isolating means are closed
  —   Prevent the isolating means from being closed when any medium-voltage compartment door of the controller is open

Control-circuit transformers should be provided with primary and secondary fuses. Instrument potential transformers should be provided with primary fuses. The control circuit voltage should be 115 V AC to permit contactor testing from an external control source. Interlocking for this test feature should be arranged so that power cannot be applied to the motor and so that the 115 V control circuit is disconnected from the normal control-circuit transformer.

Medium-voltage controllers should be wired and assembled as complete, enclosed, and freestanding units with provision for sway bracing. A means of ground connection of the enclosure by bare metal-to-metal contact with the ship's structure should be provided. The secondaries of instrument transformers should also be grounded to the ship's structure.

10.8 Motor starter comparison—direct on line (DOL), reduced voltage, and soft starter

10.8.1 DOL starter (Refer to Figure 10-13)

a) Advantage

— Simple to operate, enclosure size is small.
— Due to full voltage operation, it provides high starting torque, high accelerating torque and can attain full speed in a very short t ime.
— Commonly used onboard ship, particularly for small size motors

b) Disadvantage

— Due to across the line full voltage starting, the current requirement for breakaway torque is high, as much as 6 to 12 times the full load current on the motor. This generates a transient voltage dip in the system, which directly affects the generator response. Transient response simulation of the electrical system is often necessary for larger motors. The range of the large motor is dependent on the generator size onboard the ship. The shipboard emergency generator is usually small and regulatory body requirement is to directly connect the emergency fire pump and steering gear pump on the emergency switchboard. It is necessary to perform a starting voltage dip study for these pumps to ensure starting capability. There are multiple options available to mitigate the issue. One is to provide large size generator. Another is to provide reduced-voltage starting or solid state such as soft start.

10.8.2 Reduced voltage starter

The autotransformer type starter controls the starting voltage much below the rated voltage; the starting voltage is adjustable. A Wye-delta configuration also provides a different voltage level for starting, which must transition from the wye configuration to delta configuration. If the system transition is not acceptable for the auxiliaries, closed transition is also available. However, these reduced voltage starting systems must be thoroughly analyzed, for system performance, cost, size, complexity and so forth. Refer to Figure 10-14 through Figure 10-16 for reduced voltage starter.

Major concerns include

— When the starting voltage is reduced, the motor torque is also reduced. The torque reduction is square proportion to the voltage reduction. The torque demand must be within the reduced voltage torque generation to consider reduced voltage starting application.
— The starting current also reduces with the reduction of voltage.

10.8.3 Soft starter (refer to Figure 10-7 through Figure 10-12; Figure 10-17; Figure 10-18)

The soft starter provides adjustable voltage as well as in some cases current limiting functions during starting, so that the motor starting transient can be adjusted to provide adequate break-away torque during starting, while at the same time providing adequate starting voltage and time limitation to maintain the shipboard electrical system transient within allowable limit. This feature is very important if shipboard power generation is marginal.

The soft starter uses solid-state power electronics devices such as Scars. The power electronics devices generate harmonics and produce heat. The soft starter is usually provided with by -pass contactor, which alleviates the harmonic distortion in the electrical system and heat generation. Due to the solid-state nature of the starter, the system provides control, monitoring, diagnostic and prognostic digitally without additional cost. However, the cost of a soft starter is usually higher.

10.8.4 Frequency control starter

A frequency control starter is usually a solid-state power electronics device such as a thyristor driven device. When the frequency is reduced, the voltage changes proportionately and required torque can be maintained. Refer to the Figure 10-12, which shows four different frequencies for different speeds with constant torque. The cost of frequency control starter is usually higher, however special application requirements may dictate the use of frequency control starter. For further details of variable frequency drive large motor applications, refer to chapter 15.

Frequency control starters generate harmonic distortion in the system, which must be taken under consideration, minimized and controlled.

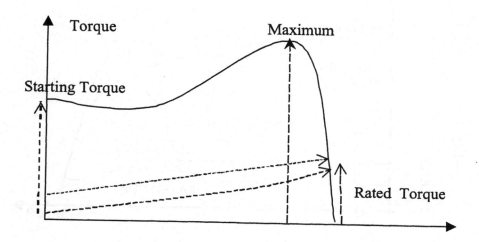

Figure 10-1: Motor Starter Diagram – Direct On Line (DOL) – Torque Characteristics

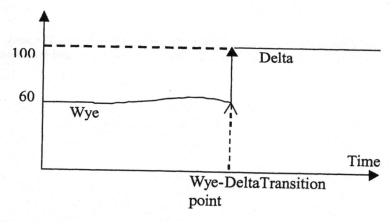

Figure 10-2: Wye-Delta Starter Diagram – Voltage and Time Characteristics

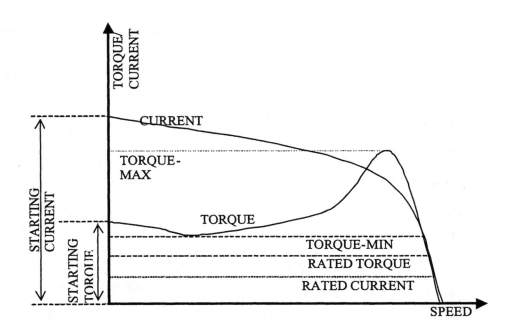

**Figure 10-3: Motor Starting Diagram—Torque/Speed and Current/Speed—
Typical for Squirrel Case Induction motor**

$$T(Torque)(Newtonmeter)(Nm) = \frac{3000 \, xP(OutputPowerInKw)}{(\pi)x(n - inRPM)}$$

Starting torque— The starting torque the breakaway torque to overcome static frictions etc. The starting torque is high which should be properly coordinated so that it does not cause adverse effect on the drive shaft and mechanical parts.

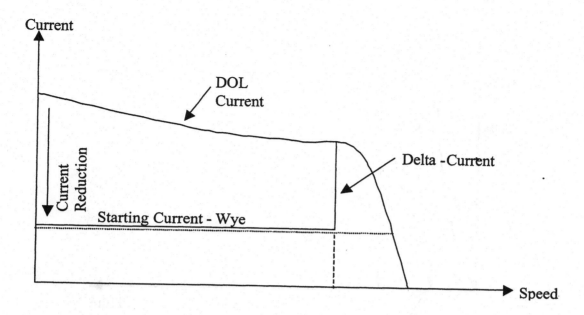

Figure 10-4: DOL and Wye-Delta Diagram—Current and RPM Characteristics

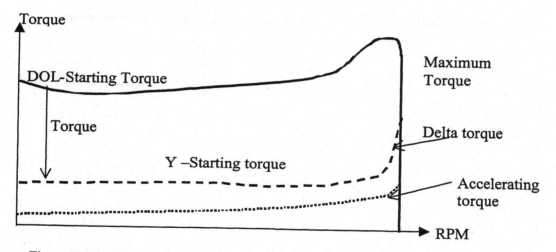

Figure 10-5: Motor Starter Diagram–Wye-Delta versus DOL – Torque Characteristics

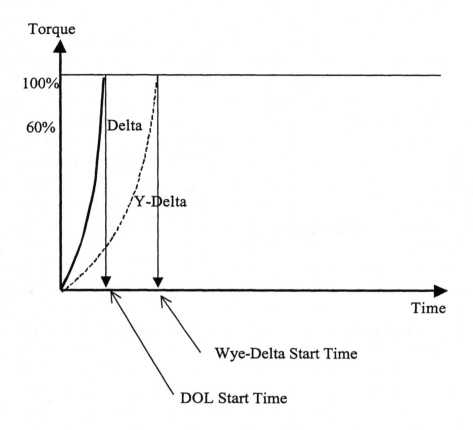

Figure 10-6: DOL versus Wye-Delta Diagram for Starting Time Characteristics

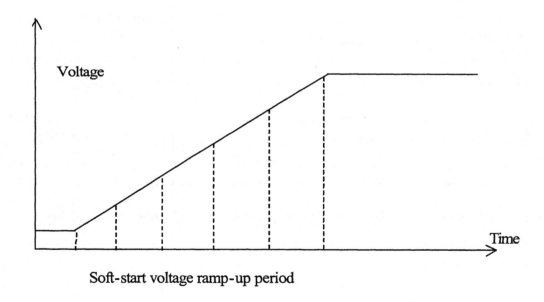

Figure 10-7: Soft Starter Voltage Ramp-up Diagram (without current limit)

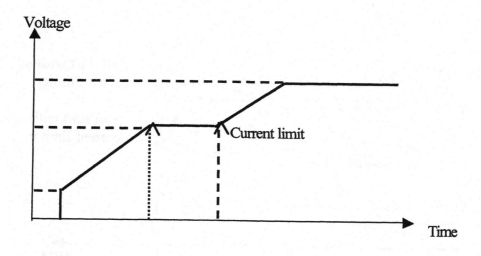

SOFT-START VOLTAGE RAMP-UP

Figure 10-8: Soft Starter Diagram —Voltage/Time Characteristics Current Ramp-Up with Current Limit

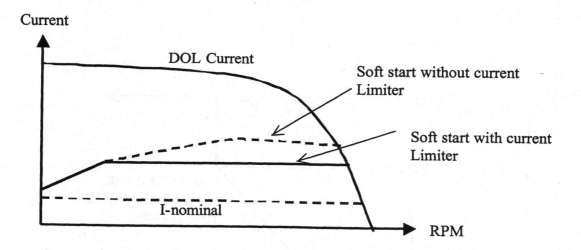

Figure 10-9: Starter Diagram—Current-Speed Characteristics—Soft Start Without Current Limit and With Current Limit (compare with DOL)

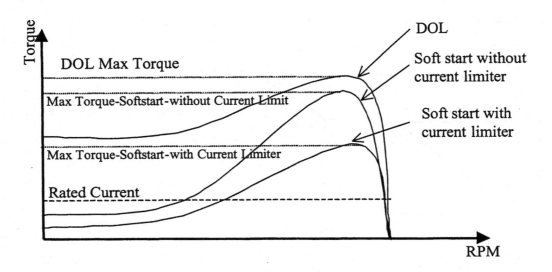

Figure 10-10: Soft Starter Diagram—Torque-RPM Relationship (with and without) Current Limiter (compare with DOL)

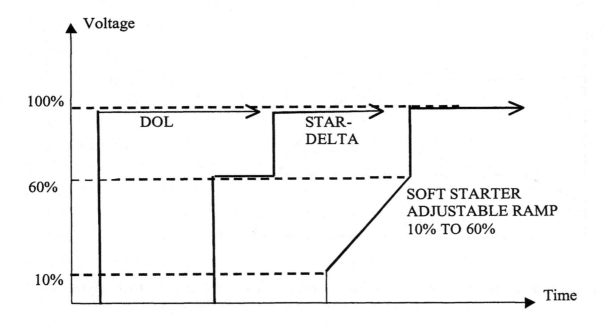

Figure 10-11: Starter Diagram—Starting Voltage—DOL, Wye-Delta, and Soft-Start Voltage Characteristics

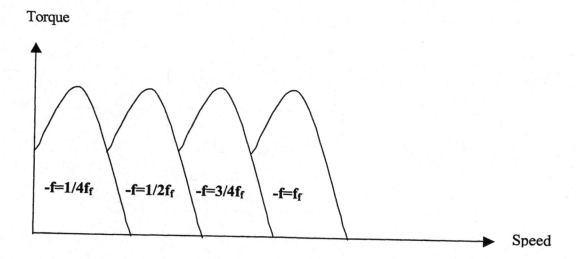

Figure 10-12: Starter Diagram—Frequency Control—Torque/Speed Characteristics

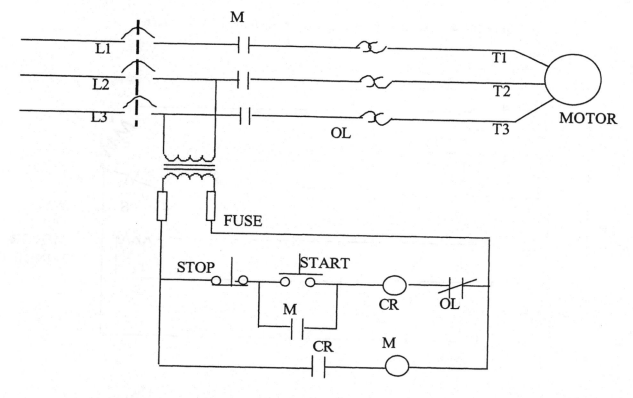

Figure 10-13: Schematic Diagram—Low-Voltage Protection (LVP) Motor Starter

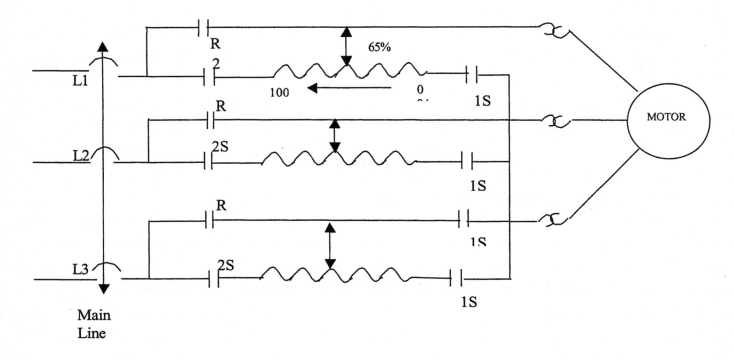

Figure 10-14: Schematic Diagram—Autotransformer Motor Starter

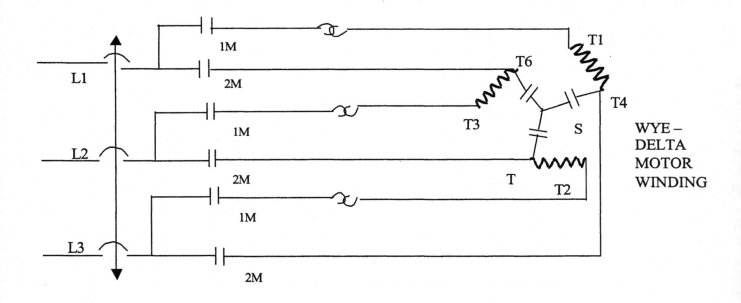

Figure 10-15: Schematic Diagram—Wye-Delta Open Transition Motor Starter

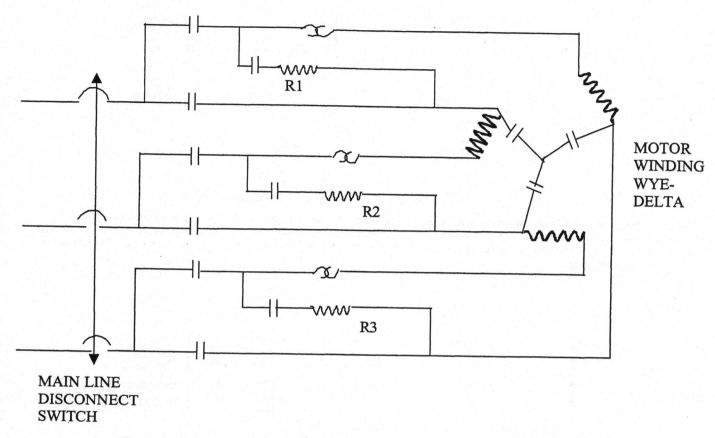

Figure 10-16: Schematic Diagram—Wye-Delta Closed Transition Motor Starter

Only the closed transition scheme is shown; for detailed schematics, refer to the manufacturer drawing.

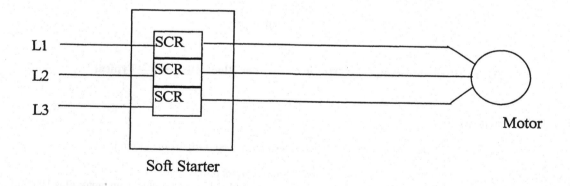

Figure 10-17: Soft Starter Block Diagram Without Bypass Switch-schematic Diagram

This is a typical block diagram only. For detail control schematics of soft starter, refer to soft starter manufacturer diagrams.

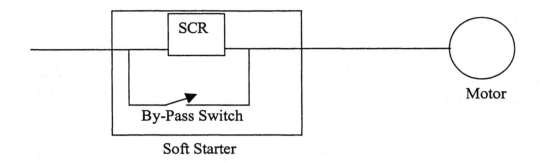

Figure 10-18: Soft Starter Block Diagram with Bypass Switch

NOTE—The bypass switch in the soft starter is to by-pass the SCR circuit when the motor attains speed. The by-pass switch is recommended due to the fact that while the SCRs are energized they produce heat. For shipboard application the generated heat must be considered for cooling. The SCRs are prone to heat rated failure. The cooling system can be a simple fan to a more complicated cooling system with heat exchanger.

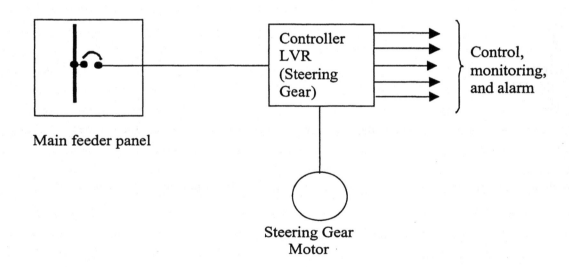

Figure 10-19: Steering Gear Motor Controller Block Diagram—Typical

NOTES (Steering gear system protection)

1—The circuit breaker (protective device) has instantaneous trip only. The instantaneous trip is for system fault protection only.
2—The instantaneous trip setting not to be less than 200% of the locked rotor current of the motor.
3—The main feeder for the steering gear motor should be rated for 125% of the full load current of the main motor plus 100% of other connected loads such as the control load.

11 Electrical power distribution and protection system—shipboard (SWBS 320)

11.1 General

A shipboard electrical system consists of power generation and distribution to meet power demand requirements. The ship service power generation and distribution voltages are generally 35 000 V, 15 000 V, 5000 V, 4160 V, 2000 V, 1000 V, 690 V, 480 V, 380 V, 230 V, 120 V, three-phase, 50 Hz or 60 Hz, 230 V, 120 V, 1 Phase, 50 Hz or 60 Hz, and DC voltages, 48 V, 24 V, 12 V. There are many other voltages and frequencies in use around the world.

	300 V	600 V (US) and 1000 V (IEC)	1000 V–2000 V	5000 V	5000 V–15 000 V
Cable	Signal, control, lighting etc.	Low-voltage power	Low-voltage power	Medium-voltage power	Medium-voltage power
Switchgear, circuit breaker, etc.	Low-voltage power	Low-voltage power	Medium-voltage power	Medium-voltage power	Medium-voltage power

Table 11-1: Low-Voltage and Medium-Voltage Classification for Commercial Ships

The shipboard power generation and distribution phase sequence is in A, B, and C and the direction of rotating equipment is also an A, B, C sequence. There are other phase conventions such as R, S, T and U, V, W. We will discuss the A, B, C phase sequence here. It is important to maintain correct phase sequence to maintain proper rotation of the rotating equipment, regardless of whether the power is from the ship service switchboard, emergency switchboard, or shore power. In order to maintain the A, B, C phase convention, the generator output terminals are marked with letters A, B, and C, the switchboard bus bars are marked with A, B, C, and distribution panel boards are marked with the A, B, C phase sequence.

There are two major further classifications, *grounded* and *ungrounded* electrical distribution systems. The National Electric Code requirement for shore based electrical installation is a grounded neutral system, while the shipboard electrical installation is generally ungrounded. However, the grounded distribution system is also used onboard ship for non-essential equipment for special safety consideration.

11.2 Grounded system

The grounded electrical system onboard ship is convenient; however, it produces high fault current, which activates protective devices such as blowing fuses or tripping breakers. A grounded electrical system requires a ground detection system to monitor ground current. The detection system provides a method of detecting ground current in amps by using an ammeter. For additional details refer to Chapter 12.

11.3 Ungrounded system

The ungrounded distribution system allows equipment to continue functioning under a single-line fault condition.

Every ungrounded electrical system, galvanically separated, is provided with a ground fault detection system. The system consists of a detection light in each phase with a test switch. The ground is detected by the brightness of the lights, which is easy to monitor. When a ground is detected in the system, the grounded circuit is isolated and the ground is lifted to clear the system.

Even in a ship with an ungrounded distribution system, some galvanically isolated systems use grounded systems such as isolated receptacles. System isolation is made by using transformers with delta primary and grounded neutral-Wye secondary.

11.4 Safety or equipment ground

Electrical equipment housings are grounded as a safety measure. Electrical safety grounding is a requirement for all shipboard electrical and non-equipment enclosures

11.5 Electrical protection system

The electrical system protective devices include full load protection, overload protection and over-current protection. Understanding the fundamentals of these protective devices and operating coordination is essential for a safe electrical distribution system. The detection of over load and protection of equipment due to overload condition and over current condition is known as coordination of protective devices. Coordination of protective devices and the fault current calculation submittal are required by the authority having approval jurisdiction. If an overload situation is prolonged, equipment may be damaged due to overheat, with serious risk of fire. The feeder cable insulation may be damaged due to current overload condition, which could cause a fire. The over current can cause serious damage to the equipment within a very short time due to very high current.

11.5.1 Protection against overload

Circuit breakers and fuses used for overload protection are required to have trip characteristics that adequately protect the system during normal and overload conditions with regard to overload capacity of each of these elements.

11.5.2 Coordination of protective devices

Protective devices are to be selected so that in the event of a fault (overload or short circuit), the protective device nearest to the fault should open first, thus eliminating the faulted portion from the system. Protective devices upstream of the fault are to be capable of carrying the fault current for a duration the short circuit current and the overload current, without opening, to allow the device nearest to the fault to open.

11.6 Generator-protective devices

Overload and short circuit protection is required for each generator in accordance with the following requirements.

11.6.1 Generator overload protection

Generators are to be protected by circuit breakers providing long-time delay (LTD) over-current protection. The LTD setting should not exceed 15% above the full-load rating of the generator. If the generator current overload is in excess of 10%, the protective device (such as circuit breaker) should trip with time delay to a maximum of 2 min in accordance with the system coordination. In some cases specific operation may not allow the breaker to trip with this coordination envelope, in which case special arrangements are made to meet the requirement.

11.6.2 Generator short circuit protection

Generator overload in excess of the LTD range of the coordination is protected by the short time rating of the breaker.

Generators are to be protected for short circuit by circuit breakers provided with short-time delay (STD) trips. For coordination with feeder circuit breakers, the short-time delay trips are to be set at a suitable current level and time, which will coordinate with the trip settings of feeder circuit breakers. Where two or more AC generators are arranged for parallel operation, each generator's circuit breaker is, in addition, provided with instantaneous short circuit trip protection.

For generators of less than 200 kW driven by diesel engines or gas turbines, which operate independently of the electrical system, consideration may be given to omission of the short-time delay trip if instantaneous and long-time trips are provided.

11.6.3 Generator thermal damage protection

Generator circuit breakers at the main and emergency switchboard are to have tripping characteristics and to be set such that they will open before the generator sustains thermal damages due to the fault current.

11.6.4 Generator reverse power protection

A reverse power protection device is to be provided for each generator arranged for parallel operation. The setting of the protective devices is to be in the range 2% to 6% of the rated power for turbines and in the range 8% to 15% of the rated power for diesel engines.

11.6.5 Generator under-voltage protection

Generators arranged for parallel operation are to be provided with means to prevent the generator circuit breaker from closing if the generator is not generating, and to open the same when the generator voltage collapses. In the case of an under voltage release provided for this purpose, the operation is to be instantaneous when preventing closure of the breaker, but is to be delayed for discrimination purposes when tripping a breaker.

11.7 Motor and motor circuit protective devices

Overload and short circuit protection is required for each motor circuit in accordance with the following requirements.

11.7.1 Motor branch circuit protection (ABS 4-8-2/9.17)

Motor branch circuits are to be protected with circuit breakers or fuses having both instantaneous and long-time delay trips or with fuses. The setting is to be such that it will permit the passage of motor starting current without tripping. Normally, the protective device is to be set in excess of the motors full load current but not to be more than the limitations given in the table below. If that rating or setting is not available, the next higher available rating or setting may be used. In cases where the motor branch circuit cable has allowable current capacity in excess of the motor full load current, the protective device setting may exceed the applicable limitation, but not to exceed 150% of the allowable current carrying capacity of the feeder cable.

When fuses are used to protect polyphase motor circuits, they are to be arranged to protect against single phasing.

11.7.2 Motor overload protection (ABS 4-8-2/9.17.2)

The overload protective devices of motors are to be compatible with the motor overload thermal characteristics, and are to be set at 100% of the motor rated current for continuous rated motor. If this is not practical, the setting may be increased to, but in no case exceeding, 125% of the motor rated current. This overload protective device may also be considered the overload protection of the motor branch circuit cable.

11.7.3 Under-voltage release protection (low-voltage release-LVR)

For motors of essential, vital, and emergency services under-voltage release is to be provided unless the automatic restart upon restoration of the normal voltage will cause hazardous conditions. Special attention is to be paid to the starting currents due to a group of motors with under-voltage- release controllers being restarted automatically upon voltage resumption after a power blackout. Means such as sequential starting are to be provided to limit excessive starting current, where necessary.

11.8 Transformer protective devices

Overload and short circuit protection is required for each transformer in accordance with the following requirements.

11.8.1 Transformer primary side protection

Each power and lighting transformer along with its feeder is to be provided with short circuit and overload protection. The protective device is to be installed on the primary side of the transformer and is to be set at 100% of the rated primary currents of the transformer. If this setting is not practicable, it may be increased to, but in no case exceed 125% of the rated primary current. The instantaneous trip setting of the protective device is not to be activated by the in-rush current (magnetizing current) of the transformer when switching into service.

11.8.2 Transformer secondary side protection

Where the secondary side of the transformer is fitted with a protective device set at not more than 125% of the rated secondary current, the transformer primary side protective device may be set at a value less than 250% of the rated primary current.

11.8.3 Transformer parallel operation

When the transformers are arranged for parallel operation, means are to be provided to disconnect the transformer from the secondary circuit. Where power can be fed into secondary windings, short-circuit protection (i.e., short-time delay trips) is to be provided in the secondary connections.

11.9 Feeder protection

Protection against short-circuit is to be provided for each non-earthed conductor (multipole protection) by means of circuit breakers, fuses or other protective devices.

11.9.1 Generator feeder protective devices

Each generator feeder's conductors are to be protected by a circuit breaker, or fuse with disconnecting switchgear, from short circuit and overload at the supply end. Fuse ratings and rating of time-delay trip elements of circuit breakers are not to exceed the rated current capacity of the feeder cables.

If the standard rating or setting of the overload protective device does not correspond to the current rating of the feeder cable, the next higher standard rating or setting may be used provided it does not exceed 150% of the allowable current carrying capacity of the feeder cable. This is allowed as the cable thermal characteristics changes much slower, expecting that the overload device will activate.

11.10 Circuit breaker fundamentals

The following sections describe some of the fundamental characteristics of circuit breakers for shipboard applications.

11.10.1 Circuit breaker rated breaking capacity

The rated breaking capacity of every protective device is not to be less than the maximum prospective short circuit current value at the point of installation. For alternating current (AC), the rated breaking capacity is not to be less than the root mean square (rms) value of the prospective short-circuit current at the point of installation. The circuit breaker is to be capable of breaking any current having an AC component not exceeding its rated breaking capacity, whatever the inherent direct current (DC) component may be at the beginning of the interruption.

11.10.2 Circuit breaker—making capacity rating

The rated making capacity of every circuit breaker which may be closed on short circuit is to be adequate for the maximum peak value of the prospective short-circuit current at the point of installation. The circuit breaker is to be capable of closing onto a current corresponding to its making capacity without opening within a time corresponding to the maximum time delay required.

11.10.3 Circuit breaker temperature rating

In general circuit breakers are calibrated for a temperature of 40 °C ambient, and the thermal rating of the breakers are good for 80% of the nameplate rating of the breaker. For marine shipboard installation, the circuit breakers should be constructed and tested to UL-489, marine supplement A. Circuit breakers for shipboard application are required to be calibrated at 50 °C and rated for 100% loading. For circuit breakers with 100% continuous rating and ambient temperature higher then 40 °C, they are to be derated by the manufacturer and certified accordingly.

For IEC compliant system, the breakers are to be constructed and tested per IEC-60947.

11.11 Vital and non-vital services (IEEE Std 45, DNV, and GL)

The following statements have been made by various regulatory bodies regarding vital versus non-vital service classifications.

11.11.1 IEEE Std 45-1977, 13.15

Vital services are normally considered to be those required for the safety of the ship and its passengers and crew. These may include propulsion, steering, navigation, firefighting, emergency lighting, and communications functions. Since the specific identification of vital services is influenced by the type of vessel and its intended service, this matter should be specified by the design agent for the particular vessel under consideration.

11.11.2 IEEE Std 45-1998, 3.10.21

The last sentence of IEEE Std 45-1977, 13.15, was changed to

The identification of all vital services in a particular vessel is generally specified by the government regulatory agencies.

11.11.3   IEEE Std 45-2002, 3.10.20, Vital services

Services normally considered to be essential for the safety of the ship and its passengers and crew. These usually include propulsion, steering, navigation, firefighting, emergency power, emergency lighting, electronics, and communications functions. The identification of all vital services in a particular vessel is generally specified by the government regulatory agencies.

11.11.4   Essential consumers per GL

Essential consumers are consumers, which are required for

— The maneuverability and safety of the ship
— The safety of passengers and crew
— The drive specific to the type of vessel on ships with a special class notation (dredger)
— The preservation of unobjectionable condition of cargo (e.g. on refrigerated cargo vessels with class KAZ)

Essential consumers for which supply from an emergency source of power is specified

11.11.5   Essential consumers per DNV

Essential equipment, which needs to be in continuous operation for maintaining the vessel's maneuverability with regard to propulsion and steering.

11.11.6   Nonessential consumers per GL

Nonessential consumers are those which, in order to maintain the source of electric power, may briefly be taken out of service, and whose temporary disconnection does not impair the maneuverability of the ship or the safety of the passengers, crew, vessel or machinery.

11.11.7   Important services per DNV

Equipment which need not be in continuous operation for maintaining the vessel's maneuverability, but which is necessary to maintain the vessel's main functions as defined in Pt1. Ch.1 Sec.2 or which according to these Rules is subject to approval when installed.

11.11.8   Nonimportant per DNV

Equipment, which is not essential or important according to the above, is classified as nonimportant.

Electrical loads are classified as Essential, Important and Non-essential. Not all international regulatory bodies recognize all three classifications. Below, we attempt to monitor different classification society (ABS, USCG, SOLAS, DNV, GL) load class recognition.

11.11.9   Vital

Safety devices and systems are considered vital per NVIC USCG 2-89. In addition to the emergency loads the vital loads may be connected to the emergency power system provided the emergency source is sized to supply these loads at 100% load factor.

11.11.10 Nonvital

The loads intended to improve the safety or survivability of the vessel in certain operating modes and which are not considered for sizing the emergency generator.

	Equipment	Essential	Nonessential	Important
1	Emergency Generator start-up	GL		
2	Emergency Lighting	GL		
3	Navigation Light	GL		
4	Radio Communications	GL		
5	Fire Detection & fire alarm	GL		
6	Emergency Fire pump	GL		
	Fire Pump			DNV
7	Steering gear	GL, DNV		
8	Auxiliary boiler	GL		
9	Bilge & ballast pump	GL		DNV
10	Boiler firing system	GL		
11	Control & regulating system for essential equipment	GL		
12	W.T. door control	GL		
13	Engine room vent fan	GL	GL	DNV
14	Foam pump	GL		
15	F.O. Transfer & Booster Pump	GL, DNV		DNV
16	Thruster	GL, DNV		DNV
17	Lighting	GL		
18	L.O. Pump	GL, DNV		
19	Main Engine auxiliary blower	GL,		
20	Startup & control air system	GL	GL	DNV
21	Windlass	GL	GL	
22	Controllable pitch propeller (CPP)	GL, DNV		

Table 11-2: Essential, Nonessential, and Important Load Classifications (GL & DNV)

11.12 Electrical system circuit designations

The electrical system circuit designations for Power, Lighting, Interior Communications, Controls, etc. are usually identified with unique circuit designations for each system, which are used throughout the ship design and development process such as drawing, documentation, equipment identification, cable identification. IEEE Std 45-2002, Annex B provides circuit designations for commercial ships. The Navy ship circuit designations are different from those of a commercial ship; the IEC standard designations are different from US commercial and Navy designations. Refer to Table 11-3 and Table 11-4 and Figure 11-9 through Figure 11-12 for circuit designations, examples, and explanations.

Circuit designation IEEE Std 45	Prefix IEEE Std 45	Circuit number	Explanation
Ship Service Power (450 V system)	P	6P-412-(3)T-250 (80 ft)	Ship service power, 450 V system, circuit number 6P412 with three 250MCM three conductor IEEE (Commercial cable) 600 V cable and cable length is 80 ft
Emergency Power (450 V System)	EP	6EP-412-(2)T-106 (50 ft)	Emergency Power, 450 V system circuit number 6P412 with three 250MCM three conductor IEEE (Commercial cable) 600 V cable and cable length is 80 ft
Propulsion Power(6600 V)	PP	3PP-6612-8KV(3)T-250 (250 ft)	Ship Propulsion power, 6600 V system, circuit number 3PP6612 with three 250MCM three conductor IEEE (Commercial cable) 8KV cable and cable length is 250 ft
Shore Power (450 V)	SP	2SP-424-T-400 (50 ft)	Ship service shore power, 450 V system, circuit number 2SP-424 with one 400MCM three conductor IEEE (Commercial cable) 600 V cable and cable length is 50 ft
Emergency Lighting (120 V)	EL	3EL-124-T-26 (150 ft)	Emergency Lighting 120 V system, circuit number 3EL-124 one 26 MCM three conductor IEEE (Commercial cable) 300 V cable and cable length is 150 ft
Telephone-Sound Powered - Engineers	C-2JV	C-2JV-15, TPS18-7 (65 ft)	Sound Powered Telephone , circuit number C-2JV with one 18 AWG Twisted Pair Shielded, seven pairs, IEEE (Commercial cable) and cable length is 65 ft

Table 11-3: Circuit Designation and Explanation (Commercial) (typical example only)

Ship service power (450 V system)	Circuit designation	Remarks
IEEE Std 45 Circuit Number (Refer to Figure 11-10)	1P-401-(7)T-250 (80 ft)	Ship Service, 460 V Power circuit number 1P-401 with seven 250MCM three conductor IEEE (Commercial) 600 V cable and cable length is 80 ft
US Navy Circuit Number (Refer to Figure 11-13)	24-4PP-(03-114-1)-T-200 (125 ft)	Ship Service Power 450 V system, Power Panel 03-114-1, circuit number 2, Navy 600 V cable, three 200MCM and cable length is 125 ft
IEC Circuit Number (Refer to Figure 11-11)	MS-11-17-2x3x120 mm^2 (75 m)	Power branch circuit 17 of power distribution panel 1 of switchboard 1 with 1000 V, 120 mm^2 three conductor cable and cable length of 75 m

Table 11-4: Circuit Designation—IEEE Std 45, US Navy, and IEC (typical example only)

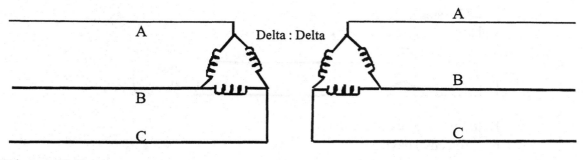

Delta : Delta

Primary Voltages:
Single Phase - Phase AB = Phase BC = Phase CA = 480 V and Three Phase = ABC =480 V

Secondary Voltages:
Single Phase -Phase AB = Phase BC = Phase CA = 120 V
Three Phase = ABC = 120 V

Figure 11-1: Delta-Delta 480 V/120 V Transformer—Ungrounded Distribution System

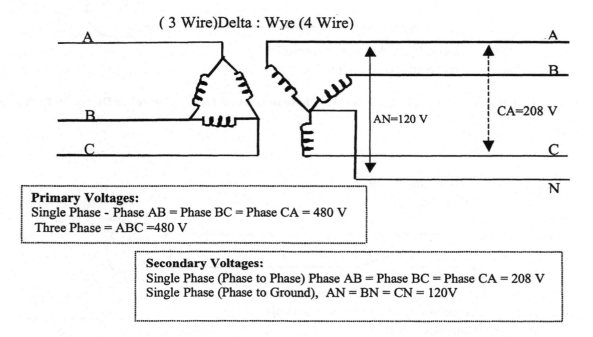

(3 Wire)Delta : Wye (4 Wire)

AN=120 V

CA=208 V

Primary Voltages:
Single Phase - Phase AB = Phase BC = Phase CA = 480 V
 Three Phase = ABC =480 V

Secondary Voltages:
Single Phase (Phase to Phase) Phase AB = Phase BC = Phase CA = 208 V
Single Phase (Phase to Ground), AN = BN = CN = 120V

Figure 11-2: Delta-Wye (4 Wire Secondary) 480 V/208 V/120 V Transformer Grounded Distribution

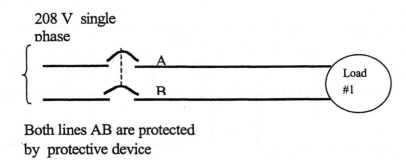

208 V single
phase

Both lines AB are protected
by protective device

Figure 11-3: Single Phase Load with Protective Device

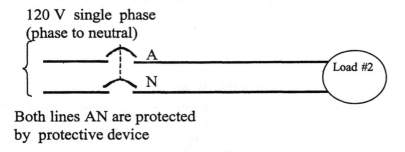

120 V single phase
(phase to neutral)

Both lines AN are protected
by protective device

Figure 11-4: Delta-Wye 480 V/208 V/120 V –AN Phase to Ground Load – Protective Device on Both Legs

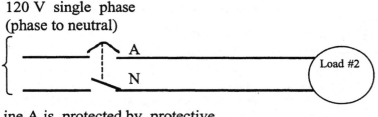

120 V single phase
(phase to neutral)

Line A is protected by protective
device and line neutral is a sample
disconnect switch

**Figure 11-5: 120 V –AN Phase to Ground Load – Line A Protected by Protective Device
and the N phase with Disconnect Switch**

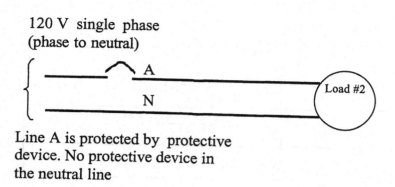

120 V single phase
(phase to neutral)

Line A is protected by protective
device. No protective device in
the neutral line

Figure 11-6: 120 V –AN Phase to Ground Load – Protective Device on Line A Only

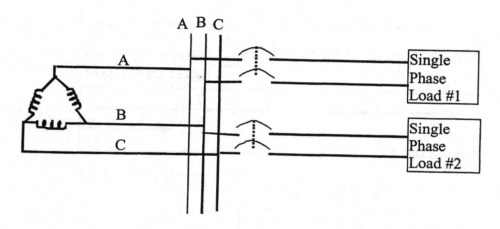

Figure 11-7: Three Phase Delta Distribution and Single Phase Load

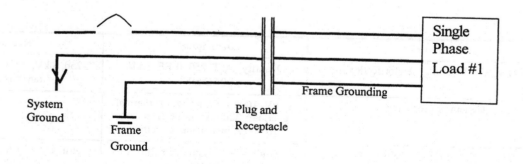

Figure 11-8: Unprotected Neutral Distribution with Equipment Safety Ground

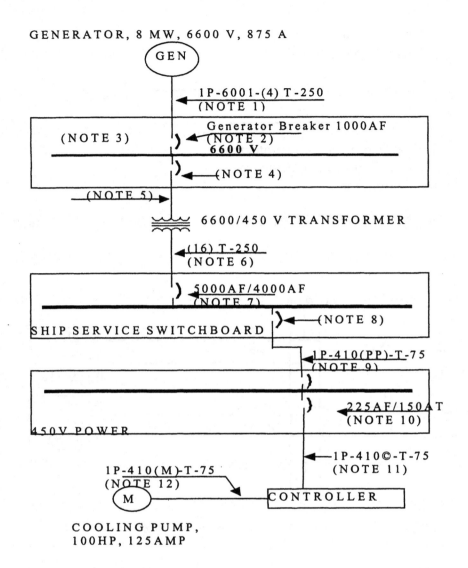

Figure 11-9: Electrical One-Line Diagram-6600 V Generation and 450 V Ship Service Distribution
(for notes refer to Table 11-5)

Notes	Items	Description	Remarks
1	6600 V, 875 A main generator feeder size	Quantity – 4, T-250, IEEE, 8 kV, 90 °C cable	T-250, 8 kV, 90 °C cable ampacity rating is 287A
2	Generator breaker size	11 kV, 1000 A frame, 900 A thermal overload, 3200 A short time delay, 10 000 A instantaneous setting	
3	6600 V/450 V transformer, 3000 kVA primary breaker type	11 kV, 400 A frame with 300 A thermal overload setting	Primary-260 A and secondary 3800 A
4	6600 V/450 V transformer, 3000 kVA primary feeder size	Quantity 1, T-168, IEEE Std 45, 8 kV, 90 °C cable	T-250 , 8 kV cable ampacity is 287 A
5	6600 V/450 V transformer, 3000 kVA secondary feeder size	Quantity 16, T-250, IEEE Std 45, 600 V, cable	3800 A is the transformer secondary current rating. The T-

Notes	Items	Description	Remarks
			250 cable rating is 259 A
6	6600 V/450 V transformer, 3000 kVA, secondary breaker type at the ship service switchboard	600 V, air frame breaker, 5000 A frame capacity with 4000 A trip element	3800 A is the transformer secondary current rating
7	Power panel feeder breaker at the ship service switchboard for a load of 500 A	600 V, 600 A frame air breaker with 600 A thermal overload trip unit	
8	Power panel feeder size	Quantity 2, T-250, IEEE Std 45, 600 V, cable	The T-250 cable rating is 259 A
9	Cooling pump feeder breaker at the panel board	600 V, 225 A frame molded case breaker with 150 A thermal overload trip unit	Pump rating 100 Hp, 125 A
10	100 HP, cooling pump motor feeder from the panel board to the motor controller	Quantity 1, T-83, IEEE Std 45, 600 V, cable	Pump rating 100 HP, 125 A. IEEE Std 45 cable -83 is rated at 131 A
11	Cooling pump motor controller overload heater	Thermal overload trip coil rating 130-133 A	Pump rating 100 Hp, 125 A
12	Cooling pump feeder from the motor controller to the 100 Hp motor	Quantity 1, T-83, IEEE Std 45, 600 V, cable	Pump rating 100 Hp, 125 A

Table 11-5: 6600 V System Device ID and Explanation (refer to Figure 11-9) 6600 V, 8 MW, 875 A, Three-phase, 60 Hz Propulsion Power Distribution System

Remarks:

1. For three conductor medium voltage cable ampacity ratings refer to Section 18. 3. 4 remarks 1, 2, 3.

2. For 600 V, 3 conductor IEEE-45 cable ampacity refer to Table 25, LSE, LSX, T/N , E, and X cable

3. The ampacity derating should be considered for higher ambient considerations

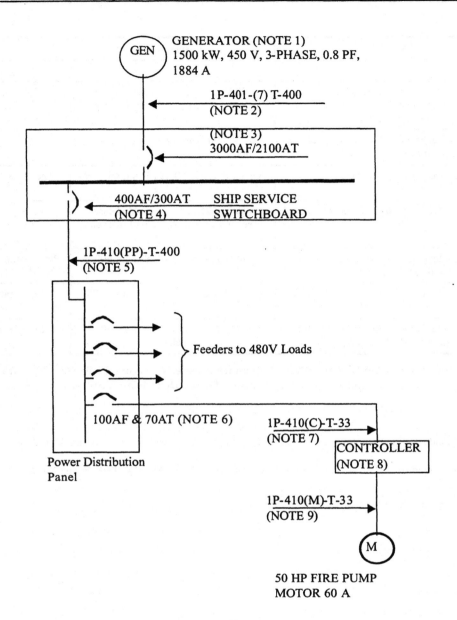

GENERATOR (NOTE 1)
1500 kW, 450 V, 3-PHASE, 0.8 PF,
1884 A

1P-401-(7) T-400
(NOTE 2)

(NOTE 3)
3000AF/2100AT

400AF/300AT   SHIP SERVICE
(NOTE 4)   SWITCHBOARD

1P-410(PP)-T-400
(NOTE 5)

Feeders to 480V Loads

100AF & 70AT (NOTE 6)

1P-410(C)-T-33
(NOTE 7)

CONTROLLER
(NOTE 8)

Power Distribution
Panel

1P-410(M)-T-33
(NOTE 9)

M

50 HP FIRE PUMP
MOTOR 60 A

**Figure 11-10: One-Line Diagram with Feeder Rating, Protection Device, and Identification—
450 V Ship Service Distribution (for notes refer to Table 11-6)**

Notes	Items	Description	Remarks
1	Generator		
2	Generator feeder size	Quantity – 7, T -400 (IEEE Std 45, 600 V, cable)	Per IEEE Std 45, Table 25 -400 is rated 342 A
3	Generator breaker type	Air breaker 3000 A frame, 600 V, 2100 A long time delay	Per IEEE Std 45, 7.5.1, generator circuit breaker long time delay trip setting should not exceed 115% of the full load amp rating of the generator

Notes	Items	Description	Remarks
4	Panel board feeder breaker at the ship service switchboard for a total 260 A load	Molded case circuit breaker , 400 A frame, 600 V, with 300 A thermal overload trip unit	The power panel feeding the 50 HP fire pump motor and other motor loads of total L 260 A
5	Power panel feeder size	Quantity –1, T -400 (IEEE Std 45, 600 V, cable)	
6	Fire pump feeder breaker at the panel board	Molded case circuit breaker, 100 A frame, 600 V, with 70 A thermal trip unit	Fire pump is 50 HP, at 60 A—full load. per IEEE Std 45, 5.9.5.3, the breaker long time delay trip should be set at 115% to 125%.
7	50 HP, fire pump motor feeder circuit breaker from the panel board to the motor controller	Quantity –1, T -33 (IEEE Std 45, 600 V, cable)	Per IEEE Std 45, Table 25 rating is 75 A
8	Thermal overload at the fire pump motor controller	Thermal overload trip coil rated at 600 V, 64 A	
9	Fire pump motor feeder from the motor controller to the 50Hp motor	Quantity-1, T -33 (IEEE Std 45, 600 V, cable)	

Table 11-6: Device Identification and Explanation (refer to Figure 11-10) Generator, 1500 kW, Three-phase, 460 V, 0.8 PF, 1884 A, Ship Service Electrical Power Distribution System

Remarks:

1. For 600 V cable ampacity, refer to IEEE 45-2002, Table 25

2. The ampacity derating may be necessary for higher ambient considerations

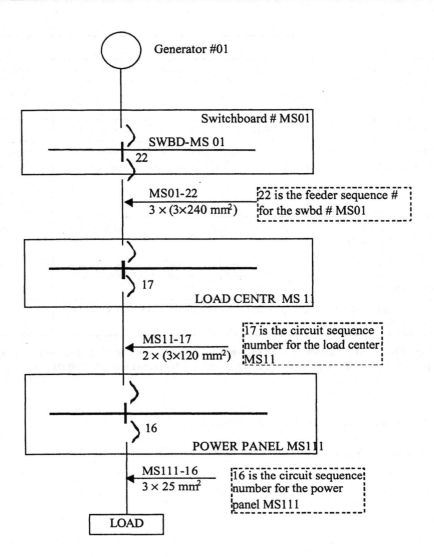

Figure 11-11: Circuit Designation by IEC Standard (Typical Power Distribution Example)

POWER PANEL # 434

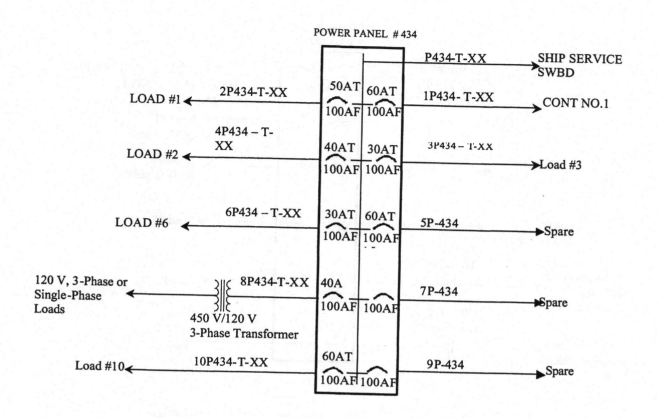

Figure 11-12: Power Panel—Commercial Ship Design

Ten-circuit power panel is shown with main feeder and individual branch circuit feeder. Each branch circuit is shown with circuit breaker frame size and trip rating.

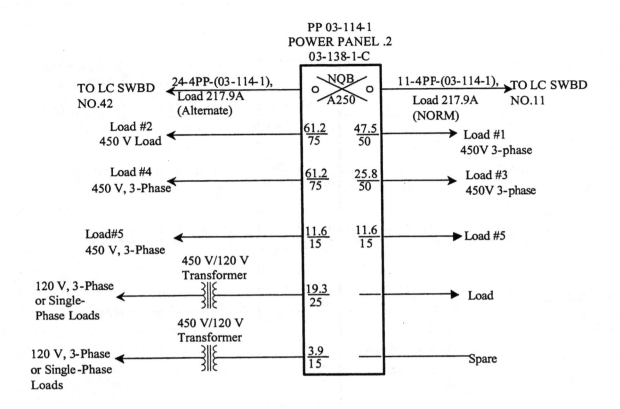

Figure 11-13: Power Panel Configuration for US Navy Ship Design

The figure above is typical with integral ABT for normal and stand-by power supply.
(The Panel board distribution circuit breakers are shown with full load ampere in each branch circuit and the circuit breaker trip rating.)

12 Grounding system—shipboard electrical distribution (SWBS 320)

12.1 General

There are different types of grounding systems used for shipboard electrical distribution system. These include the ungrounded distribution system, solidly grounded distribution system, and high or low resistance grounding system. The following are the most commonly used grounding systems :

a) Ungrounded distribution system (low voltage)

An ungrounded electrical distribution system (Insulated neutral) is preferred by many over a grounded distribution system to reduce the possibility of loss of critical ship service loads due to ground fault. In case of a single-phase ground fault, equipment will continue to run, while a ground detection lamp brightness difference indicates the ground fault condition in ungrounded distribution system. The ground fault clearing actions are, locate the fault by switching on/off to detect the grounded circuit and clear the fault, change over to alternate feeder to ensure continuity of equipment operation, or turn off the service of the grounded circuit and provide alternate standby service. In single phase ground situation equipment will run as it will not trip the protective device. Ungrounded system can be safe, if there is no phase to ground path, however there may be a leakage ground path through the cable insulation and equipment creating the possibility of electric shock due to inadvertent contact of energized equipment. Additionally, the ungrounded system may be dangerous in single phase to ground fault situation as the remaining active phases , each phase to ground may have voltage. Therefore it should not be assumed that electrical live parts, such as cable can be touched on the ungrounded distribution system.

b) Grounded distribution system (solidly grounded or earthed system)

There are special cases when a solidly grounded electrical system is used onboard ship for safety when using portable electric equipment such as electrical hand tools. This type of solidly grounded system is required and generally accomplished by the use of galvanic isolation such as a delta-wye transformer.

c) High resistance grounded distribution system

The high resistance grounded system is generally used for high power and medium voltage and above. The resistance value of the system is determined to minimize the circulation of ground fault current. In this system the ground fault current is continuously monitored and managed for service continuity and equipment protection.

12.2 Ungrounded electrical installation—ground detection system (refer to Figure 12-1)

In this sort of installation, there are three sets of ground detection lamps used for ground detection, one set for each phase. The lamps are grounded to a common neutral point with a normally closed switch by connecting in a WYE configuration. The lamps are usually provided with integral transformer to provide 12V, and 25watt configuration. Normally the system line voltage is equal to the phase-to-phase voltage divided by the square root of three. However, if one phase is grounded, the lamp in that phase will go weaker . The switch is provided to test the system. The switch opens the line to provide phase-to-phase voltage at the lamp, turning the lamp to full brightness.

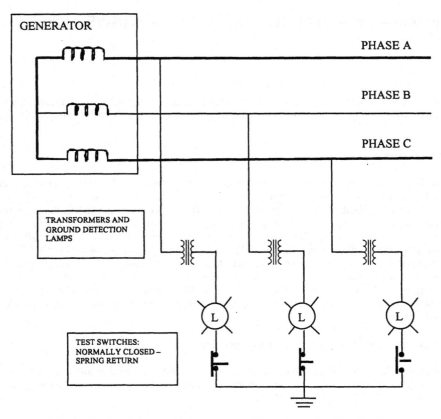

Figure 12-1: Ungrounded System – Ground Detection System

12.3 Grounded installation—ground detection devices (refer to Figure 12-2)

A neutral-grounded, four wire, dual voltage, ground detection system consists of an ammeter, test switch, transducer and a current transformer (CT). The CT is connected to the neutral line with a transducer and ammeter to measure ground current scale of 0 to 10 A.

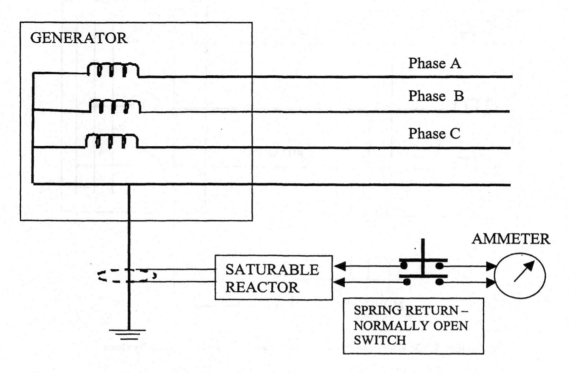

Figure 12-2: Grounded System– Ground Detection System

12.4 Ungrounded installation—insulation monitoring system (refer to Figure 12-3)

The insulation monitoring system continuously monitors the insulation level of the system. The insulation monitoring system is capable of providing remote indication, monitoring by indication and alarm.

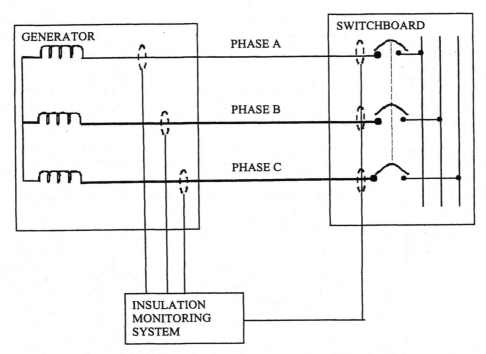

Figure 12-3: Ungrounded System – Insulation Monitoring System

12.5 Generator and its feeder protection—differential protection

Differential protection for medium-voltage generators and feeders is required. For a 480 V system, large generators, above 1500 kVA, differential fault protection is required. In other 480 V systems, with generators of any size, fault protection should be considered. (*Note*: All of the aforementioned protection mechanisms are to be provided on the generator side of the circuit breaker.) This protection is for the generator and the generator feeder cable. The differential protection sets the system at a specific value, and if the system reaches that value, it automatically trips the feeder breaker. Furthermore, if provision is made, it can disable the generator excitation system as well.

The generator specification must include the requirement of the generator side CTs to be installed and terminated inside the generator enclosures, making sure that CTs and termination points are accessible. The switchboard specification must include the requirements of the generator differential protection, including the switchboard CTs.

12.6 NVIC 2-89, Ground detection system (extract only)

3.2 Equipment Ground, Grounded Systems, and Ground Detection

Ground Detection General

Grounds can be a source of fire and electric shock. In an ungrounded system, a single ground has no appreciable effect on current flow. However, if low resistance grounds occur on conductors of different potentials, very large currents can result. In grounded system, a single low impedance ground can result in large fault currents. To provide for the detection of grounds, the regulations require that ground detection means be provided for each electric propulsion system, each ship's service power system, each lighting system, and each power or lighting system that is isolated from the ship's service power and lighting system by transformers, motor generator sets, or other device. This indication need not be part of the main switchboard but should be co-located with the switchboard (i.e. at the engineering control console adjacent to the main switchboard).

Ground Detection Ungrounded System

The indication may be accomplished by a single bank of lights with a switch that selects the power system to be tested, or by a set of ground detector lights for each system monitored. In an ungrounded three-phase system, ground detection lamps are used. The ground lamps are connected in a "wye" configuration with the common point grounded. A normally closed switch is provided in the ground connection.

If no ground is present on the system, each lamp will see one-half of the phase-to-phase voltage and will be illuminated at equal intensity. If line "A" is grounded by a low impedance ground, the lamp connected to line "A" will be shunted out and the lamp will be dark. The other two lamps will be energized at phase-to-phase voltage and will be brighter than usual. If a low resistance ground occurs on any line, the lamp connected to that line will be dimmed slightly and the other two lamps will brighten slightly The switch is provided to aid in detecting high impedance grounds that produce only a slight voltage shift. When the switch opens the ground connection, the voltage across each lamp returns to normal (phase voltage) and each lamp will have the same intensity. This provides a means to observe contrast between normal voltage and voltages that have shifted slightly. Lamp wattages of between 5 W and 25 W when operating at one-half phase-to-phase voltage (without a ground present) have been found to perform adequately, giving a viewer adequate illumination contrast for high impedance grounds. Should a solid ground occur, the lamps will still be within their rating and will not be damaged. For lesser grounds, the lumen output of the lamps will vary approximately proportional to the cube of the voltage. This exponential change in lamp brightness (increasing in two and decreasing in one) provides the necessary contrast.

Ground Detection—Grounded System

On grounded dual voltage systems, an ammeter is used for ground detection. This ammeter is connected in series with the connection between the neutral and the vessel ground. To provide for the detection of high impedance grounds with correspondingly low ground currents, the regulations specify an ammeter scale of 0 to 10 A. However, the meter must be able to withstand, without damage, much higher ground currents, typically around 500 A. This feature is usually provided by the use of a special transducer such as a saturable reactor in the meter circuit. Some ammeters use a non-linear scale to provide for ease in detecting movement at low current values.

Other types of solid-state devices are becoming available that can provide ground detection. They should not be prohibited, but should be evaluated to determine that they are functionally equivalent to the lights and ammeters historically used. Some systems also include a visual and/or audible alarm at a preset level of ground current.

12.7 IEEE Std 45 recommendations for grounding electrical systems and equipment

In IEEE Std 45-1998, the grounding electrical system and equipment was in 11.39, and in IEEE Std 45-2002, the same is in 5.9.6. There are additions and deletions in 2002 edition. Both version are provided for ease of understanding the changes.

12.7.1 IEEE Std 45-1998, 11.39.1, System grounding—general (changed in 2002; see Section 13.7.2 in this Handbook)

Low-voltage (600 V or less) distribution systems should be insulated (ungrounded) to the maximum extent possible to ensure continuity of service under all operating conditions. Ungrounded distribution systems should have all current-carrying conductors, including source of power and all connected loads, completely insulated from ground throughout the system. The ungrounded system should have provisions for continuous ground fault monitoring.

When a solidly grounded distribution system is used, the neutral conductor should be full-sized to preclude overheating due to harmonic distortion from non-linear loads. (See Clause 33 for restrictions in hazardous locations.)

Medium-voltage systems may be grounded through resistance or reactance. For further guidance on these systems, refer to the regulatory agencies and the classification societies.

The secondary winding of each instrument and control transformer should be grounded at the enclosure.

12.7.2 IEEE Std 45-2002, 5.9.6.1, System grounding—general (changed from IEEE Std 45-1998, 11.39.1)

Methods of grounding low-voltage (600 V or less) power distribution system should be determined considering the following:

a) Grounded systems reduce the potential for transient over voltage.
b) To the maximum extent possible, system deign should allow for continuity of service under single-line-to-ground fault conditions, particularly in distribution systems supplying critical ship's service loads.
c) Systems should be designed to minimize the magnitude of ground fault currents flowing in the hull structure.

To satisfy these criteria, it is recommended that systems be designed per one of the following grounding philosophies:

— Ungrounded, with all current-carrying conductors completely insulated from ground throughout the system. Ungrounded systems should have provisions for continuous ground fault monitoring.
— High resistance grounded such that single-line-to-ground faults are limited to 5 A, maximum. High-resistance grounded systems should have provisions for continuous ground fault monitoring. In addition, cables or raceways containing power conductors should be provided with equipment grounding conductors sized in accordance with NEC Table 250-122 to minimize the possibility of ground fault currents flowing in the hull structure.
— Solidly grounded. Solidly grounded designs should be limited to systems supplying non-critical loads, such as normal lighting, galley circuits, and so on. When a solidly grounded distribution system is used, the neutral conductor should be full-sized to preclude overheating due to harmonic distortion from nonlinear loads. (See Clause 33 [not quoted in this Handbook] for restrictions in hazardous locations.) In addition, cables or raceways containing power conductors should be provided with equipment grounding conductors sized in accordance with NEC Table 250-122 to minimize the possibility of ground fault currents flowing in the hull structure.

Medium-voltage systems may be grounded through resistance or reactance. For further guidance on these systems, refer to the regulatory agencies and the classification societies.

The secondary winding of each instrument and control transformer should be grounded at the enclosure.

12.7.3 IEEE Std 45-2002, 5.9.6.2, Grounding points (bold indicates change from IEEE Std 45-1998, 11.39.1.1)

For each separately derived system or part of a system that is desired to be operated as a grounded system, a system ground connection should be provided. Separate grounded and ungrounded systems may be provided in a ship if motor-generator sets or transformers with independent primary and secondary windings isolate the two parts. For 115-V single-phase circuits feeding isolated convenience receptacles, the transformer secondary supplying the 115-V circuit may be solidly grounded and GFCI interrupters installed for the convenience receptacles.

The system grounding point should be selected on the following basis:

— Three-phase system: At the neutral
— Single-phase, three-wire system: At the phase midpoint
— Single-phase, two-wire system: At either power conductor

The grounding connection to the system or parts of a system should be made as close to the source of power as possible rather than at the load ends of the system. There should be only one point of connection to ground on any **separately derived** grounded system. **Where there are multiple sources on a separately derived system, an isolated ground bus shall be provided with only one connection to ground.**

12.7.4 IEEE Std 45-2002, 5.9.6.3, Grounding arrangements (bold indicates change from IEEE Std 45-1998, 11.39.1.2)

Grounded low-voltage systems (600 V AC or less) should be solidly grounded or high-resistance grounded. However, a low impedance system may be used where the maximum fault current of a line-to-ground fault is sufficiently higher than a maximum three-phase fault to necessitate the use of circuit breakers with increased short-circuit interrupting capabilities, or would exceed the magnitude of current for which the generator windings are braced. If an impedance in the ground connection is required, a value of impedance should be selected that will result in the line-to-ground fault current being equal to or less than the three-phase fault current.

The connection to the hull should be made to a suitable structural frame or longitudinal girder. In the case of large-capacity systems, the connection should be made to both a frame and a girder. Grounding devices should have an insulation class suitable for the line-to-line voltage rating of the system. The thermal capabilities of any external impedance should be coordinated with the characteristics of the over-current protective device for the power source to ensure that ground-fault currents will be interrupted before the thermal limit of the grounding device is exceeded.

For grounded systems, a disconnect feature should be provided for each generator ground connection to permit checking the insulation resistance of the generator-to-ground, before the generator is connected to the bus.

12.7.5 IEEE Std 45-1998, 11.39.1. 4, Neutral grounding

Each propulsion, power, lighting, or distribution system having a neutral bus or conductor shall have the neutral grounded at a single point. The neutral of a dual-voltage system (i.e., three-phase four-wire AC, three-wire DC, or single-phase three-wire, AC) should be solidly grounded at or directly adjacent to the generator switchboard. Where the grounded system includes a power source, such as a transformer, the single-point ground connection should be made at or directly adjacent to the switchboard or distribution panel for the power source.

12.7.6 IEEE Std 45-2002, 5.9.6.6, Generation and distribution system grounding

The neutral of each grounded generation and distribution system should be grounded at the generator switchboard, except the neutral of an emergency power generation system should be grounded with no direct ground connection at the emergency switchboard. The emergency switchboard neutral bus should be permanently connected to the neutral bus on the ship service switchboard, and no switch, circuit breaker, or fuse should be in the neutral conductor of the bus-tie feeder connecting the emergency switchboard to the main switchboard.

The ground connection should be accessible for checking the insulation resistance of the generator to ground before the generator is connected to the bus.

Fuses should not, and circuit breakers need not, be provided for the neutral of a circuit. The grounded conductor of a circuit should not be disconnected by a switch or circuit breaker unless the ungrounded conductors are simultaneously disconnected.

Medium-voltage transformer primary neutrals should not be grounded except when all generators are disconnected and power is being supplied from shore.

12.7.7 IEEE Std 45-2002, 5.9.6.7, Tank vessel grounded distribution systems (same as IEEE Std 45-1998, 11.39.1.10)

Distribution systems on tank vessels of less than 1000 V, line-to-line, should not be grounded. If the voltage of the distribution system on a tank vessel is 1000 V or more, line-to-line, and the distribution system is grounded, any resulting current should not flow through hazardous locations.

12.7.8 IEEE Std 45-2002, 5.9.6.8, Sizing of neutral grounding conductors

In an AC system, the conductor that connects the system neutral to the single-point ground should be equal in capacity to the largest generator conductor supplying the system or equivalent for paralleled generators.

12.7.9 IEEE Std 45-2002, 5.9.7.1, Ground detection—general

Means to continuously monitor and indicate the state of the insulation-to-ground should be provided for electric propulsion systems and integrated electric plants; ship service and emergency power systems; lighting systems; and power or lighting distribution systems that are isolated from the ship service or emergency power and lighting system by transformers, motor-generators, or other devices.

For insulated (ungrounded) distribution systems, a device or devices should be installed that continuously monitor and display the insulation level and give audible and visual alarms in case of abnormal conditions.

When a ship is designed for operation with an unattended machinery space, ground detection alarms should be connected to the machinery control monitoring and alarm system.

Ground indicators should be located at the ship service switchboard for the normal power, normal lighting, and emergency lighting systems; at the emergency switchboard for emergency power and lighting, and at the main power switchboard for integrated propulsion systems. All ground indicators should be readily accessible.

12.7.10   IEEE Std 45-1998, 11.39.2, Ground detection (extract only)

For insulated (ungrounded) distribution systems, a device or devices should be installed that continuously monitor and display the insulation level and give audible and visual alarms in case of abnormal conditions. These devices should be in addition to ground detection lights.

12.7.11   IEEE Std 45-1998, 11.39.1.2, Ground detection lamps on ungrounded systems (revised by IEEE Std 45-2002; see Section 13.7.12 of this Handbook)

Ground detection for each ungrounded system should have a monitoring and display system, which has a lamp for each phase that is connected between the phase and ground. This lamp should operate at more than 5 W and less than 24 W when at one-half voltage in the absence of a ground. The monitoring and display system should also have a normally closed, spring return-to-normal switch between the lamps and the ground connection.

12.7.12   IEEE Std 45-2002, 5.9.7.2, Ground detection lamps on ungrounded systems

Ground detection for each ungrounded system should have a monitoring and display system that has a lamp for each phase that is connected between the phase and the ground. This lamp should operate at more than 5 W and less than 24 W when at one-half voltage in the absence of a ground. The monitoring and display system should also have a normally closed, spring return-to-normal switch between the lamps and the ground connection. If lamps and continuous ground monitoring utilizing superimposed DC voltage are installed, the test switch should give priority to continuous ground monitoring and should be utilized only to determine which phase has the ground fault by switching the lamp in. With continuous ground monitoring tied into the alarm and monitoring system, consideration will be given to alternate individual indication of phase to ground fault. If lamps with low impedance are utilized, the continuous ground monitoring is reading ground fault equivalent to the impedance of the lamps, which are directly connected to the ground.

Where continuous ground monitoring systems are utilized on systems where nonlinear loads (e.g., adjustable speed drives) are present, the ground monitoring system must be able to function properly.

12.7.13   IEEE Std 45-2002, 5.9.7.3, Ground detection on grounded neutral AC systems (bold indicates change from IEEE Std 45-1998, 11.39.2.4)

Ground detection for each AC system that has a grounded neutral should have an ammeter and ammeter switch that can withstand the maximum available fault current without damage. The ammeter should indicate the current in the ground connection and should have a scale that accurately, and with clear definition, indicates current in the 0-A to 10-A range. The ammeter switch should be the spring return-to-on type.

The ammeter and current transformer should both be of such a design that ground-fault currents do not damage them. Where the ammeter is located in a remote enclosure from the current transformer, a suitable protective device should be provided to prevent high voltage in the event of an open circuit. A short-circuiting switch should be connected in parallel with the protective device for manually short-circuiting the remote part of the current transformer.

For high resistance grounded system, an indicating ammeter or voltmeter should be provided to indicate ground current flow.

12.7.14   IEEE Std 45-2002, 5.9.7.4, Equipment grounding

Exposed non-current-carrying metal parts of fixed equipment that may become energized because of any condition for which the arrangement and method of installation does not ensure positive grounding, should be permanently grounded through separate conductors or grounding straps, securely attached, and protected against damage.

The metal case of each instrument, relay, meter, and instrument transformer should be grounded.

Instrument and control transformer enclosures should be grounded to the ship structure.

Each receptacle outlet that operates at 55 V or more should have a grounding pole. However, this requirement does not apply to lamp bases, shades, reflectors, or guards supported on lamp holders or lighting fittings constructed of or shrouded in non-conducting material. Grounding poles are also not required on portable appliances that have double insulation, or portable appliances that are protected by isolating transformers. Grounding poles are not required on bearing housings that are insulated in order to prevent the circulation of current in the bearings. Grounding poles are not required on apparatus supplied at not more than 55 V.

12.7.15   IEEE Std 45-2002, 5.9.7.5, Equipment grounding methods (same as 1998-11.39.3)

All non-current-carrying metallic parts of electrical equipment should be effectively grounded by the following methods. Metal frames or enclosures of apparatus should be fixed to, and be in metallic contact with, the ship's structure, provided that the surfaces in contact are clean and free from rust, scale, or paint when installed and are firmly bolted together. Alternatively, they should be connected to the hull either directly by ground strap or, for portable equipment, via the grounding terminal of a receptacle outlet. A reading of 0.1 ohm (DC resistance) or less should be achieved between an equipment enclosure and an adjacent structural ground potential point.

Metallic cable sheaths or armor should not be solely relied upon for achieving equipment grounding. The metallic sheaths and armor should be grounded by means of connectors, or cable glands approved, listed, or labeled for the purpose and designed to ensure an effective ground connection. The stuffing tube should be firmly attached to, and be in effective electrical contact with, a grounded metal structure. Conduits should be grounded by being screwed into a grounded metallic enclosure, or by nuts on both sides of the wall of a grounded metallic enclosure where contact surfaces are clean and free from rust, scale, or paint.

As an alternative to the methods described in the above paragraph, armor and conduit may be grounded by means of clamps or clips of corrosion-resistant metal, making effective contact with the sheath or armor and grounded metal. All joints in metallic conduits and ducts and in metallic sheaths of cables that are used for ground continuity should be solidly made and protected against corrosion.

Every grounding conductor should be of copper or other corrosion-resistant material and should be securely installed and, where necessary, protected against damage and electrolytic corrosion.

On wood and composite ships, a continuous-ground conductor should be installed to facilitate the grounding of non-current-carrying exposed metal parts. The ground conductor should terminate at a copper plate of area not less than 0.2 m² [2 ft²] fixed to the keel below the light waterline in a location that is fully immersed under all conditions of heel.

Every ground connection to the ship structure, or on wood and composite ships to the continuous ground conductor, should be made in an accessible position and should be secured by a screw or connector of brass or other corrosion-resistant material used solely for that purpose.

All armor or other metal coverings of cable should be electrically continuous throughout the entire length and should be effectively grounded to the hull of the ship at both ends, except for branch circuits (final sub-circuits), which may be grounded at the supply end only. The metallic braid or sheath should be terminated at the stuffing tube or connector where the cable enters the enclosure and should be in good electrical contact with the enclosure.

Methods of securing aluminum superstructures to the steel hull of a ship often include insulation to prevent galvanic corrosion between these materials. In such cases, a separate bonding connection should be provided between the superstructure and the hull. The connection should be made in a manner that minimizes galvanic corrosion and permits periodic inspection.

(Same as 1998-11.39.3.1)

12.7.16   IEEE Std 45-2002, 5.9.7.6, Grounding of portable equipment (same as IEEE Std 45-1998, 11.39.3.2)

Portable electrical equipment energized from the ship's electrical system should have all exposed metal parts grounded. This should be accomplished by an additional conductor (green) in the portable cable and a grounding device in the attachment plug and receptacle. Further safety can be provided by the use of an isolating transformer. Double insulated portable electrical equipment need not have exposed metal parts grounded.

Rating or setting of automatic overcurrent device in circuit ahead of equipment, conduit, etc. not exceeding (A) (NEC TABLE-250-122)[a]	Copper conductor size in AWG or kcmil (Circular Mils) (NEC TABLE-250-122)	As an example Single conductor, X-type, 90 °C rated and 45 °C ambient ampacity for IEEE Std 45-2002, Table 25
15	14 (4110)	34
20	12 (6530)	43
30	10 (10 400)	54
40	10 (10 400)	54
60	10 (10 400)	54
100	8 (16 500)	68
200	6 (26 300)	88
300	4 (41 700)	118
400	3 (52 600)	134
500	2 (66 400)	156
600	1 (83 700)	180
800	1/0 (106 000)	207
1000	2/0 (133 000)	240
1200	3/0 (168 000)	278
1600	4/0 (212 000)	324
2000	250 (250 000)	359
2500	350 (350 000)	446
3000	400 (400 000)	489
4000	500 (500 000)	560

NOTES

1—NEC Table 250-122 also states "Note: Where necessary to comply with Section 250-2(d), the equipment grounding conductor shall be sized larger than this table."

2—NEC 250-2(d) **Performance of fault path**— The fault current path shall be permanent and electrically continuous, shall be capable of safely carrying the maximum fault likely to be imposed on it, and shall have sufficiently low impedance to facilitate the operation of overcurrent devices under fault conditions. The earth shall not be used as the sole equipment grounding conductor or fault current path.

[a]NEC Table 250-122 For copper conductor only—Minimum size equipment grounding conductors for grounding raceway and equipment

Table 12-1: Grounding Conductors

13 Electric propulsion system—shipboard (SWBS 235)

13.1 General

The conventional propulsion plant consists of prime movers, such as steam engine, diesel engine, or turbines, which drive the propulsion to drive the ship. The major equipment between the prime mover and the propeller are the reduction gear, main propulsion shaft, propulsion shaft bearings, stern shaft, stern tube bearing, and the stern tube seal. The propulsion drive system is designed to support maximum speed requirements of the ship, as well as all other maneuvering speeds of the ship.

The propellers are classified in two major categories: fixed pitch propeller (FPP) and controllable pitch propeller (CPP). The controllable pitch propeller (CPP) is supported by an hydraulic system, which controls the pitch of the propeller, in turn controlling the propeller thrust. The main propulsion shaft can be long, supported by a number of main shaft bearings. The main propulsion prime mover, main shaft, and stern tube alignment is a major task and requires precision work. The hydraulic system of the controllable pitch propeller is a major maintenance item. The mechanical parts of the propeller blade pitch mechanism are not accessible for maintenance, so any repairs must be done in drydock. Otherwise, the direct drive system is simple, and overall efficiency of the system is very good.

The FPP system propulsion thrust control is provided directly by the prime mover control. The reversing function is performed by stopping the prime mover and reversing it, or by a reverse gear system.

Propulsion shaft misalignment contributes to major vibration in the ship.

The integrated electric propulsion and electrical power systems with electric drive have gained popularity in commercial and military shipbuilding industry over direct drive propulsion. The propeller is driven by an electric motor. The direct drive propulsion system requires the engine is to be connected directly to the propeller through propulsion shaft and stern shaft. For this reason the direct propulsion engine is known as the prime mover. The electric propulsion system engines are not prime movers, but are known as diesel generators. The integrated electric propulsion drive electric motor is directly coupled to the propeller with a stern shaft; alternatively the electric motor may be in the same housing as the propeller. The engines drive the generator at synchronous speed, generating electric power. The electric motor is usually installed very close to the propeller. This arrangement totally eliminates the need for a stern shaft and stern tube bearing system.

Demand for power density plays heavily in the selection of diesel engines versus turbines. There are no predefined requisites for the location of these generators, because the generators do not have any mechanical connection to the propeller. The designers can select any number of generators, and their locations, within the machinery space boundary of generators similar to the ship service generators.

Integrated propulsion electric drive motors can be DC, or AC, synchronous or induction type. 50 or 60 Hz power is converted to variable frequency/voltage by variable frequency drives such as cyclo-converter drive, load commutating inverter, and pulse width modulation (PWM).

Overall fuel economy plays a vital role in commercial and military ship application. Due to the fact that multiple generators are dedicated for power generation, the system can easily be configured for maximum utilization of a combination of generators for fuel efficiency.

Electrical system configuration, power flow management, fault management are significant factors for electric power generation and distribution. In addition, variable speed drive contributes to harmonic distortion, which is detrimental to over all ship service system performance.

Integrated electric plant is promising for military ships, which provides highly flexible electrical power generation and distribution within and around damage control zones with adequate redundancy contributing to survivability.

All electric ship conceptions provide open options of selecting steam, high pressure air and hydraulic auxiliary equipment, which are usually complex and high maintenance items. A contaminated hydraulic system is a costly maintenance item, while electric equipment maintenance requirement is very reasonable. Power generation, distribution, and power management contributes to high load on the prime movers, which contributes to fuel efficiency and less maintenance; this in turn contributes to reduced manning.

13.2 IEEE Std 45-2002 recommendations for electric propulsion and maneuvering systems

IEEE Std 45-2002 makes several recommendations with regards to the propulsion and maneuvering systems, as briefly summarized below.

13.2.1 IEEE Std 45-2002, 31.1, Electric propulsion and maneuvering systems—scope

The application of power electronics technology to large motor drive systems has resulted in the successful installation of shipboard electric propulsion systems that derive input power from a central, fixed-frequency power generation plant that also provides power to the ship service loads. These integrated electric power systems require careful consideration of the physical location requirements of system equipment, cable protection, and control devices as well as special attention to power distribution and load management to ensure an uninterrupted supply of power to vital systems.

There are currently a variety of electric propulsion concepts, such as, but not limited to, fixed speed controllable pitch propeller and variable speed propeller with fixed or controllable pitch. Variable speed propellers are usually supplied from static power converters, such as DC/DC drives, DC drives with thyristor converters, AC drives with pulse width modulated (PWM) converters, load commutated inverters (LCI), and cycloconverters. Each type may give different considerations to network and mechanical design of the system.

This clause provides recommendations covering general specifications, testing, installation, operation, and maintenance of electric propulsion systems. Although these recommendations relate specifically to the electric propulsion equipment, they also address mechanical equipment where required for the successful functioning of the entire system.

13.2.2 IEEE Std 45-2002, 31.3.1, System requirements—general

The design of an integrated electric power system should consider the power required to support ship service loads and propulsion loads under a variety of operating conditions, with optimum usage of the installed and running generator sets.

In order to prevent excessive torsional stresses and vibrations, careful consideration should be given to coordination of the mass constants, elasticity constants, and electrical characteristics of the system. The entire system includes prime movers, generators, converters, exciters, motors, foundations, slip -couplings, gearing, shafting, and propellers.

The normal torque available from the propulsion motors for maneuvering should be adequate to permit the vessel to be stopped or reversed, when the vessel is traveling at its maximum service speed, in a time that is based on the estimated torque-speed characteristics of the propeller during maneuvering and on other necessary ship design characteristics, as determined from hull model testing. Adequate torque margin should also be provided in AC propulsion systems to guard against the motor pulling out of synchronism during rough weather and on a multiple screw vessel, when turning. This margin should be based on information related to propeller and ship characteristics, as well as the propulsion drive's characteristics.

The electric propulsion system may be utilized by a dynamic positioning system. The applicable class and regulatory requirements for dynamic positioning regarding power system and thrust devices should then apply for the propulsion system in addition to the requirements in this clause.

Systems having two or more propulsion generators, two or more propulsion drives, or two or more motors on one propeller shaft should be so arranged that any unit may be taken out of service and disconnected electrically, without affecting the other unit.

13.2.3 IEEE Std 45-2002, 31.4, Prime movers for integrated power and propulsion plants

Prime movers, such as diesel engines, gas turbines, or steam turbines, for the generators in integrated electric power systems shall be capable of starting under dead ship conditions in accordance with requirements of the authority having jurisdiction. Where the speed control of the propeller requires speed variation of the prime mover, the governor should be provided with means for local manual control as well as for remote control.

The prime mover rated power, in conjunction with its overload and the large block load acceptance capabilities, should be adequate to supply the power needed during transitional changes in operating conditions of the electrical equipment due to maneuvering, sea, and weather conditions. Special attention should be paid to the correct application of diesel engines equipped with exhaust gas-driven turbochargers to ensure that sudden load application does not result in a momentary speed reduction in excess of limits specified in Table 4 [see 5.4.1 in this Handbook].

When maneuvering from full propeller speed ahead to full propeller speed astern with the ship making full way ahead, the prime mover should be capable of absorbing a proportion of the regenerated power without tripping from overspeed when the propulsion converter is of a regenerative type. Determination of the regenerated power capability of the prime mover should be coordinated with the propulsion drive system. The setting of the overspeed trip device should automatically shut down the unit when the speed exceeds the designed maximum service speed by more than 15%. The amount of the regenerated power to be absorbed should be agreed to by the electrical and mechanical machinery manufacturers to prevent overspeeding.

Electronic governors controlling the speed of a propulsion unit should have a backup mechanical fly-ball governor actuator. The mechanical governor should automatically assume control of the engine in the event of electronic governor failure. Alternatively, consideration would be given to a system, in which the electronic governors would have two power supplies, one of which should be a battery. Upon failure of the normal supply, the governor should be automatically transferred to the alternative battery power supply. An audible and visual alarm should be provided in the main machinery control area to indicate that the governor has transferred to the battery supply. The alternative battery supply should be arranged for trickle charge to ensure that the battery is always in a fully charged state. An audible and visual alarm should be provided to indicate the loss of power to the trickle charging circuit. Each governor should be protected separately so that a failure in one governor will not cause failure in other governors. The normal electronic governor power supply should be derived from the generator output power or the excitation permanent magnet alternator. The prime mover should also have a separate overspeed device to prevent runaway upon governor failure.

13.3 Salient features of electric propulsion per IEEE Std 45-2002

Torque margin should be provided in the propulsion system to guard against the motor pulling out of synchronism during rough weather and, on a multiple screw vessel, when turning. This margin should be based on information related to propeller and ship characteristics, as well as the propulsion drive's characteristics. Again, propulsion system dynamic analysis, with steady state and transient mathematical modeling and simulation, is necessary at the concept design phase as well as the detailed design phase.

The prime mover-rated power, in conjunction with its overload and large block load acceptance capabilities, should be adequate to supply the power needed during transitional changes in operating conditions of the electrical equipment due to maneuvering, sea, and weather conditions. Special attention should be paid to the correct application of diesel engines equipped with exhaust gas-driven turbochargers to ensure that sudden load application does not result in a momentary speed reduction in excess of the specified limits.

When maneuvering from full propeller speed ahead to full propeller speed astern with the ship making full way ahead, the prime mover should be capable of absorbing a proportion of the regenerated power without tripping from over-speed when the propulsion converter is of a regenerative type. Determination of the regenerated power capability of the prime mover should be coordinated with the propulsion drive system.

Propulsion system electrical steady state and transient mathematical modeling and simulation is necessary at the concept design phase as well as the detailed design phase. This analysis shall be performed for fault current analysis, electrical load flow analysis, and total harmonic analysis.

Mechanical mathematical modeling, electrical modeling and simulation, and vessel speed torque and powering characteristics must be closely analyzed to guarantee that the vessel's intended operational requirements are met.

13.4 Justification of selecting electric propulsion system

The following classes of ships are considered for specific service requirement in view of integrated electric propulsion system.

13.4.1 Commercial ships

Electric propulsion provides less vibration and less noise, compared to direct propulsion. Both features directly contribute to the comfort of the ship's crews--specifically cruise ship passengers. Due to multiple engine generator configuration, it is easier to program an electrical power generating set for optimum engine loading. Optimization of engine loading contributes to overall fuel economy and cost savings. With multiple diesel generator sets, it is easier to perform maintenance to the engines under almost all operating conditions.

The electric propulsion ship build cycle is simpler. The direct drive engines are the first ones to be procured and installed, as those engines are at the bottom of the engine room, and engines are to be in line with the propeller shaft line. The direct drive engines are required to be purchased first as the manufacturing time is longer and installation requirement is earlier. In comparison, the propulsion drive engines' installation location is not as critical as that of the direct drive engines. Due to multiple diesel generators, the engine ratings are smaller, and easier to handle for installation. Engines can be installed a deck higher, allowing for a more relaxed procurement time span. The propulsion motor installation is much easier than that of the engines; and the motors can easily be transported and placed in the propulsion motor room. The podded propulsion drive (Figure 13-) makes it very simple for installation on board ship, as the pod consists of the propulsion motor and the propeller. The pod can be installed late in the build cycle, after completing the engine room work and after completing the aft part of the ship. For ships requiring frequent port service, it is simple to produce required power during port in and port out maneuvering with variable speed electric drive propulsion system.

13.4.2 Military ships

A military ship's electrical propulsion provides additional features such as:

— Enhancement of the vibration- and noise-related ship signature characteristics, due to inherently reduced equipment-generated vibration and noise. Multiple and dispersed electrical propulsors, such as propulsion motors and propellers, enhance the operational capability in case of damage to the vessel.

— Multiple and dispersed electrical power generation and distribution enhances electrical redundancy. Power distribution is easier to set up for ring bus configuration and/or zonal distribution for better redundancy in view of zonal battle damage. A completely automated electrical propulsion power management system contributes to overall fuel efficiency.

13.5 Electric propulsion system dynamic analysis

Electric propulsion plant dynamic analysis should be performed to simulate and predict transient performance of the electric propulsion system onboard ship. Transient performance is observed during the stopping function, maneuvering function, etc. The dynamic analysis is performed in order to:

— Support propulsion plant detailed engineering, design and development.
— Aid integration of mechanical and electrical propulsion equipment.
— Determine adequacy of control system algorithms employed to control and to protect the propulsion plant and to ensure that the rate of load application on the prime movers / generator sets does not exceed the prime mover performance limit.
— Determine the dynamic braking characteristics of the electrical propulsion system and if there is a requirement of dynamic braking resistor to absorb and control regenerative power during crash maneuver
— Coordinate with the prime mover manufacturer to establish the regenerative power absorption capability of the prime mover and match with ship's electrical loading to establish system performance prediction.

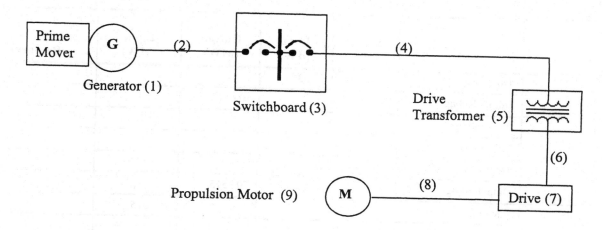

Figure 13-1: Block Diagram for Electrical Propulsion Train with Typical Efficiency (see notes in table below)

			Effi-ciency (%)	Loss (MW)	Output (MW)	Input (MW)	Remarks
1	Generator	Input rating				11.49	Generator input power is equal to the engine output power
		Efficiency	96				
		Total loss		0.442			
		Output rating			11.0494		
2	Generator cable	Input rating				11.0494	
		Efficiency	99.5				
		Total loss		0.055			
		Output rating			10.9944		
3	Switchgear	Input rating				10.9944	
		Efficiency	99				
		Total loss		0.11			
		Output rating			10.9834		
4	Switchgear cable	Input rating				10.9834	
		Efficiency	99.5				
		Total loss		0.0546			
		Output rating			10.9288		
5	Transformer	Input rating				10.9288	
		Efficiency	98				
		Total loss		0.214			
		Output Rating			10.7148		
6	Transformer Cable	Input Rating				10.7148	
		Efficiency	99.5				

			Effi-ciency (%)	Loss (MW)	Output (MW)	Input (MW)	Remarks
		Total Loss		0.0533			
		Output Rating			10.6615		
7	Drive	Input rating				10.6615	
		Efficiency	97				
		Total loss		0.31			
		Output rating			10.3515		
8	Drive cable	Input rating				10.3515	
		Efficiency	99.5				
		Total Loss		0.0515			
		Output rating			10.3		
9	Propulsion motor	Input rating				10.3	
		Efficiency	97				
		Total loss		0.3			
		Output rating			10		
For system overall electrical efficiency is 10 MW/11.49 MW = 87% of power generation. This is for estimation only.							

Table 13-1: Electric Propulsion Pant Power Calculation (refer to Figure 13-1; estimate only)

13.6 ABS propulsion redundancy—direct prime mover driven propulsion system

The following figures are to show conventional redundant propulsion systems, directly driven by a prime mover such as diesel engines, gas turbines etc.

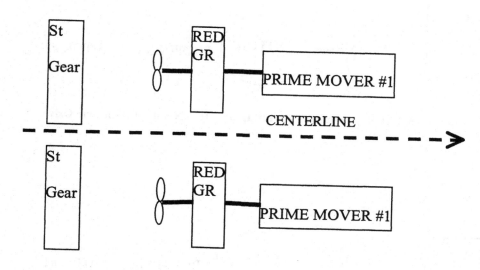

Figure 13-2: ABS Propulsion Redundancy "R2" Notation—Direct Prime Mover Driven Propulsion System

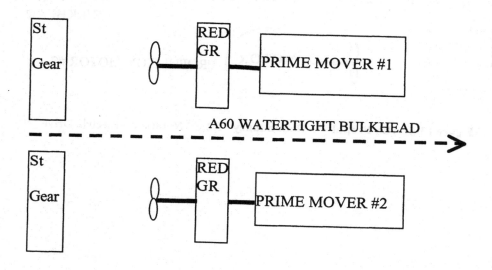

Figure 13-3: ABS Propulsion Redundancy "R2-S+" Notation

13.7 ABS propulsion redundancy—electric motor driven propulsion system

Figure 13-5 and Figure 13-6 are for an electric drive system, where the propulsion is driven by electric motor. Both configurations meet the ABS redundancy requirement. Figure 13- is for ABS R2-S classification, and Figure 13- is for ABS R2-S+ classification.

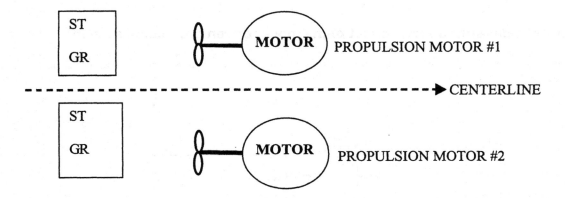

Figure 13-4: ABS Propulsion Redundancy "R2" Notation—Electric Drive

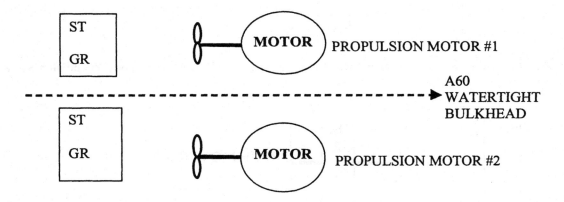

Figure 13-5: ABS Propulsion Redundancy "R2-S" Notation—Electric Drive

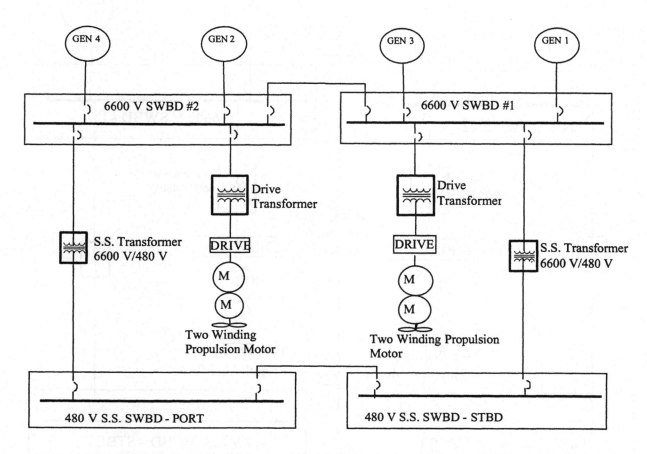

Figure 13-6: Electrical One-Line Diagram—Electrical Propulsion System with ABS R2-S Redundancy

The figure above shows a typical electrical one-line diagram for ABS class redundancy requirement. This power generation and distribution including the electric propulsion system meets the ABS R2-S classification.

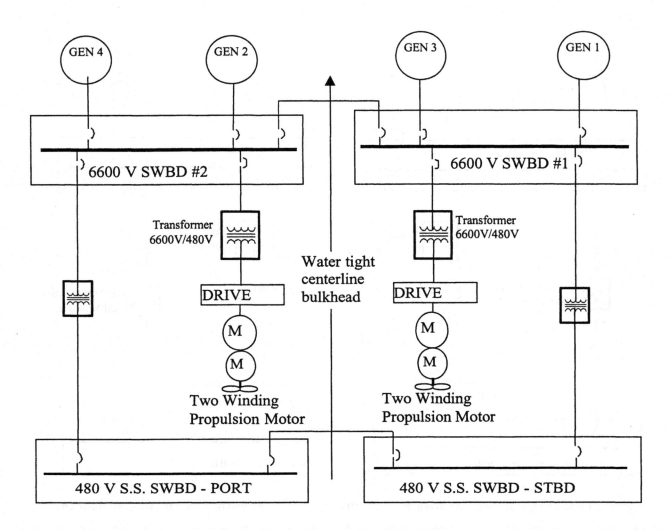

Figure 13-7: Electrical One-Line Diagram—Electrical Propulsion System with ABS R2-S+ Redundancy

Figure 13-7 is a typical electrical one-line diagram in compliance with the ABS redundancy requirement of R2-S+ classification. In this configuration the ship service power is through a 6600 V/480 V transformer. The effect of harmonics must be thoroughly analyzed for this configuration.

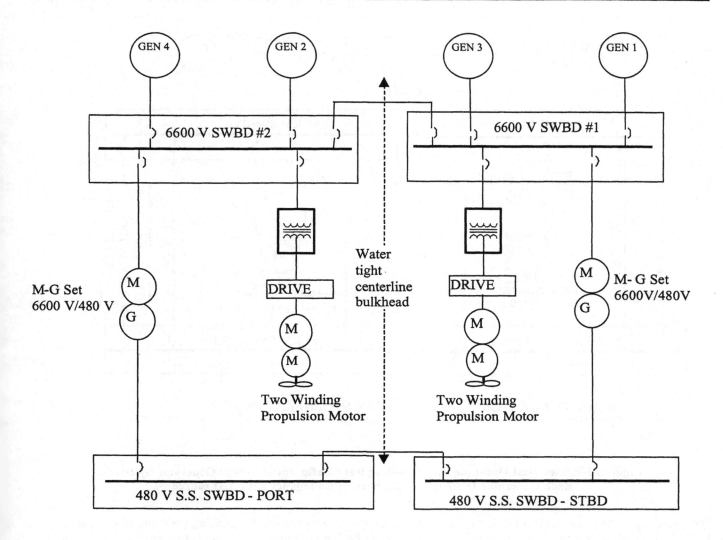

Figure 13-8: Electrical One-Line Diagram-Electrical Propulsion System with ABS R2 -S+ Redundancy and Motor-Generator Set

Figure 13-8 is a typical electrical one-line diagram in compliance with the ABS redundancy requirement of R2-S+ classification. In this configuration, the ship service power is through a 6600 V/480 V motor-generator set. The effect of harmonics is managed by this configuration, supplying clean power to the ship service switchboard.

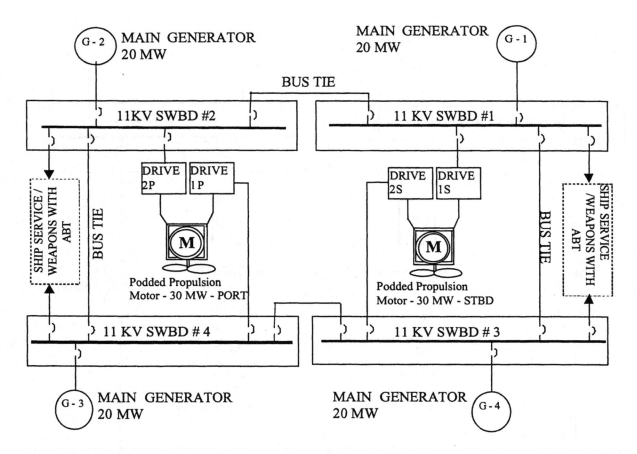

**Figure 13-9: Electrical One-Line Diagram-Ring bus Configuration—Four Generator System;
Each Generator Designated Ship Service/Emergency with ABT source**

The figure above is an electrical propulsion system configuration with power generation and ring bus distribution system. The electric propulsion is shown as a podded propulsion system. Each propulsion motor is of the two winding type. Each winding is powered from a dedicated drive and each drive is powered from a dedicated switchboard.

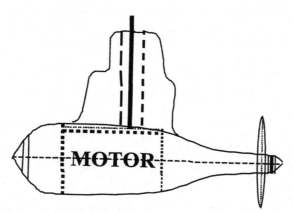

Figure 13-10: Podded Propulsion Motor – Sample Drawing (not to scale)

A podded propulsion motor is a highly compact assembly, ready to be installed.
The unit consists of the following:
1. Propulsion Motor
2. Motor housing
3. Propeller
4. Mounting block assembly
5. Service duct
6. Slip-ring unit

14 Electrical system redundancy requirement—shipboard

14.1 General

Shipboard redundancy of electrical power generation, distribution, and controls is a requirement. However, the redundancy requirement varies with the application criticality of the ship and ship's systems. The normal operating mode as well as the failure mode of shipboard systems leads to the requirement of redundant auxiliaries to ensure operational continuity of vital auxiliary equipment and systems. Redundant equipment, system installations and service continuity are required by regulatory bodies. The commercial ship redundancy requirement is different from that of the military ship. The commercial ship's redundancy requirement is for backup support of power and equipment in case of failure due to operational malfunction of equipment, fire damage, and flood damage. The military ship's redundancy consideration is for back-up support for wear and tear-related failures of equipment, as well as the survival requirement of the battle damaged equipment, systems, and / or compartments.

As an example, the steering gear system is vital for steering the ship. The regulatory body requirement for a one-rudder ship is to provide redundant steering gear hydraulics and steering gear motors. The electric power to one of the motors is required to be supplied from the emergency switchboard, as a redundant support measure. The steering gear motor controllers are of the low-voltage release type so that, during power failure and immediate resumption of ship service power, the steering gear system will be the first one to get back on line. If the main power system does not resume power within the allotted time, the emergency generator will come on line and take over the steering gear operation, which is connected to the emergency switchboard. This switchover will happen within 45 seconds. The steering gear feeders must have maximum port and starboard separation, so that only one feeder is exposed to local fire damage. The steering gear feeder circuit breaker does *not* have overload thermal trip protection, which guarantees that the steering system will not trip due to a local overload situation. The steering gear system will have complete control from the local station at the steering gear room to assure the system control is available if all remote controls fail. These are the redundancy requirements for single rudder steering gear system. The requirement for a dual rudder system is full duplication of a single rudder system.

14.2 Commercial ship automation redundancy requirement (computerized system)

IEEE Std 45-2002's statements on the subject of redundancy are summarized below.

14.2.1 IEEE Std 45-2002, 9.3, Control system design—general

When an automatic machinery control system is provided, the control and monitoring system should be designed and installed to ensure safe and effective operation, to at least the same level as could be obtained by skilled watch-keeping personnel. The system as installed should be capable of meeting all load and operational demands under steady state and maneuvering conditions without the need for manual adjustment or manipulation. Therefore, the system design should ensure the following:

— Unsafe effects of failure of automatic systems or remote control systems are minimized by design, and failures are limited to failsafe states.

— Prompt alerting of an appropriate crew member, either directly or by reliable instrumentation and alarms, of machinery failure, flooding, or fire.

— An alternative means to operate the vessel safely and to counteract the effects of the machinery failure, fire, or flooding.

— Indication at an attended control station of the status of operation of the equipment controlled from that location.

Computer-based or computer-assisted systems are to be designed to ensure operational capability upon loss of any processing component of the system that may cause an unsafe operating condition of the plant. Unless backup, hardwired safety systems, and emergency shutdown systems are fitted, all of a computer-based system's equipment or peripherals should be provided in duplicate and should automatically transfer duty functions to the corresponding standby system/equipment upon failure of an on-line equipment or system. Such action should be alarmed at the remote station. Additionally, particular attention should be paid to the following criteria in the design and installation of automatic or centralized control systems:

a) Maintainability: The capability of keeping the control system in the designed state of operation.

delaying the operation of the vessel.

c)   Standardization: The types of components used. Items that are readily available should be utilized wherever possible.

d)   Operational capability: The capability of keeping display and control device movements as simple as possible. Direction of control device movements should be parallel to the axis of the display they control.

e)   Reliability: The probability that a system will function within its design limits for a specified period of time.

f)   Accessibility: The ability to maintain components and subsystems without removal of, or interference with, other components and subsystems.

The control system should be designed to incorporate hierarchical degrees of automation, starting from local manual control to full unattended automatic operation. The system should be designed such that loss of automatic features automatically shifts the level of control to the next lower step. The design of the system should be such that the transfer to the next lower step does not change the status of the plant or the commanded order from that of the higher level. Additionally, the system should incorporate a feature that allows the operator to set the level of control desired from the main control station.

14.2.2 IEEE Std 45-2002, 9.5.1, Machinery control—general

The centralized control system should support real-time monitoring and control of ship propulsion, electrical, auxiliary, and steering functions.

When provided for essential machinery functions, standby machines should incorporate automatic changeover features. The changeover function should be alarmed. Vital control, safety, and alarm systems should automatically transfer power sources to backup systems upon failure of the operating power source. The system should be designed to ensure the changeover operation does not further degrade the plant condition by inadvertently compounding the situation.

Effective means should be provided to allow propulsion units to be operated under all sailing conditions, including maneuvering. The speed, direction of thrust and, if applicable, the pitch of the propeller, should be fully controllable from the navigating bridge. This includes all thrust conditions, from that associated with maximum controllable open-water astern speed to that associated with the maximum controllable ahead speed.

Control functions from a console may be designed for either remote manual or automatic control. Vital systems that are automatically or remotely controlled should be provided with the following:

a)   An effective primary control system.

b)   A manual alternate control system. A method should exist that provides for alternative positive control of the vital equipment.

c)   A safety control system. Methodologies should be incorporated into the control system either in hardware or software that preclude the unsafe operation of the equipment. Unsafe operation is considered to be operation that will cause severe and permanent damage of the equipment, or risk to operating personnel.

d)   Instrumentation to monitor system parameters necessary for the safe and effective operation of the system. The instrumentation installed should be sufficient to provide a skilled worker with an adequate ability to assess the status, operational performance, and health of the equipment under control. Human factors and ergonometric considerations should be included in the instrumentation design.

e)   An alarm system. A method should be provided to alert the operator to abnormal and dangerous operating conditions.

Provision should be made for independent manual control in the event of a loss of an element of, or the entire centralized control system. The provision should include methods and means of overriding the automatic controls and interlocks. The instrumentation and control provided should be capable of supporting the manual operation for indefinite periods of time.

14.2.3 IEEE Std 45-2002, 9.5.2, Control hierarchy

For ships with more than one control station, a decreasing authority should be assigned according to the following control station locations:

a) Local controls at the controlled equipment
b) Control station(s), in the machinery spaces, closest to the controlled equipment [local control station(s)]
c) Remote control station(s) outside of machinery spaces
d) Navigating bridge (or bridgewing) control station

The control station of higher authority should be designed to include a supervisory means for transferring control from a station of lower authority at all times, and to block any unauthorized request from any station of lower authority. A station of higher authority must be capable of overriding and operating independently of all stations of lower authority. The overriding action when executed should be alarmed at the remote location affected. Transfer of control from one station to another, except for the specific case of override by a station of higher authority, is to be possible only with acknowledgment by the receiving station.

The control function at a control station of higher authority should not depend on the proper functioning of the control station of the lower authority. The control functions from only one station should, except for emergency control actions, initiate a command signal to controlled equipment. Transfer of control between stations should be accomplished smoothly and control of the controlled equipment should be maintained during the transfer. Failure of a control function at a station of lower authority should automatically and smoothly transfer the control to the station of the next higher authority. This transfer should generate an alarm at both stations. Appropriate indications with discriminatory intelligence should be provided at each station, except locally at the controlled equipment, to identify the station in control and any loss of control.

14.3 ABS Commercial ship power generation and distribution redundancy requirement

The ABS redundancy requirement is to ensure that both ACC and ACCU ship class propulsion systems with their auxiliaries and the steering gear systems, are reliable and available to reduce the risk to personnel, the ship, other ships and other structures. The ABS guidelines must be followed during the design phase to fulfill these requirements. The following redundancy classification notation is extracted from the ABS redundancy guide of 1997:

— **R1:** A vessel fitted with multiple propulsion machines but only a singe propulsor and steering system will be assigned the class notation R1

— **R2:** A vessel fitted with multiple propulsion machines and also multiple propulsors and steering system (hence, multiple propulsion systems) will be assigned the class notation R2

— **R1-S:** A vessel fitted with only a single propulsor but having the propulsion machines arranged in separate spaces such that a fire or flood in one space would not affect the propulsion machines in the other spaces will be assigned the class notation R1-S

— **R2-S:** A vessel fitted with multiple propulsors which has the propulsion machines and propulsors, and associated steering systems arranged in separate spaces such that a fire or flood in one space would not affect the propulsion machines and propulsors, and associated steering systems in the other spaces.

— **+ (Plus Symbol):** The mark + will be affixed to the end of any of the above class notations (e.g. R1+, R2-s+) to denote that the vessel's propulsion capability is such that, upon a single failure, propulsive power can be maintained or immediately restored to the extent necessary to withstand adverse weather conditions without drifting.

14.4 Military ship power generation and distribution redundancy requirement

Per the general specification (1995) for military ship construction, SWBS 320c (Systems reliability), two sources of power shall be installed to ensure continuity of service to selected ship electrical systems and equipment. One source of power, the normal source, shall be from a ship service switchboard; the other source shall be from either an emergency switchboard or from another ship service switchboard in a different watertight subdivision from the normal source. Those functions which cannot be handled by the emergency plant shall be supplied by normal and alternate ship service sources of power, where two or more ship service switchboards are installed. For ships with only one ship service switchboard fed by two ship service generators, where the switchboard is provided with a circuit breaker to isolate the generator feeds to two separate sections of the board, each section shall be considered a separate source of power. The vital systems to be provided with an emergency or alternate source of power shall be determined from the characteristics for each ship or its assigned mission with the fleet. In providing emergency power for loads, Category I has the highest priority. If sufficient emergency generator capacity cannot be provided for all loads in Categories I through III, the main machinery plant shall be arranged to provide independent operation of the ship service generators in split-plant mode so that two sources of ship service power can be provided. Any electrical support function required of any system or equipment listed below shall receive the same back-up power supply reliability and shall be supplied from the same power sources.

Category I (highest priority)

In all cases, when emergency power is provided an emergency switchboard shall supply loads associated with **"Emergency Ship Control"** which includes the following:

Fire and foam pumps, Auxiliaries to support the emergency generator prime mover such as Booster pumps, Compartment ventilation motors, Fuel transfer pumps, Starting air pumps, Emergency communications, Emergency lighting, Interior communications, Navigation machinery space circle W ventilation, Steering gear auxiliaries, and Navigation radar VHF bridge-to-bridge radio.

Vital propulsion auxiliaries required for cold starting the ship's plant and necessary for machinery protection, including as a minimum Emergency or standby lubricating oil service pumps, Feed transfer pumps, Forced draft blowers (low speed), Fuel service pumps.

Propulsion plant reliability

The electrical distribution system shall have the capability of restoring the propulsion plant after loss of all ship service power at sea. Redundant auxiliaries shall be segregated and shall be supplied from two separate and independent power distribution panels. Each panel shall have a different normal and alternate source of power.

The following propulsion plant auxiliaries shall be connected to power distribution panels having two sources of power:

Auxiliary machinery cooling water pump, excitation power unit (electric propulsion), forced draft blower, fuel service pump, lube oil service pump, main and auxiliary vacuum pumps, main circulating pump, main condensate pump, main feed booster pump.

In addition, on multiple propulsion plant installations with only electrically-driven pumps:

One fuel service pump, one lube oil service pump, one main condensate pump, and one main feed booster pump of each propulsion plant shall be connected to a two-source power distribution panel of a nonassociated plant.

Bus transfer equipment applications

Automatic bus transfer (ABT) units shall be installed for selection between two power supplies for the following: fire and foam pumps, emergency lighting, fire extinguishing auxiliaries and controls, IC switchboard and panels, pumps associated with the main and auxiliary machinery plant, having low voltage release (LVR) control, and steering power panel.

Manual bus transfer (MBT) units shall be installed for all other loads that require two sources of power. Loads which have LVR or Low Voltage Protection (LVP) control features other than fire fighting systems may be combined and supplied from an ABT unit. Bus transfer units shall be located adjacent to the panel or equipment they supply.

14.5 ABS redundancy classifications

14.5.1 Prime mover driven propulsion redundancy notation "R1"

Figure 14-1 below shows a single engine room with two redundant prime movers, one reduction gear and one steering gear system in accordance with ABS classification notation "R1."

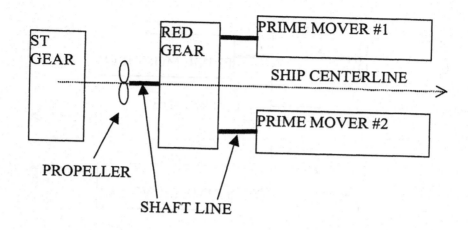

Figure 14-1: ABS Propulsion Redundancy "R1" Notation

14.5.2 Prime mover driven propulsion redundancy notation "R1-S" (Figure 14-2)

Figure 14-2 shows two separate engine rooms, separated by an A-60 watertight bulkhead. One prime mover is in each engine room, providing completely isolated power to the propeller through a common reduction gear. In this configuration, there is one steering gear. This configuration is ABS redundancy classification notation "R1-S."

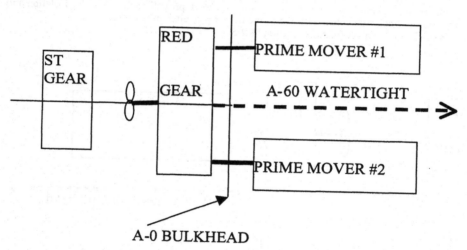

Figure 14-2: ABS Propulsion Redundancy "R1-S" Notation

The propulsion arrangement is fully redundant. There are two propulsion systems, each fully independent. Both prime movers are in the same machinery space and same watertight compartment. This arrangement meets ABS R2-S redundancy notation.

14.5.3 Prime Mover Driven Propulsion Redundancy Notation "R2-S" (Figure 14-3)

Figure 14-3 shows shows an engine room consisting of two prime movers. Each prime mover is providing completely isolated power to each propeller. In this configuration, there are two steering gears providing redundant prime mover, propulsion and steering gear. As both engines are in the same engine room, there is no fire separation for the engines. This configuration is for ABS redundancy classification notation "R2-S."

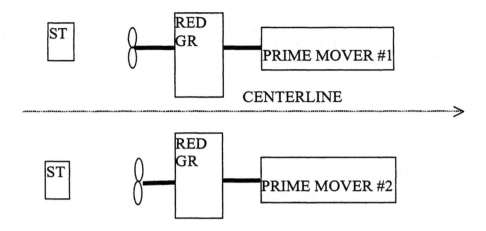

Figure 14-3: ABS Propulsion Redundancy "R2-S" Notation

The propulsion arrangement is fully redundant. There are two propulsion systems, which are fully independent. Each prime mover is in its own watertight compartment. This arrangement meets ABS R2-S+ redundancy notation. If one engine room is flooded or is otherwise de-commissioned, the other engine room will not be affected. Any one of the prime movers can deliver ship's take-home powe r.

14.5.4 Prime mover driven propulsion redundancy notation "R2-S⁺" (Figure 14-4)

Figure 14-4 shows two engine rooms separated by an A-60 watertight bulkhead, providing complete fire isolation and completely redundant propulsion configuration. Each side consists of one prime mover propeller, and steering gear.
This configuration is for ABS redundancy classification notation "R2-S."

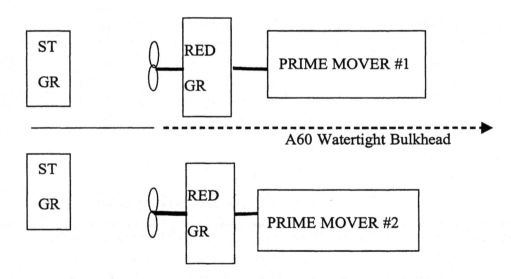

Figure 14-4: ABS Propulsion Redundancy "R2-S+" Notation

14.5.5 Electric motor driven propulsion redundancy notation "R2" (Figure 14-5)

Figure 14-5 shows two redundant electric propulsion plants in one machinery space. Each electric propulsion motor is dedicated to a propeller supported by a dedicated steering gear system.

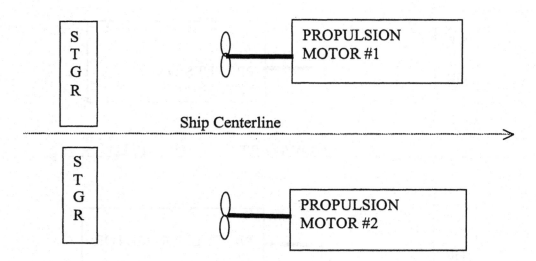

Figure 14-5: ABS Propulsion Redundancy "R2" Notation for Electric Drive

For R2-S ABS class redundant propulsion configuration, each propulsion motor is in a separate engine room, separated by an A60 watertight bulkhead.

14.5.6 Electric motor driven propulsion redundancy notation "R2-S" (Figure 14-6)

Figure 14-6, below, shows two redundant electric propulsion plants; each plant is in a separate machinery space separated by A-60 watertight bulkhead. Any casualty such as fire, flood, electric fault in one machinery space will have no impact on the other machinery room.

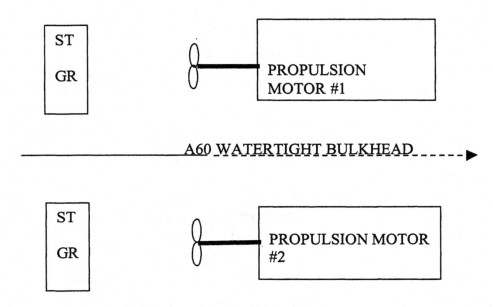

Figure 14-6: ABS Propulsion Redundancy "R2-S" Notation for Electric Drive

14.5.7 Electrical One Line Diagram- Electric Motor Driven Propulsion System with R2 Redundancy (Figure 14-7)

In Figure 14-7, power generation consists of four main generators and one emergency generator. Each main generator is rated for 6 MW, three-phase, 60 Hz, 6.6 kV for a total of 24 MW power generation. Two propulsion motors are each rated at 10 mW for a total 20 MW of propulsion power. Two redundant ship service transformer or motor generator sets are shown at 2000 kVA ship service power. Two main switchboards are shown, each to support two main generators. Two 480 V ship service switchboards are provided to support ship service loads. This design is a complete redundant design; however, there is no physical fire or flooding zone separation. This design qualifies for ABS class R2 notation.

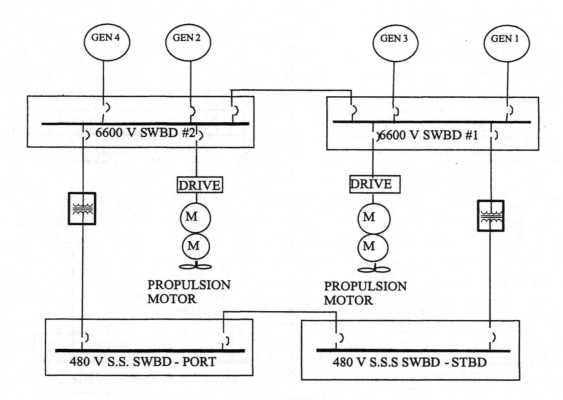

Figure 14-7: Electrical One-Line Diagram—Electrical Propulsion System with ABS R2 Redundancy

14.5.8 Electrical One Line Diagram - Electric Motor Driven Propulsion System with "R2-S⁺" Redundancy with S.S. Transformer (Figure 14-8)

In Figure 14-8, the power generation consists of four main generators, one auxiliary generator and no emergency generator. Two main generators are each rated at 8 MW and two main generators, each rated at 2 MW, provide a total of 20 MW power generation. Two propulsion motors, each rated at 8 MW, give a total of 16 MW propulsion power. Two redundant ship service transformer sets are shown at 2500 kVA ship service power. Two main switchboards are shown, each to support two main generators. Two 480 V ship service switchboards support ship service loads. One auxiliary generator rated at 1200 kW, 480 V is dedicated for ship service loads. This design is a complete redundant design, although there is no physical fire or flooding zone separation. In addition, generator configuration must have dead ship start capability, as there is no emergency generator. This design qualifies for ABS class R2 notation.

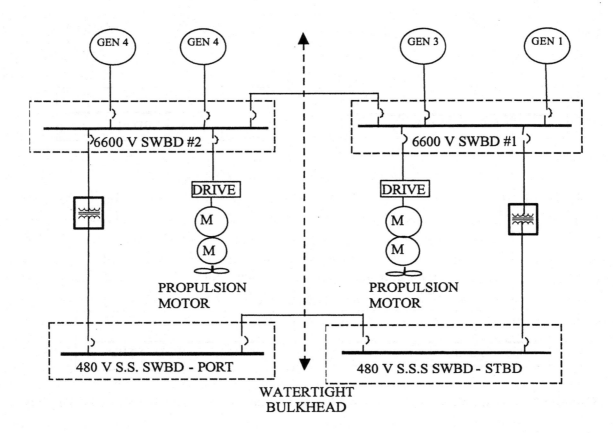

Figure 14-8: Electrical One-Line Diagram-Electrical Propulsion System with ABS R2 -S+ Redundancy with S.S. Transformer

14.5.9 Electrical one-line—electrical propulsion system with R2 redundancy with MG set (Figure 14-9)

In Figure 14-9, there are two 6600 V switchboards, each with two main generators. The ship service switchboards are supplied by dedicated motor-generator sets. In this configuration, the ship service power is clean. This arrangement qualifies as an ABS R2-S redundancy configuration.

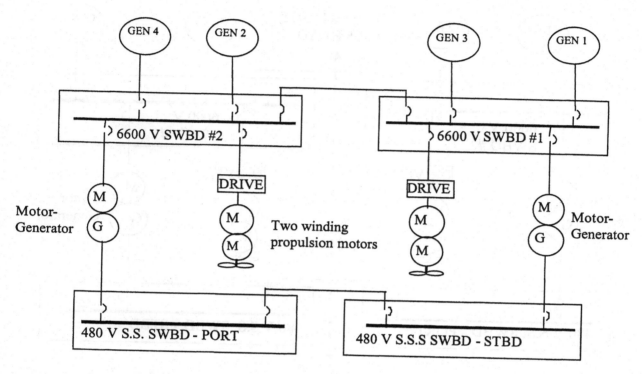

Figure 14-9: Electrical One-Line Diagram—Electrical Propulsion System with ABS R2 Redundancy with MG set

NOTES on the figure above:

1. Each propulsion motor is of the two-winding type. Each winding can run independently.
2. Ship Service power is supplied with the motor generator set. With this arrangement, the ship service power is clean, which meets Mil-Std 1399 power requirements.

14.5.9.1 Electrical one-line—electrical propulsion system with "R2-S+" redundancy with MG set (Figure 14-10)

In Figure 14-10, the ABS "R2 – S $^{+}$" classification requirement is to separate each engine room with an A60 class bulkhead.

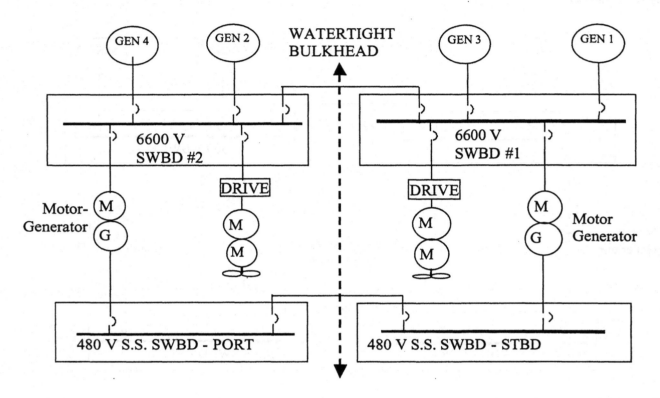

Figure 14-10: Electrical One-Line Diagram-Electrical Propulsion System with ABS R2-S+ Redundancy with MG set

This arrangement has the following features:

1. Each propulsion motor is of the two-winding type. Each winding can run independently.

2. Ship Service power is supplied with the motor generator set. In this arrangement the ship service power is clean, which meets Mil-Std 1399 power requirements.

3. This is a fully redundant electric propulsion system. Each propulsion system is within its own watertight boundary.

14.5.11   Propulsion redundancy configuration – high-level view (Figure 14-11)

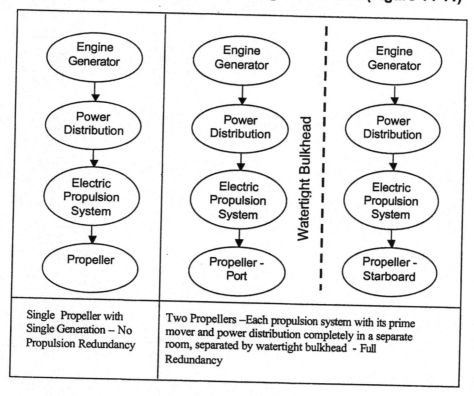

Figure 14-11: Propulsion Redundancy Configuration

For additional redundancy details refer to Section 2. For steering gear redundancy configuration details refer to Section 21.

15 Electric propulsion drive system—shipboard (SWBS 235)

15.1 General

Variable-speed electric propulsion drives are increasingly being considered for use in shipboard propulsion systems. Some of the reasons for this include:

— Multiple generators can be dedicated to generate the required power
— The controllable pitch propulsion system can easily be replaced with adjustable speed propulsion motor
— There is no need for inline engines as generating sets can be installed in any convenient location
— The propulsion motor start up response is much faster then the conventional propulsion system
— It is easier to provide redundant equipment in electric propulsion

The increased high power handling capabilities, fast switching speed, ease of power electronics control , ease of drive control, and low cost of modern semiconductor devices made converter drives affordable for shipboard propulsion drive application and many other shipboard applications. In order to understand the application of these power semiconductors, it is essential to understand the characteristics of those devices. The choice of power electronics devices such as diodes, thyristors, GTOs, IGBTs and IGCTs depend on the application in respect to their properties. Some of the properties to consider include

— Voltage and current rating determine the power handling capability
— Power required by the device control circuitry
— Device conduction loss is directly related to on-state voltage and resistance
— Switching time relates to energy loss
— The temperature coefficient and thermal management of the device

Several different types of drives are shown in this section. For typical power converter drive schematics, refer to the figures as follows: Figure 15-7 is a Load commutating inverter (Current Source) for synchronous motor, Figure 15-8 is a Pulse width modulation (PWM) drive (Voltage source) for induction motor, Figure 15-9 is a Load commutating inverter (LCI) (Current source) for induction motor, and Figure 15-10 is a Cycloconverter for synchronous motor.

15.2 IEEE Std 45-2002, 31.8, Propulsion power conversion equipment

Italics indicate slight modifications for clarification.

The propulsion power conversion equipment should fulfill applicable requirements, class rules, and regulations and comply with either IEC 60146, 61136-1, -2, -3, or design guidelines of IEEE Std 519-1992 to the maximum extent practicable. The *converter* equipment should be designed for continuous operation within the maximum ambient and cooling *media* temperatures (if *liquid*-cooled) without reduction of the equipment's rated performance criteria.

If drives are fitted with forced-ventilation or *liquid* cooling, means should be provided to monitor the cooling system. In the case of a cooling system failure, the current should be reduced automatically to avoid overheating, or the converter may shut down *(safety)* if necessary. Failure of the cooling system or automatic reduction of current should be indicated by an audible and visual alarm in the *control console*. The alarm signal can be generated by a reduction in the flow of liquid coolant, by the loss of the electric supply to the ventilation fans, or by an increase in temperature of semiconductor's heat sink or equivalent alternatives.

Drive enclosures and other parts subject to corrosion should be made of corrosion-resistant material or of a material rendered corrosion resistant.

The static power converters should be mounted in such a manner that they may be removed without dismantling the unit.

Propulsion drive enclosures shall have a protection equivalent to switchboard requirements (See Clause 8 of IEEE Std 45-2002

Whenever power converters for propulsion are applied to integrated electric plants, the drive system should be designed to maintain and operate with the power quality of the electric plant. The effects of disturbances, both to the integrated power system and to other motor drive converters, should be regarded in the design. Attention should be paid to the power quality impact of the following:

a) Multiple drives connected to the same main power system.

b) Commutation reactance, which, if insufficient, may result in voltage distortion adversely affecting other power consumers on the distribution system. Unsuitable matching of the relation between the power generation system's sub-transient reactance and the propulsion drive commutation impedance may result in production of harmonic values beyond the power quality limits.

c) Harmonic distortion can cause overheating of other elements of the distribution system and improper operation of other *ship service* power consumers.

d) Adverse effects of voltage and frequency variations in regenerating mode.

e) Conducted and radiated electromagnetic interference and the introduction of high-frequency noise to adjacent sensitive circuits and control devices. Special consideration should be given for the *shipboard power conversion equipment* installation, filtering, and cabling to prevent electromagnetic interference.

The following propulsion drive unit protection should be provided:

— Drive over-voltage protection by suitable devices applied to prevent damage.

— Semiconductor elements in static power converters should have short-circuit protection, unless they are rated for the full short-circuit available at the point of application. When fuses are used for protection, blown fuse indication should be provided.

— Load limiting control to ensure that the permissible operating current of semiconductor elements cannot be exceeded during normal operation.

The propulsion converter should be equipped with a blackout prevention function ensuring that it does not cause overload of the power generation system while ensuring power available to essential ship service loads. This should be effective in normal operations and after a fault in the power system, e.g., loss of one generator.

The propulsion converter should withstand or restart automatically after a short loss of power supply, e.g., after a cleared short-circuit in the power distribution system, if the propulsion unit is necessary to maintain the required maneuverability in the intended operation.

When maneuvering requires regenerative braking of the propulsion unit, the converter should either be equipped with load dumping resistors or regenerative line supply. Dimensioning of the dumping resistors and regenerative line supply should ensure safe dynamic operation of the vessel under all specified conditions, including crash-stop maneuver. Special attention should be made to the ability to regenerate power to the power system without causing excessive voltage and frequency variations, and the propulsion converters with regenerative line supply should provide means to limit the amount of regenerated power to a level that can be absorbed by the line network, as shown in Table 35.

Table 35—Temperature measurement points—propulsion conversion equipment

Sensor	Location	Function	Comment
Cooling media temperature	Cooling media	Warning + trip	In hot spot[a]
Cooling air temperature	Cold air	Warning	If water cooled[b]
Water leakage indicator	Heat exchanger	Warning	If water cooled[c]

[a]Minimum one + one spare. 3-wire PT100 or equivalent. Warning + trip limit to be advised by vendor. Cooling air temperature cold air warning if water-cooled.

[b] Minimum one + one spare per cooling circuit. 3-wire PT100 or equivalent. Warning limit to be advised by vendor. Water leakage indicator heat exchanger warning if water-cooled.

[c] Passive level switch.

15.3 Power converter—semiconductor—power solid-state device fundamentals

Various types of semiconductor switching devices are used with power variable drive converters. The following semiconductor power devices will be discussed for general understanding:

— Diodes
— Silicon-controlled rectifier (SCR) or thyristor
— Thyristor—gate turn-off (GTO)
— Controllable Power Electronic switches : Insulated gate bipolar transistor (IGBT), Integrated Gate Commutated Thyristor (IGCT), etc

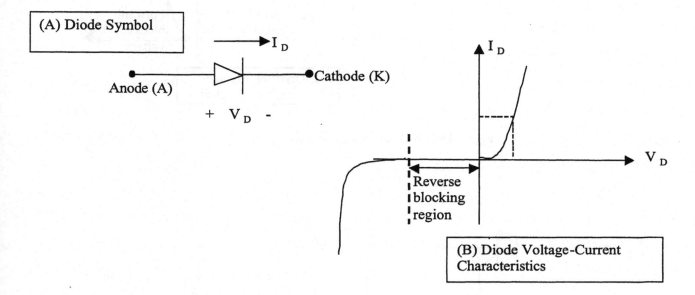

Figure 15-1: Diode (example only)

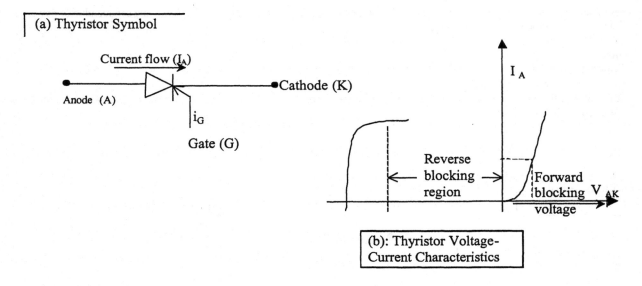

Figure 15-2: SCR or Thyristor (example only)

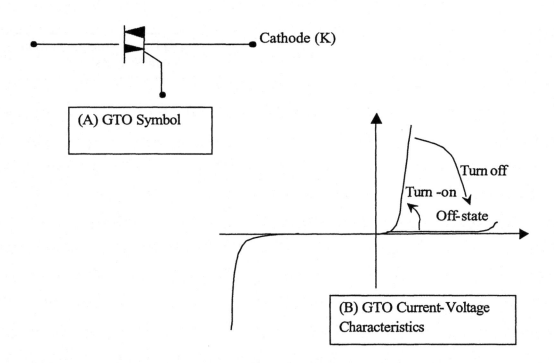

Figure 15-3: Gate Turn-Off (GTO) Thyristor (example only)

A Gate turn-off (GTO) thyristor can be switched on and off. The GTO requires large negative current to turn OFF.

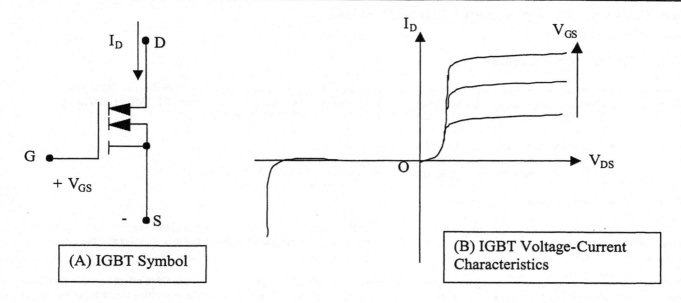

Figure 15-4: IGBT – Integrated Gate Bipolar Transistor (example only)

IGBT switching ON-OFF can be done up to 300 kHz.

15.4 Power converter fundamentals (refer to Figure 15-5)

Power converters are used in a three-phase two-way rectifier or six-pulse bridge, converting the three phase AC voltage to a controllable DC voltage by means of six semiconductor devices, V1 through V6.

V1, V3, and V5 form a positive bridge, providing positive voltage V+. V2, V4, and V6 provide negative voltage output V-. One semiconductor device in the positive bridge and one in the negative bridge simultaneously conduct current while others are blocked, so that the DC motor is connected between the phases that are conducting. The shifting of current from one semiconductor to another is called commutation.

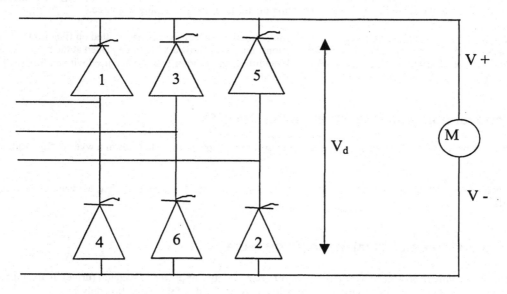

Figure 15-5: Power Converter Schematics (example only)

15.5 Power electronics converter EMI and RFI issues

15.5.1 EMI/RFI fundamentals

A power electronics driven system, such as a variable frequency drive (VFD), generates radio frequency interference (RFI) in the range of 0.5 MHz to 1.7 MHz, (example only) and electromagnetic interference frequencies (EMI) in the range of 1.7 MHz to 30 MHz (example only). The high frequency-related EMI and RFI generation is caused by several factors:

- high carrier frequencies of the pulse-width modulation

- associated short rise times of power electronics (IGBT output devices for example)

- reflected waves from the motor terminals

EMI is also produced by the harmonics, which are generated by carrier frequencies, rise times and reflected waves. Reflected waves are caused by the capacitive effects of long motor leads and the resulting impedance mismatch between the motor cables and the motor windings.

The EMI/RFI will travel to the motor along the motor leads and will be transmitted to ground via the capacitive effect between the motor windings and the motor frame, the capacitive effect between the line conductors and bond wire, and the capacitive effect between the line conductors and conduit. The EMI/RFI will then seek a return path to the source, that is, to the input of the VFD unit.

15.5.2 Power conversion unit—solutions to power cable EMI/RFI problems

Due to the "skin effect" of high frequencies passing on a wire, the bond wire will appear as a high impedance path to those frequencies. For this reason, cable with both a metallic shield and an outer insulating jacket is preferred. An EMI/RFI filter at the drive input terminals will provide a return path for noise that will effectively reduce the distance of noise travel along the ground grid and keep the noise away from the upstream transformer and incoming power lines. A isolation transformer will also provide a return path that will keep the noise away from the upstream transformer and incoming power lines.

15.5.3 Power conversion unit—solutions to signal and EMI/RFI problems

Reduction of noise on signaling circuits may be achieved by using shielded cable. It is generally preferable to ground the shield at the source of the signal rather than at the receiving end, thus preventing the need of noise returning to source.

The use of optical signal isolators will also reduce noise on signaling circuits. Control signals transmitted on fiber optic cables will prevent noise problems caused by long runs and, in some applications, may also allow a VFD to be placed at the motor location, thus eliminating both long motor lead cables and the need for filters. In addition, short motor leads, when used with an EMI/RFI filter or drive isolation transformer, will also shorten the return path along the system.

15.5.4 Power conversion unit—measurement of EMI/RFI

Common mode noise may be measured by connecting three one meg-ohm resistors to the line leads in a wye configuration and an oscilloscope.

It is possible for a standing wave at the motor terminals of a 480 V AC system to be 4 to 5 times higher. Measure voltage standing wave at the motor terminals.

15.6 Power conversion unit (drive) comparison table

The following table gives a comparison of the characteristics of different varieties of power conversion (drive). It should be noted that the calculated data are based on given parameters. (The data given below is typical and is for information only.)

	Cycloconverter	Current source inverter	Voltage source inverter
Topology	Direct conversion without DC link capacitors or inductors	Thyristor rectifier and inverter with DC current link	Diode rectifier with DC voltage link and switching inverters
Semiconductors	Thyristors	Thyristors	Diodes (rectifier) IGCT, IGBT (inverter)
Motor type	Synchronous	Synchronous	Synchronous / Induction
Start-up amps	negligible (transformer inrush)	negligible (transformer inrush)	negligible (transformer inrush)
Start-up torque transients	negligible	Pulsating up to 10% of nominal speed	negligible
Power consumption, low thrust	negligible	negligible	negligible
Power Factor - full load	> 0.76	> 0.85	> 0.95
Power factor variation with load (cos Φ)	0 .. 0.76 (prop. speed)	0 .. 0.85 (prop. speed)	> 0.95 ($\cong$ constant)
Dynamic response (power)	Slow	Slower	Ultra fast (<50ms)
Torque ripple	Smooth	Pulsating	Smoothest
Zero-thrust crossing	Smooth	Pulsating, discontinuous	Smoothest
Stationary torque capability at zero speed	Approx 100%	Approx 100%	Approx 70%
Motor matching required	Some	Yes	No
Generator power factor (typical)	0.7	0.75	0.8
Harmonics	Difficult to filter, solve by system engineering	Easier to filter	Very easy to filter

Table 15-1: Power Converter Unit (drive) Comparison

15.7 Electric drive—power transmission loss calculation form (example only)

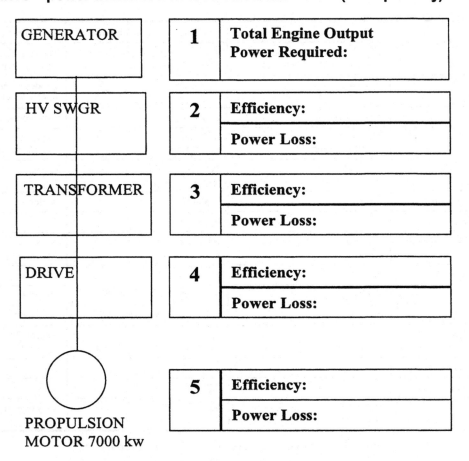

Figure 15-6: Electric Drive—Power Transmission Loss Calculation Form (example only)

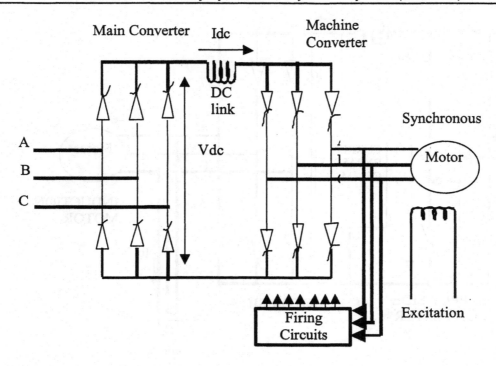

Figure 15-7: Load Commutating Inverter (LCI) Drive for Synchronous Motor (current source) (example only)

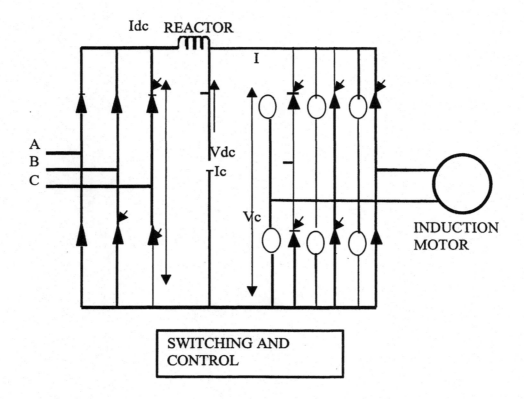

Figure 15-8: Pulse Width Modulation (PWM) Drive (Voltage source) for Induction Motor (example only)

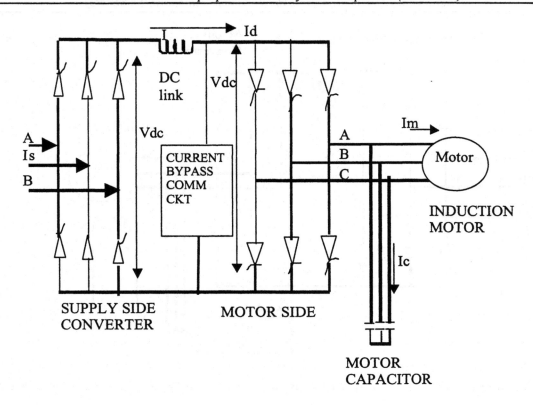

Figure 15-9: Load Commutating Inverter (LCI) Drive (Current source) for Induction Motor (example only)

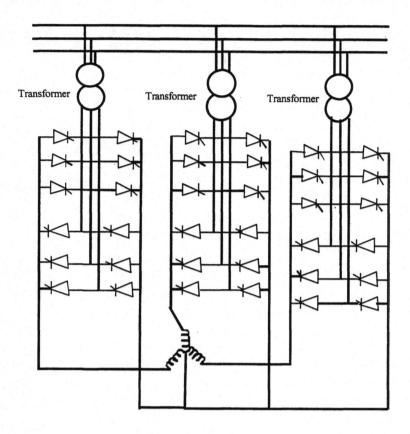

Figure 15-10: Cycloconverter For Synchronous Motor (example only)

16 Electric propulsion motor – shipboard (SWBS 235)

16.1 General

The traditional shipboard propulsion system uses a prime mover-driven shaft-mounted propeller, where the propeller control is performed by the control of the prime mover. Recently, however, the propeller is being mounted onto the electrical motor shaft directly, and control of the propeller is performed by the control of the motor. This is known as an electric propulsion system. A shipboard direct drive propulsion system consists of electric propulsion motor(s), propulsion motor drive equipment such adjustable frequency drives, and associated voltage conversion transformers. The propulsion motor must be able to respond to the propeller dynamic operating behavior under all operational conditions of the ship. The following operating conditions should be considered:

— Normal full load cruising power
— Normal variable cruising power
— Crash stop
— Crash ahead
— Crash astern— Ship running at full speed ahead to full power astern
— Ballard loading— Propulsion motor loading (torque) characteristics from stop to full power ahead with the ship at zero speed.
— Propeller stalling
— Regenerative power
  — Propeller, propeller shaft, and propulsion motor rotational energy
  — Propeller hydrodynamic force caused by the ship's forward motion through the water
— Regenerative Power management during crash astern
— Dynamic Braking characteristics— to absorb and control regenerative power during crash ahead to crash astern maneuvers
— Short time overload rating
— Long time overload rating
— Dual armature winding for redundancy
— Propeller generated vibration withstand capability with thrust bearing
— Induction motor for propulsion
— Synchronous motor for propulsion
— Synchronous motor excitation details
— Propulsion motor for propeller thrust
— Propulsion motor shaft must meet propeller shafting requirement
— Propulsion motor anti-condensation heater
— Propulsion motor cooling requirement
— Propulsion motor bearing lubrication system
— Propulsion motor protection
— Propulsion motor turning gear (The turning gear is for maintenance purpose. The propulsion motor turning speed can be 1 revolution in 5 min to 10 min) During turning the propulsion motor turning the lube oil jacking pump must be running for bearing lubrication
— Remote monitoring of the cooling system leak detector, stator winding temperature, Lube Oil inlet and outlet temperature, air inlet and outlet temperature.
— Internal lighting for motor inspection
— Slip ring viewing window and light
— Propulsion motor power cable voltage rating (2.5 to 3 times the operating voltage due to variable frequency drive application)

The propulsion motor speed variation from minimum to maximum and maximum to minimum is required to be bumpless.

16.2 IEEE Std 45-2002, 31.7.1, Propulsion motors—general (extract)

Propulsion motors should be of substantial and rugged construction, and motor construction should be in accordance with Clause 13 of IEEE Std 45-2002, with additional requirements described in this clause.

Static converter fed propulsion motors should be designed and rated for starting and operation in an adjustable speed drive without any degradation of the insulation, degradation of commutation, and derating. All windings should utilize Class F or H insulation systems that resist moisture, oil vapor, and salt air. In lieu of detailed calculation of additional losses, e.g., due to harmonic currents, the windings should utilize a 55 °C temperature rise over the ambient temperature.

Propulsion motors should be enclosed and ventilated and should be provided with forced ventilation when required by the service. The exhaust air should be discharged through ducts from the motor enclosure. These ducts should be arranged to prevent the entrance of water or foreign material. Ventilation may be provided by the recirculation of air through a closed or partially closed system employing water coolers. Where the coolers are of sufficient capacity to provide 40 °C cooling air at the maximum condition, allowable temperature rises should be based on this ambient temperature. In this case, extreme care should be taken to prevent the water from entering the motor via leaking cooler tubes. The air entering the motors should be filtered to minimize the entrance of oil vapor and foreign material. Means should be provided for totally enclosing the motor when not in use if forced air-cooling is provided by external (in the weather) ventilation ducts. Abnormal brush wear and slip ring maintenance may occur in motors containing silicon materials. Silicon should not be used in any form that can release vapor in enclosed motor interiors.

Air ducts should be provided with high-temperature alarms, dampers, and means of access for inspection. Dampers need not be provided for recirculating systems.

Means should be provided to prevent circulating currents from passing between the motor shaft journal and the bearings. The lubrication of propulsion motor bearings and shafting should be effective at all normal speeds from continuous creep speeds to full speed, ahead, and astern. The shafts and bearings should be self-lubricated. They should not be damaged by slow rotation, under any operational temperature conditions, either when electrical power is applied to the motor or when the propellers may induce rotation.

Pressure or gravity lubrication systems, if used, should be fitted with a low oil pressure alarm and provided with an alternative means of lubrication, such as an automatically operated standby pump, an automatic gravity supply reservoir, or oil rings.

If the insulation is not comprising self-extinguishing material, provisions for a fire extinguishing system suitable for fires in electrical equipment should be provided for motors that are enclosed or in which the air gap is not directly exposed.

Effective means, such as electric heating, should be provided to prevent the accumulation of moisture from condensation when motors are idle.

Propulsion motors should be provided with means for obtaining the temperatures of the stationary windings, as shown in Table 34 of IEEE Std 45-2002. The temperatures are to be displayed at a convenient location such as the main control console. A remote audible alarm actuated when thermal limits are exceeded should be provided. For machines with a heat exchanger type closed circuit cooling method, either the flow of primary and secondary coolants or the winding temperatures should be monitored and alarmed. Liquid coolant leakage detection should be provided and alarmed if applicable. Such detectors should be provided for all sizes of propulsion motors and not limited to machines of certain sizes and rating.

16.2.1 IEEE Std 45-2002, 31.7.2, Propulsion motor excitation

Each propulsion motor exciter should be supplied by a separate feeder.

There should be no overload protection in excitation circuits that would cause the opening of the circuit.

Static excitation power supplies for propulsion motors may be incorporated in the propulsion drive cabinets for the associated motor or may be in separate, freestanding cabinets in the drive or motor room. The standby propulsion exciter power supply should be physically and electrically isolated from the main excitation power supplies and should incorporate an output transfer switch to apply excitation power to the main propulsion systems.

Motor drive should be stopped and should reduce the motor voltage to zero after opening of the field circuit. In constant voltage systems with two or more independently controlled motors in parallel on the same generator, the motor circuit breaker should be tripped when the excitation circuit is opened by a switch or contactor.

16.2.2 IEEE Std 45-2002, 31.7.3, AC propulsion motors

Motors should be polyphase with a voltage between phases not exceeding 13 800 V.

The rating of the propulsion motor should consider the power needed to fulfill thrust requirements both for bollard pull conditions (dynamic positioning or maneuvering) as well as for sailing conditions. Propulsion motor torque, power, and speed characteristics should conform to the propeller characteristics provided by the propeller vendor.

16.3 IEEE Std 45-2002, 31.13.1, Podded propulsion—general

Podded propulsion systems consist of an electric motor placed in a housing located under the hull of the vessel. The electrical motor is directly connected to one or two propellers. The pod is steerable or fixed. A steerable podded propulsion unit constitutes both propulsion machinery and the steering system.

The requirements of the IEC standards, especially the shipbuilding related standards of the serial IEC 60092, have to be fulfilled.

16.3.1 IEEE Std 45-2002, 31.13.2.1, Pod unit—electric motor

The electric motor can be an induction (asynchronous) motor, a synchronous motor, or a permanent magnet motor. If the vessel is equipped with only one podded propulsor, the motor should consist of at least two electrically independent winding systems.

The electric motor should be designed in accordance to 31.7 of IEEE Std 45-2002.

16.3.2 IEEE Std 45-2002, 31.13.2.2, Power transmission system

Power transmission between rotatable and fixed unit components shall either by slip ring assembly or by flexible cable supply power continuously. If slip ring units are used, it shall be possible to install provisions for detecting sparking. The brushes shall be visible for inspection during operation. The degree of protection shall be at least NEMA 4x or equivalent IP enclosure (see Annex C of IEEE Std 45-2002. IP44. Flexible connection for fluids shall be avoided. Any damage to the fluid connection shall not lead to liquid penetration to essential electrical equipment.

16.3.3 IEEE Std 45-2002, 31.13.2.3, Steering system *(italics in original)*

The steering gear system should be of electrohydraulic or electrical type. If the pod is used as a steering device, class and regulatory requirements for steering systems shall apply for the pod steering system and the term *rudder angle* should be interpreted as *pod azimuth angle*.

The podded propulsion unit should be provided with a dual redundant steering system. If more than one azimuthing pod is provided, each shall have a fully independent steering system. The steering system should be capable of moving, stopping, and holding the pod unit at any desired angle within design limits. Requirements for steering gear (Clause 32 of IEEE Std 45-2002) apply as far as possible.

16.3.4 IEEE Std 45-2002, 31.13.2.4, Shaft, bearing, sealing systems, and the propellers

Provisions for continuously or regularly monitoring shall be installed. If water leakage in shaft sealing may penetrate into the vessel's hull, the water shall be collected in a bilge system equipped with pumps.

16.3.5 IEEE Std 45-2002, 31.13.2.5, Ventilation and cooling unit

Ventilation and cooling systems should ensure that under maximum operating conditions; the electrical motor will not exceed the maximum temperature limits. The systems should be designed so that a single failure of the ventilation and cooling system will not lead to loss of propulsion power. The ventilation systems shall be monitored for proper operation. The air flow into the electrical motor should be filtered, and other measures be taken to avoid accidental intake of foreign bodies.

16.3.6 IEEE Std 45-2002, 31.13.2.6, Auxiliary systems

All essential auxiliary systems shall be placed as close to the pod unit as possible. They shall be monitored and supplied from at least two different sources of electrical power for redundancy purposes.

16.4 Motor excitation system

Figure 16-1 through Figure 16-3 show sample propulsion motor excitation systems: separately excited, compound excitation and series excited varieties.

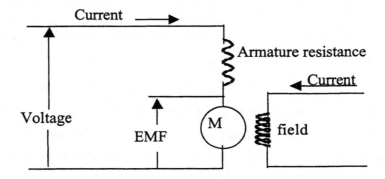

Figure 16-1: DC Motor Excitation (Separately Excited)

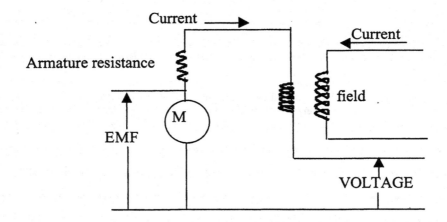

Figure 16-2: DC Motor Excitation (Compound Excitation)

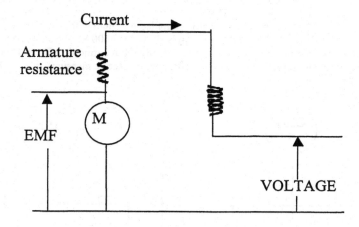

Figure 16-3: DC Motor Excitation (Series Excitation)

16.4.1 Redundant propulsion systems

Figure 16-4 and Figure 16-5 are sample propulsion systems with two electric propulsion motors. These two figures are provided to show redundant propulsion systems in compliance with ABS redundancy requirements.

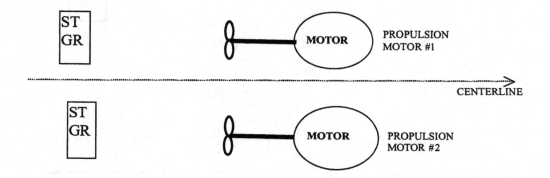

Figure 16-4: ABS Propulsion Redundancy "R2" Notation—Electric Drive

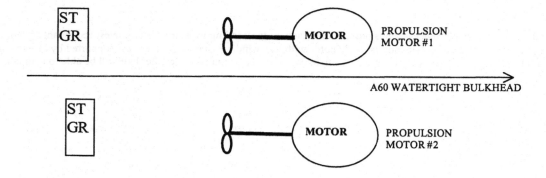

Figure 16-5: ABS Propulsion Redundancy "R2-S" Notation—Electric Drive

For R2-S ABS-class redundant propulsion configuration, each propulsion motor is in a separate engine room, separated by an A60 watertight bulkhead.

Figure 16-6 is a sample electrical one line diagram for an electric propulsion system with two motors. Each motor is shown as a two winding motor to establish redundancy. The power generation is shown as 6600 V. The ship service uses transformers to provide 480 V distribution power. An electric propulsion system suffers inherently from harmonic effects. This arrangement with ship service transformer may not guarantee clean ship service distribution. Additional corrective measures, such as harmonic filtering, may be necessary depending on the type of drive used.

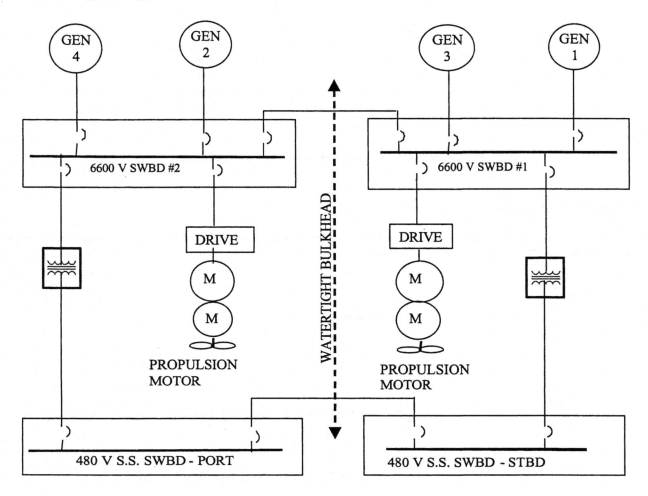

Figure 16-6: Electrical One-Line Diagram—Electrical Propulsion System with ABS "R2-S+" Redundancy with Ship Service Transformer

In general, dedicated and independent drive system is provided for each motor winding to establish completely redundant system for each side of the propulsion. For both sides two motors, each motor with two windings and each winding is supported by its dedicated drive so that if one motor winding fails or one drive system fails in each side the propulsion the other half will be able to maintain propulsion power.

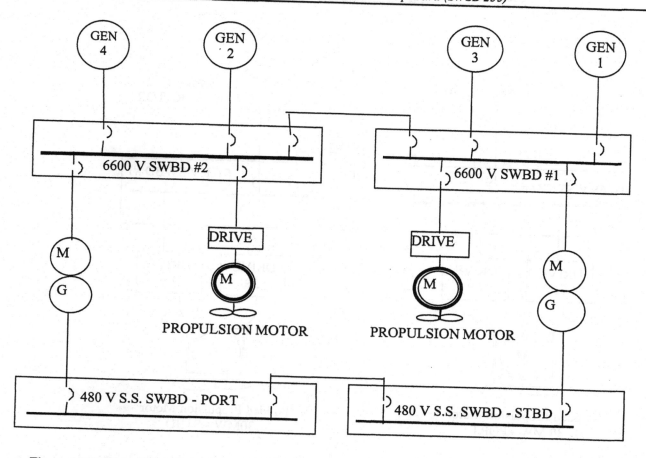

Figure 16-7: Electrical One-Line Diagram—Electrical Propulsion System with ABS "R2" Redundancy with Motor-Generator Set

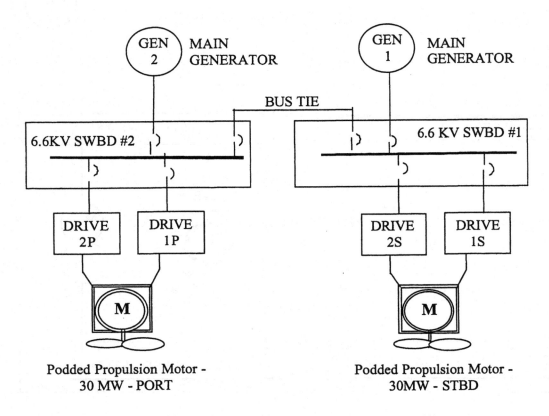

Figure 16-8: Electrical One-Line Diagram—Podded Propulsion system

Figure 16-8 is a sample electrical one line diagram for an electric propulsion system with two podded electric propulsion motors. The power generation is shown as 6600 V.

17 Transformer—shipboard (SWBS 314)

17.1 General

Inductance is one of the passive, two terminal elements in electric circuits, the simple model of which is a current-carrying coil of electric wire. The change in current in the coil produces a change in the magnetic field, which in turn induces voltage in the coil. The same analogy is applied to two or more coils of electric wire, which are so situated that the magnetic field produced by one the current in one coil induces voltage in the other coil. Such a device is called an electric transformer.

The purpose of a transformer in shipboard electrical systems is to isolate the different parts of the electric power distribution system into several partitions, normally in order to obtain different voltage levels such as ship service transformers 450 V to 208 V or 120 V, 6600 V to 450 V to 450 V to 208 V to 120 V.

Transformers are also used to isolate the different parts of the electric power distribution system into several partitions, normally in order to obtain a phase shift. Phase-shifting transformers can be used to feed frequency converters, e.g. for variable speed propulsion drives, in order to reduce the injection of distorted currents into the electric power network by canceling the most dominant harmonic currents. This reduces the voltage distortion for generators and other consumers. The transformers also have a damping effect on high frequency conductor emitted noise, especially if the transformer is equipped with a grounded copper shield between primary and secondary windings.

There are numerous different transformer designs in use, but the most common types for shipboard application are the air insulated dry type and the resin insulated (cast or wound) type. Regulations, and ambient conditions, govern the type of transformer selected.

Physically, the transformer is normally built as three-phase units, with three-phase primary coils and three-phase secondary coils around a common magnetic core. The magnetic iron core constitutes a closed path for magnetic flux, normally with three vertical legs and two horizontal yokes; one at the bottom and one at the top. The inner winding constitutes the low-voltage or secondary windings, and the outer is the primary or high-voltage winding. The ratio of primary to secondary windings gives the transformation ratio. The coils may be connected as a (Wye) Y-connection or (Delta) Δ-connection. The connection may be different on primary and secondary sides, and in such transformers, not only the voltage amplitude will be converted, but there will also be introduced a phase shift between the primary and secondary voltages.

Transformers may be designed according to NEMA standards or IEC standards. For converter transformers, it is essential that the design accounts for the additional thermal losses due to the high content of harmonic currents.

The shipboard power distribution transformers for US application are the following:

— Ship service power transformers HV /480 V
— Ship service power transformers 480 V/120 V
— Propulsion drive power transformers (voltage rating varies)
— Emergency service power transformers 480 V/120 V

The shipboard power distribution transformers for IEC application are the following:

— Ship service power transformers HV /400 V
— Ship service power transformers 400 V/230 V
— Propulsion drive power transformers (voltage rating varies)
— Emergency service power transformers 400 V/230 V

Per IEEE Std C57.12-01, the power transformer kVA shall be the output that can be delivered for the time specified at a rated secondary voltage and rated frequency without exceeding the specified temperature rise limitation under prescribed condition of test, and within the limitation of established test.

In other conventions, the transformer kVA rating may be the incoming power, where the delivered power is the rated incoming power minus the internal power loss and minus the reactive power consumption in the transformer.

17.2 Transformer inrush current

When a transformer is energized, a high inrush current occurs due to the transient saturation of the core material. The transient current inrush decays in a few seconds, but often the duration is long enough for protective devices to take effect—for example, breakers may trip.

17.3 Transformer cooling requirements

The ship service transformer and ship service emergency transformer are required to be naturally air-cooled to maintain the rated temperature rise. The heat-generating element is the generally high transformer wattage loss (no load and load). The selection of transformer location is very important to ensure a natural flow of air around the transformer. Otherwise the transformer ambient temperature will rise and contribute to overheating the transformer.

Single-phase transformer (kVA)	Three-phase transformer (kVA)
1	15
3	30
5	45
7.5	75
10	112.5
15	150
25	225
37.5	300
50	500
75	750
100	1000

Table 17-1: Preferred Continuous kVA of Single-Winding and Three-Winding Transformer (refer to IEEE Std C57.12.01)

17.4 IEEE Std 45-2002 recommendations for transformers

The following section calls out the most salient recommendations of IEEE Std 45 regarding transformers.

17.4.1 IEEE Std 45-2002, 12.1, Transformers—General

Dry type transformers should have copper windings, be air cooled by natural circulation, and have a drip proof enclosure as a minimum. Where used for essential services and located in areas where sprinkler heads or spraying devices for fire prevention are fitted, they should be enclosed so that water cannot cause malfunction.

In cases where capacity, space, or other restrictions warrant, transformers may be of the immersed (nonflammable liquid), self-cooled, or other suitable type. Immersed type transformers should be suitable for operation at 40° inclinations without leakage and provided with a liquid level gauge to give indication of the level of liquid. Drip tray(s) or other suitable arrangements should be provided for collecting liquid leakage.

All transformers should be capable of withstanding the thermal and mechanical effects of a short-circuit at the terminals of any winding for 2 s without damage. Foil-wound transformers constructed of conductors that are uncoated should be vacuum impregnated. Transformers should comply with ANSI C57 or IEC 60726, as applicable to the type, size, application, and voltage rating of the units installed.

17.4.2 IEEE Std 45-2002, 12.3, Type, number, and rating of transformers

The number and rating of transformers supplying services and systems essential to the safety or propulsion of the ship should have sufficient capacity to ensure the operation of those services and systems even when one transformer is out of service. Transformers should be either the three-phase type or the single-phase type, suitable for connection in a three-phase bank, with a Class B temperature rise. All distribution and control transformers should have isolated primary and secondary windings. Transformers with electrostatic shielding between windings should be used in distribution systems containing nonlinear load devices. Autotransformers should be used only for reduced voltage motor starting or other suitable special applications.

17.4.3 IEEE Std 45-2002, 12.5, Parallel operation of transformers

Transformers for parallel operation should have coupling groups and voltage regulation characteristics that are compatible. The actual current of each transformer operating in parallel should not differ from its proportional share of the load by more than 10% of full load current. A means of isolating the secondary connections should be provided.

17.4.4 IEEE Std 45-2002, 12.6, Temperature rise for transformers

The limits of temperature rise in a 40 °C ambient should be in accordance with Table 16. Transformers should also be designed to operate in an ambient temperature of 50 °C without exceeding the recommended total hot spot temperature, provided the output kilovoltampere at rated voltage does not exceed 90% of the rated capacity of the transformer with Class A insulation and 94% of the rated capacity of a transformer with Class B insulation.

Table 16—Transformers and DC balance coils temperature rise

	Copper temperature rise by resistance (°C)				Hottest spot temperature rise (°C)			
Class of insulation								
Part	A	B	F	H	A	B	F	H
Insulated Windings	55	80	115	150	65	110	145	180

NOTE—Metallic parts in contact with or adjacent to insulation shall not attain a temperature in excess of that allowed for the hottest spot copper temperature adjacent to that insulation.

17.4.5 IEEE Std 45-2002, 12.7, Terminals and connections of transformers

Provision should be made to permit the ready connection of external cables to the primary and secondary leads in an enclosed space of adequate size to prevent overheating. Terminals should be readily accessible for inspection and maintenance.

Single-phase transformers should permit the use of single conductor cables, without undesirable inductive heating effects, when interconnecting three single-phase transformers in a three-phase bank.

Transformers should be furnished with a permanently attached diagram plate showing the leads and internal connections and their markings and the voltages obtainable with the various connections.

17.4.6 IEEE Std 45-2002, 31.6, Propulsion drive transformers

Static converter propulsion transformers should be designed and rated for starting and operation in an adjustable speed drive without any degradation of the insulation, degradation of commutation, and derating. All windings should utilize Class F or H insulation systems that resist moisture, oil vapor, and salt air. In lieu of detailed calculation of additional losses, e.g., due to harmonic currents, the windings should utilize a 55 °C temperature rise over the ambient temperature for liquid-filled transformers.

This clause is applicable as additional requirements to Clause 12 of IEEE Std 45-2002. See Section 18 of this Handbook for additional details on transformers whose main purpose is to feed main and excitation power to propulsion converters or auto transformers intended for continuous operation in relation to propulsion converters, e.g., for regenerative breaking. For auxiliary supply transformers, starting transformers, and so on, Clause 12 of IEEE Std 45-2002 should apply.

Propulsion transformers should be evaluated where there is a need for, either or in combination,

—   Adjusting voltage level from distribution system to propulsion converter voltage
—   Reducing harmonic distortion (e.g., 12-pulse configuration)
—   Reducing conductive born noise (EMI), using transformer with conductive shielding between primary and secondary
—   Separate motor drive circuit's from distribution system's grounding philosophy

Where these criteria are fulfilled without the use of transformer, the propulsion transformer will normally be omitted.

Rating determination of the propulsion transformer should include

—   Continuous operation at 105% of maximum continuous load of actual converter
—   Overload duty cycle as specified for propulsion converter, if any
—   Harmonic losses
—   Transformer rating should be according to IEC 60726, or applicable IEEE Std C57™ series standards

Enclosure should be protected against corrosion. Propulsion transformers with enclosure should be equipped with heaters, suitable to keep it dry when not connected. The heaters should be automatically connected when main supply is disconnected. A light indicator should indicate when heater power is connected.

When using air to water heat exchanger, double-tube water pipes should be used, of a material corrosion resistant to the cooling medium. Special attention should also be made to the minimum distances in any direction to ensure proper cooling and inspection/ maintenance work.

Each part of the transformer enclosure should be grounded by a ground strap, or equivalent, to protective earth. Induced currents in enclosure or clamping structure during energizing or in normal operation should not cause arcing.

Propulsion transformers should have a copper shield between primary and secondary windings for reducing conductor borne noise and to reduce the risk for creepage and flashover between primary and secondary windings. This may be omitted when converter is equipped with EMI filters at line supply, but then an overvoltage limiting device should be installed on the outputs of a step-down transformer.

The primary windings should be equipped with full capacity taps at 2.5% and 5% above and below normal voltage taps.

Sufficient space should be provided for termination of the required number of incoming and outgoing cables. Cables should be supported at a distance of a maximum of 600 mm.

Access panels for inspection or cable termination work should be bolted or locked. Locking of hatches that can be opened without a tool or key, e.g. handles, is not acceptable. The transformer should be clearly marked on enclosure and all access panels by warning signs: If greater than 1000 V, "Danger—High Voltage <voltage level>" or if lower voltages, "Danger—<voltage level>".

Terminals for instrumentation should be separated from power supplies, in a separate or in a two-compartment junction box with EMI suppressing device. The propulsion transformers should be equipped with the following instrumentation:

Propulsion transformers should be provided with means for obtaining the temperatures of the windings, as shown in Table 33. The temperatures are to be displayed at a convenient location such as the main control console. A remote audible alarm actuated when thermal limits are exceeded should be provided. For transformers with a heat exchanger type closed circuit cooling method, either the flow of primary and secondary coolants or the winding temperatures should be monitored and alarmed. Liquid coolant leakage detection should be provided and alarmed if applicable. Such detectors should be provided for all sizes of propulsion transformers and not limited to units of certain sizes and rating.

Table 33, Temperature measurement points—propulsion transformer

Sensor	Location	Function	Comment
Winding temperature	Secondary windings	Warning + trip	In hot spot [a]
Cooling air temperature	Cold air	Warning	If water cooled [b]
Water leakage indicator	Heat exchanger	Warning	If water cooled [c]
Pressure switch	In transformer house	Trip	If liquid filled

[a] Minimum one + one spare. 3-wire PT100 or equivalent. Warning + trip limit to be advised by vendor.
[b] Minimum one + one spare per cooling circuit. 3-wire PT100 or equivalent. Warning limit to be advised by vendor.
[c] Passive level switch.

17.5 ABS-2002 additional rules for transformers

In addition to the IEEE recommendations, ABS has rules for transformers, which are summarized in this section.

17.5.1 ABS-2002, 4-8-5, 3.3.5, Number and capacity of transformers

The number and capacity of transformers is to be sufficient, under seagoing conditions, with any three-phase transformer or any one transformer of three single phase transformer bank out of service to carry those electrical loads for essential service and for minimum comfortable conditions of habitability. For this purpose, and for the purpose of immediate continuity of supply, the provision of a single-phase transformer carried onboard as a spare for a three phase transformer bank or ∇ -∇ connection by two remaining single-phase transformers, is not acceptable.

17.5.2 ABS-2002, 4-8-5, 3.5.2, Protection of power transformers

If the total connected load of all outgoing circuits of the power transformer secondary side exceeds the rated load, an overload protection or an overload alarm is to be fitted.

Each high-voltage transformer intended to supply power to the low-voltage ship service switchboard is to be protected in accordance with ABS-2002-4-8-2/9.19. In addition, the following means for protecting the transformers or the electric distribution system are to be provided:

17.5.3 ABS-2002, 4-8-5, 3.5.2(a), Coordinated trips of protective devices

Discriminative tripping is to be provided for the following (see ABS-2002-4-8-2/9.7):

i)   between the primary side protective device of the transformer and the feeder protective devices on the low-voltage ship service switchboard, or
ii) between the secondary side protective device of the transformer, if fitted, and the feeder protective devices on the low-voltage ship service switchboard.

17.5.4 ABS-2002, 4-8-5, 3.5.2(b), Load shedding arrangement

Where the power is supplied through a single set of three-phase transformer to a low-voltage ship service switchboard, automatic load shedding arrangements are to be provided when the total load connected to the low-voltage ship service switchboard exceeds the rated capacity of the transformer. See ABS-2002-4-8-1/5.1.5 and 4-8-2/9.9.

17.5.5 ABS-2002, 4-8-5, 3.5.2(c), Protection from electrical disturbance

Means or arrangements are to be provided for protecting the transformers from voltage transients generated within the system due to circuit conditions, such as high-frequency current interruption and current suppression (chopping) as the result of switching, vacuum cartridge circuit breaker operation, or thyristor-switching.

An analysis or data for the estimated voltage transients is to be submitted to show that the insulation of the transformer is capable of withstanding the estimated voltage transients. See ABS-2002-4-8-5/3.7.5(b).

17.5.6 ABS-2002, 4-8-5, 3.5.2(d), Detection of phase-to-phase internal faults

For three-phase transformers of 100 kVA or more, means for detecting a phase-to-phase internal fault are to be provided. The detection of the phase-to-phase internal fault is to activate an alarm at the manned control station or to automatically disconnect the transformer from the high-voltage power distribution network.

17.5.7 ABS-2002, 4-8-5, 3.5.2(e), Protection from earth-faults

Where a Y-neutral of three-phase transformer windings is earthed, means for detecting an earth-fault are to be provided. The detection of the earth fault is to activate an alarm at the manned control station or to automatically disconnect the transformer from the high-voltage power distribution network.

17.5.8 ABS-2002, 4-8-5, 3.5.2(f), Transformers arranged in parallel

When transformers are connected in parallel, tripping of the protective devices at the primary side is to automatically trip the switch or protective devices connected at the secondary side.

17.5.9 ABS-2002, 4-8-5, 3.5.3, Voltage transformer for control and instrumentation

Voltage transformers are to be protected against short-circuit by fuses on the primary and secondary sides. Special consideration will be given to omitting fuses on the primary side or to fitting automatic circuit breaker on the secondary side instead of fuses.

17.5.10 ABS-2002, 4-8-5, 3.5.5, Overvoltage protection—transformer

Lower voltage systems supplied through transformers from high-voltage systems are to be protected against over-voltage due to loss of insulation between primary and secondary windings. Direct earthing of the lower voltage system or appropriate neutral voltage limiters may be fitted. Special consideration will be given to the use of an earthed screen between the primary and secondary windings of high-voltage transformers.

17.5.11   ABS-2002, 4-8-5, 3.7.5(a) and (b), Application and Plan

a) ABS -2002, 4-8-5. 3.7.5(a): Application:

Provisions of ABS-2002-4-8-5/3.7.5 are applicable to power transformers for essential services. See also ABS-2002-4-8-1/7.3.3. Items 4-8-5/3.7.5(c) and 4-8-5/3.7.5(d) are applicable to transformers of the dry type only. These requirements are not applicable to transformers intended for the following services:

— Instrument transformers
— Transformers for static converters
— Starting transformers

a) ABS -2002, 4-8-5. 3.7.5(b): Plans:

Plans. In addition to the details required in ABS-2002-4-8-3/7, the applicable standard of construction and the rated withstanding voltage of the insulation are also to be submitted for review.

17.5.12   ABS-2002, 4-8-5, 3.7.5(c), Enclosure

Transformers are to have a degree of protection in accordance with 4-8-3/Table 2 but not less than IP44. However, when installed in spaces accessible to qualified personnel only, the degree of protection may be reduced to IP2X. For transformers not contained in enclosures, see 4-8-5/3.11.

17.5.13   ABS-2002, 4-8-5, 3.7.5(e), Testing

Three-phase transformers or three-phase bank transformers of 100 kVa and above are to be tested in the presence of the Surveyor. The test items are to be in accordance with the standard applicable to the transformer. In addition, the tests required in ABS-2002-4-8-3/7.3.5 are also to be carried out in the presence of the Surveyor for each individual transformer. Transformers of less than 100 kVA will be accepted subject to a satisfactory performance test conducted to the satisfaction of the Surveyor after installation. Specific requirements are applicable for the following tests:

i) In the dielectric strength test, the short duration power frequency withstand voltage to be applied is to follow the standard applicable to the transformer but not less than the estimated voltage transient generated within the system. If the short duration power frequency withstand voltage is not specified in the applicable standard, IEC 60076-3 is to be referred to. For the voltage transient, see ABS-2002-4-8-5/3.5.2(c).

ii) The induced over-voltage withstand test (layer test) is also to be carried out in accordance with the standard applicable to the transformers in the presence of the Surveyor. This test is intended to verify the power-frequency withstand strength along the winding under test and between its phase (strength between turns and between layers in the windings). If the induced over-voltage withstand test is not specified in the applicable standard, IEC 60076-3 is to be referred to.

17.6 Figures and examples

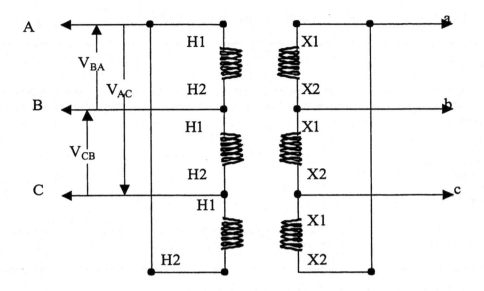

Figure 17-1a: Transformer Delta-Delta Connection

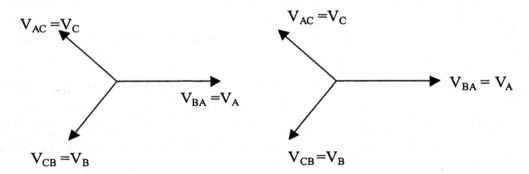

Figure 17-1b: Transformer Delta-Delta Connection – Vector Diagram

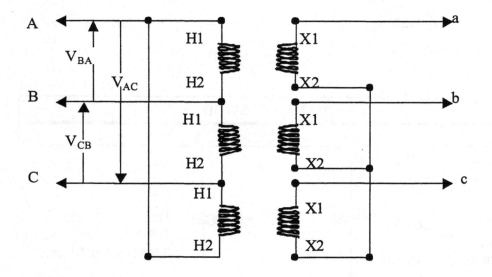

Figure 17-2a: Transformer Delta-Wye Connection

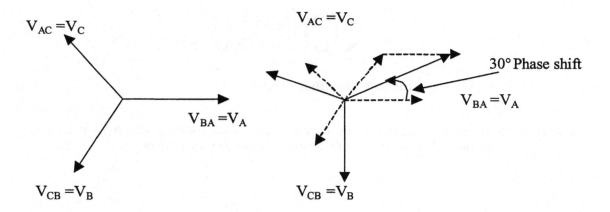

Figure 17-2b: Transformer Delta-Wye Connection – Vector Diagram

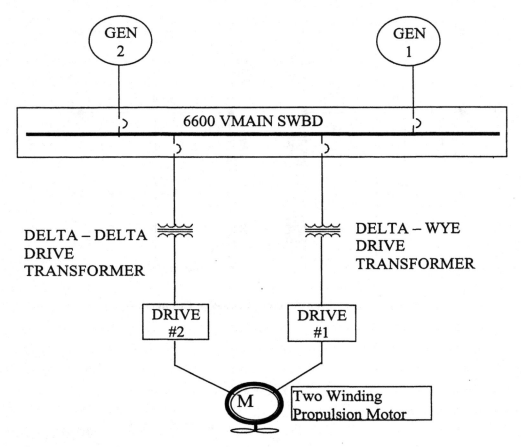

Electric Propulsion drive

Figure 17-3: Electrical One-Line Diagram-6600 V Electric Propulsion Plant with Delta-Delta and Delta-Wye Drive Transformer for Harmonic Cancellation

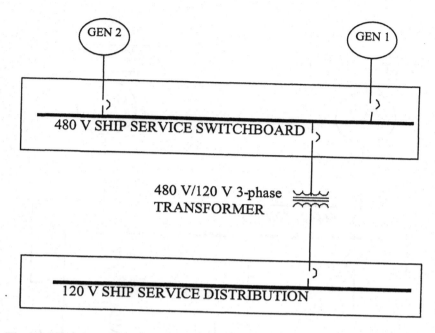

Figure 17-4: Electrical One-Line Diagram-Ship Service Power Distribution with One Ship Service Transformer

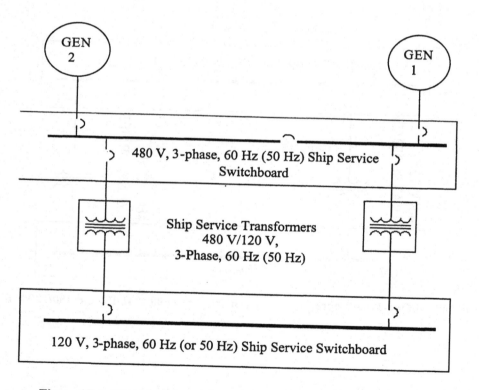

Figure 17-5: Electrical One-Line Diagram-Ship Service Distribution with 480 V/120 V Two Transformers for 120 V Distribution

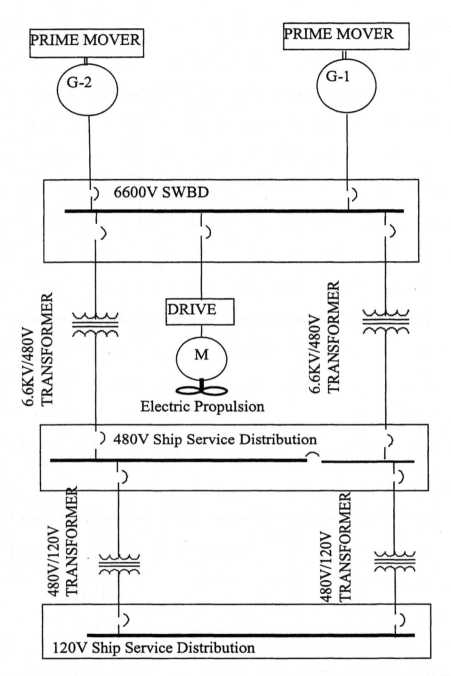

Figure 17-6: Electrical One-Line Diagram – Transformers with 6600 V/480 V and 480/120 V Distributions

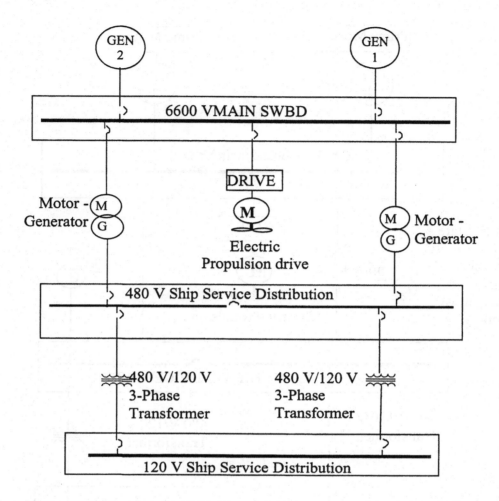

Figure 17-7: Electrical One-Line Diagram-6600 V Electric Propulsion Distribution and 480 V Ship Service Distribution with Motor Generator

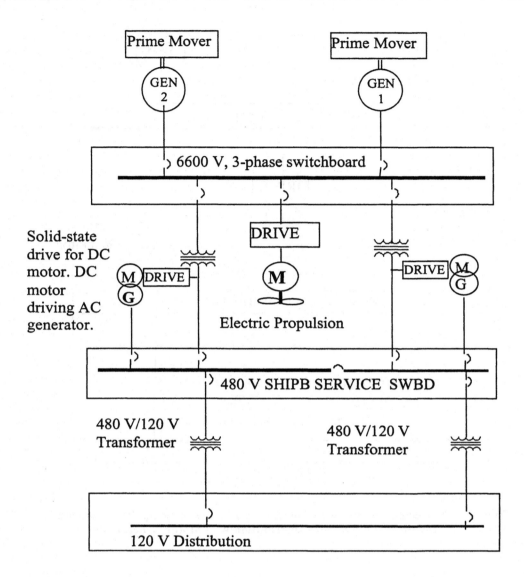

Figure 17-8: Electrical One-Line Diagram- 6600 V Electric Propulsion/distribution, 480 V Ship Service Distribution with Ship Service Transformer or Solid-State Drive DC Motor and AC Generator for Clean Power

18 Marine electrical cable (SWBS 304)

18.1 General

In IEEE Std 45-1998, Clause 8 was for cable construction, Clause 9 was for cable application, and Clause 10 was for cable installation. During the development process of IEEE Std 45-2002, it was decided to remove the cable construction clause from IEEE Std 45 and dedicate a new standard strictly for cable construction, IEEE Std 1580-2001. The IEEE Std 45-2002 clauses were rearranged for performance requirements with Clause 23 for identifying commercial cable and Navy cable, Clause 24 for cable application, and Clause 25 for cable installation. Additionally, IEEE Std 45-2002 acknowledges the IEC cable construction and performance requirements. For IEC cable, appropriate IEC 60092-350 series standards should be consulted. For US Navy cable construction and performance specifications, Mil Specifications, such as Mil-C-24643, Mil-C-24640, Mil-C-915, should be consulted.

18.2 Cable size calculation fundamentals in different rules and regulations

Cable requirements must be properly examined to ensure that the cable selection is within the ampacity limit and voltage drop limit for the intended application. The following list shows some of the many requirements by different regulatory bodies.

— ABS allows cable selection for 100% full load amperes.
— USCG and ACP allow cable selection 125% full load amperes (refer to 0 and chapter 20 for further details).
— For IEEE cable use IEEE ampacity only. Note that cable ampacity may be different from one manufacturer to another manufacturer.
— For IEC cable use, cable ampacity given in the IEC ampacity table should be used.
— IEC allows pulling cable in bundle. For cable installation in bundle, ampacity derating may apply.
— IEEE requirement is to run cable in hanger and tiers, which allows uniform heat dissipation.
— If cable is pulled in an ambient other than 40 °C, i.e. the ambient of 45 °C or 50 °C, ampacity derating should be used as recommended.
— For overhead cable runs in locations such as auxiliary machinery space and under continuous deck, proper ambient temperature must be considered for potential heat pockets.
— Cables of different ambient ratings should not be run in the same wireway. However, if the same wireway is used, the lowest ambient rated ampacity should be used for all cables.
— Circuit breakers are usually tested in the factory for 40 °C ambient and tested with cable ambient rating from 60 °C to 75 °C. If the breakers are connected to higher ambient rated cable, the breakers may be subjected to higher temperature contributing to nuisance tripping, and malfunction.
— Breaker and cable installation requirement should be thoroughly examined. If the feeder cable is directly connected to the breaker lugs and ambient is restricted, the breaker continuous rating must be evaluated. This is more so with the panel mounted breakers as the cable is directly connected to the lug or the breakers.
— The switchboard-mounted breakers are not usually direct contact with the cable as the cable gets connected to the bus. The bus spacing is usually engineered with sufficient spacing for heat dissipation. Therefore, switchboard circuit breaker derating requirement may be different from the panel board mounted breakers.

18.3 IEEE Std 45-2002 recommendations for cable performance

The most salient recommendations of IEEE Std 45-2002 in regards to cable performance are summarized below.

18.3.1 IEEE Std 45-2002, 24.8, Ampacities for cables (extract)

... Current-carrying capacities should be adjusted as noted to suit the ambient temperature in which the cable is installed if it differs from 45 °C. Cable ampacities in Table 25 of IEEE Std 45-2002 [see 19.3.3 of this Handbook] are for single banked installations. Double-banked cables should be derated in accordance with Note 6 of Table 25.

Conductors should be sized to limit conductor operating temperatures at the termination device to those designated for the termination devices involved. For listed devices, unless marked with higher temperature limits, the terminals of devices rated 100 A or less typically are limited to operating temperatures of 60 °C. Devices rated in excess of 100 A typically are limited to 75 °C. In selecting circuit conductors, the designer shall assure that the actual conductor temperature does not exceed the temperature rating of the terminal device. The derating required for motor circuits and continuous loads on devices such as circuit breakers that limits the actual current allowed in circuit wiring can be considered when determining conductor operating temperature. Other factors such as ambient temperature within enclosures and the single conductor configuration of most terminations also can be taken into account when determining the actual conductor temperatures attainable.

Other segments of the cable run where different thermal conditions exist from those at the termination point will require separate derating considerations. The lowest ampacity calculated for any 3-m section in the cable run would determine the cable size.

The ampacities for cable types manufactured and tested in accordance with U.S. Navy military specifications MIL-C-24643A, MIL-C-24640A, and MIL-DTL-915G are to be in accordance with MIL-HDBK-299. Ampacities at 45 °C are to be determined by dividing the 40 °C ampacities in MIL-HDBK-299 by the 40 °C factors contained in Note 5 to Table 25 for the appropriate insulating material. For double-banked installations, the values for U.S. Navy military specification cables are to be multiplied by 0.8, in accordance with Note 6 to Table 25.

18.3.2 IEEE Std 45-2002, 24.10

Where armored cable is used, a sheath may be added over the armor for corrosion protection.

18.3.3 IEEE Std 45-2002, 24.8, Table 25, Distribution, control and signal cable

Table 25—Distribution, control, and signal cables—single-banked, maximum current-carrying capacity (types T, T/N, E, X, S, LSE, LSX, and P @ 45 °C ambient)

| AWG/kcmil | Cross sectional | | Single conductor | | | Two conductor | | | Three conductor | | |
	mm²	Circular Mils	T	LSE LSX T/N E, X	S, P	T	LSE LSX T/N E, X	S, P	T	LSE LSX T/N E, X	S, P
			75 °C	90 °C	100 °C	75 °C	90 °C	100 °C	75 °C	90 °C	100 °C
20	0.517	1020	9	11	12	8	9	10	6	8	9
18	0.821	1620	13	15	16	11	13	14	9	11	12
16	1.31	2580	18	21	23	15	18	19	13	15	16
-	1.5	2960	20	24	26	17	20	22	14	17	18
15	1.65	3157	21	26	28	18	22	23	15	18	19
14	2.08	4110	28	34	37	24	29	31	20	24	25

12	3.3 1	6530	35	43	45	31	36	40	24	29	31
10	5.26	10 400	45	54	58	38	46	49	32	38	41
8	8.37	16 500	56	68	72	49	60	64	41	48	52
7	10.5	20 800	65	77	84	59	72	78	48	59	63
6	13.3	26 300	73	88	96	66	79	85	54	65	70
5	16.8	33 100	84	100	109	78	92	101	64	75	82
4	21.2	41 700	97	118	128	84	101	110	70	83	92
3	26.7	52 600	112	134	146	102	121	132	83	99	108
2	33.6	66 400	129	156	169	115	137	149	93	111	122
1	42.4	83 700	150	180	194	134	161	174	110	131	143
1/0	53.5	106 000	174	207	227	153	183	199	126	150	164
2/0	67.4	133 000	202	240	262	187	233	242	145	173	188
3/0	85.0	168 000	231	278	300	205	245	265	168	201	218
4/0	107.2	212 000	271	324	351	237	284	307	194	232	252
250 kcmil	126.7	250 000	300	359	389	264	316	344	217	259	282
262 kcmil	133.1	262 600	314	378	407	278	333	358	228	273	294
300 kcmil	152	300 000	345	412	449	296	354	385	242	290	316
313 kcmil	158.7	313 100	351	423	455	303	363	391	249	298	321
350 kcmil	177.3	350 000	372	446	485	324	387	421	265	317	344
373 kcmil	189.4	373 700	393	474	516	339	406	442	277	332	361
400 kcmil	203	400 000	410	489	533	351	419	455	286	342	371
444 kcmil	225.2	444 400	453	546	588	391	468	504	319	382	411
500 kcmil	253.3	500 000	469	560	609	401	479	520	329	393	428
535 kcmil	271.3	535 000	485	579	630	415	496	538	340	407	443
600 kcmil	304	600 000	521	623	678	450	539	585	368	440	478
646 kcmil	327.6	646 000	557	671	731	485	581	632	396	474	516
750 kcmil	380	750 000	605	723	786	503	602	656	413	494	537
777 kcmil	394.2	777 000	627	755	822	525	629	684	431	516	562
1000 kcmil	506.7	1 000 000	723	867	939	601	721	834	493	592	641
1111 kcmil	563.1	1 111 000	767	942	1025	637	784	854	523	644	701
1250 kcmil	635	1 250 000	824	990	1072	—	—	—	—	—	—
1500 kcmil	761	1 500 000	917	1100	1195	—	—	—	—	—	—
2000 kcmil	1013	2 000 000	1076	1292	1400	—	—	—	—	—	—

Ampacity adjustment factors for more than three conductors in a cable with no load diversity: Percent of values in Table 25 for Three Conductor Cable as adjusted.

Number of Conductors for ambient temperature, if necessary

    4 through 6: 80
    7 through 9: 70
    10 through 20: 50

    21 through 30: 45
    31 through 40: 40
    41 through 60: 35

NOTES

1—Current ratings are for AC or DC.

2—For service voltage 1001 V to 5000 V, Type T, T/N, LSE, and LSX should not be used.

3—Current-carrying capacity of four-conductor cables where one conductor does not act as a normal current-carrying conductor (e.g., grounded neutral or grounding conductor), is the same as three-conductor cables listed in Table 27 of IEEE Std 45-2002

4—Table 25 is based on an ambient temperature of 45 °C and maximum conductor temperature not exceeding: 75 °C for type T insulated cables, 90 °C for types T/N, X, E, LSE, and LSX insulated cables, and 100 °C for types S and P insulated cables.

5—If ambient temperatures differ from 45 °C, the values shown in Table 25 of IEEE Std 45-2002
should be multiplied by the following factors:

Ambient Temperature	Type T insulated cable	Type T/N, X, E, LSE, LSX insulated cable	Type S and P insulated cable
30 °C	1.22	1.15	1.13
40 °C	1.08	1.05	1.04
50 °C	0.91	0.94	0.95
60 °C	-	0.82	0.85
70 °C	-	0.74	0.74

6—The current-carrying capacities in this table are for marine installations with cables arranged in a single bank per hanger and are 85% of the ICEA calculated values [see Note 7]. Double banking of distribution-type cables should be avoided. For those instances where cable must be double banked, the current-carrying capacities in Table 25 should be multiplied by 0.8.

7—The ICEA calculated current capacities of these cables are based on cables installed in free air, that is, at least one cable diameter spacing between adjacent cables. See IEEE Std 835-1994.

8—For cables with maintained spacing of at least 1 cable diameter apart, the values from this table may be divided by 0.85.

18.3.4 IEEE Std 45-2002, 24.8 Ampacities for medium-voltage power cable

Remarks:

1. Table 27 is the ampacity table for medium-voltage power cable, copper conductor-single-banked (single-layered), maximum current-carrying capacity based on 45 °C ambient, shields grounded on one end (Open-circuited shields). The Table 27 Notes are provided here for easy reference.

2. Caution: Table 28 is the ampacity table for medium-voltage power cable, three copper conductors-single-banked (single-layered), maximum current-carrying capacity based on 45 °C ambient. This table does not provide correct ampacity. Use the ampacity table of API-RP-14F standard table 3 for medium voltage copper three-conductor cable.

3. Table 29 is the ampacity table for medium-voltage power cable, copper conductor-triplexed or triangular configuration single-banked (single-layered), maximum current-carrying capacity based on 45 °C ambient.

The allowable ampacities are based on the conductor temperature rise in a given ambient. When selecting conductor sizes and insulation ratings, consideration must be given to the following:

a) The actual conductor operating temperature must be compatible with the connected equipment, especially at the connection points.

b) Conductor selection should be coordinated with circuit and system overcurrent and short-circuit protection to avoid cable damage during through-fault conditions. See ICEA P32-382 to determine conductor short-circuit withstand current.

c) Current-carrying capacity of four-conductor cables, in which one conductor is not a current-carrying phase conductor (e.g., neutral or grounding conductor), is the same as three-conductor cables.

d) If ambient temperatures differ from 45 °C, cable ampacities should be multiplied by the following factors:

Conductor temperature	Ambient temperature						
	30 °C	40 °C	45 °C	50 °C	55 °C	60 °C	70 °C
90 °C	1.10	1.05	1.00	0.94	0.90	0.82	0.67
105 °C	1.08	1.04	1.00	0.96	0.92	0.86	0.76

e) The current-carrying capacities are for cable installations with cables arranged in a single bank per hanger and are 85% of the calculated free air values. Double banking of medium voltage cables is not recommended.

f) The calculated current capacities are based on cables installed in free air, that is, at least one cable diameter spacing between adjacent cables. See IEEE Std 835-1994.

g) For cables with maintained spacing of at least 1 cable diameter apart, the ampacities may be increased by dividing by 0.85.

h) If more than one circuit of parallel runs of the same circuit are installed, there should be a maintained spacing of 2.15 times one conductor diameter between each triangular configuration group. Otherwise cables are considered double-banked.

AWG/kcmil	Cross sectional		Single Conductor Cable					
	mm²	Circular Mils	Up to 8kV Shielded	Up to 8kV Shielded	8001- to 15kV Shielded	8001- to 15kV Shielded	8001- to 15kV Shielded	8001- to 15kV Shielded
			90 °C	105 °C	90 °C	105 °C	90 °C	105 °C
6	13,30	26,240	92	106	-	-	-	-
4	21,15	41,740	121	135	-	-	-	-
2	33,62	66,360	159	187	164	187	-	-
1	42,40	83,690	184	216	151	216	192	216
1/0	53,50	105,600	212	245	217	242	220	245
2/0	67,44	133,100	244	284	250	284	250	284
3/0	85,02	167,800	281	327	288	327	288	327
4/0	107,20	211,600	325	375	332	375	332	375
250	126,70	250,000	360	413	366	413	366	413
263	13.3,10	262,600	371	425	377	425	376	425
313	158,60	313,100	413	473	418	471	416	471
350	177,30	350,000	444	508	448	505	446	505
373	189,30	373,700	460	526	464	523	462	523
444	225,20	444, 400	510	581	514	580	512	580
500	253,30	500,000	549	625	554	625	551	625
535	271,20	535,3000	570	648	574	648	570	648
646	327,50	646,400	635	720	638	720	632	720
750	380,00	750, 000	697	788	697	788	689	788
777	394,00	777,700	709	802	709	802	701	802
1,000	506,70	1,000, 000	805	913	808	913	798	913

Table 18-1: IEEE-45-2002, Table 29—Ampacity for Medium Voltage Power Cable, Copper Conductor – Triplexed or Triangular Configuration (Single-Layered), Maximum Current-Carrying Capacity Based on 45 °C ambient – Single Conductor

AWG/kcmil	Cross sectional		Three Conductor Cable			Three Conductor Cable			
	mm²	Circular Mils	Up to 5kV Nonshielded	Up to 8kV Shielded	Up to 8kV Shielded	8001- to 15kV Shielded	8001- to 15kV Shielded	8001- to 15kV Shielded	8001- to 15kV Shielded
			90 °C	90 °C	105 °C	90 °C	105 °C	90 °C	105 °C
8	8.37	16,510	48	-	-	-	-	-	-
6	13,30	26,240	64	75	85	-	-	-	-
4	21,15	41,740	84	99	112	-	-	-	-
2	33,62	66,360	112	129	146	133	150	-	-
1	42,40	83,690	130	149	168	151	170	149	172
1/0	53,50	105,600	151	171	193	174	196	174	196
2/0	67,44	133,100	174	197	222	199	225	198	225
3/0	85,02	167,800	202	226	255	229	259	230	257
4/0	107,20	211,600	232	260	294	263	297	262	294
250	126,70	250,000	258	287	324	291	329	291	327
263	13.3,10	262,600	266	296	334	299	338	299	336
313	158,60	313,100	296	328	370	331	374	329	373
350	177,30	350,000	319	352	397	355	401	351	400
373	189,30	373,700	330	365	412	367	414	363	414
444	225,20	444, 400	365	387	437	388	438	402	470
500	253,30	500,000	393	434	490	434	490	432	490
535	271,20	535,3000	407	449	507	449	507	447	507
646	327,50	646,400	453	496	560	497	561	496	559
750	380,00	750, 000	496	541	611	542	612	541	609
777	394,00	777,700	504	550	621	550	621	550	619
1,000	506,70	1,000, 000	571	622	702	623	703	622	703

Table 18-2: API-14F- Table 3—Ampacity for Three-Conductor Medium Voltage Power Cable, 2001 Volts to 35kV, Copper Conductor Single-Banked (Single-Layered), Maximum Current-Carrying Capacity Based on 45 °C ambient

18.3.5 IEEE Std 45 25.5-2002, Cable support and retention

Multiple cables should be supported in metal hangers or trays, arranged as far as practicable to permit painting of the adjacent structure without undue disturbance of the installation. Distribution cables grouped in a single hanger should be limited to single banking, except under the limitations in Table 25 [see 19.3.3 of this Handbook], Note 6 or Note 8. Control and signal cables should preferably be single banked, but may be double banked with other signal and control cables.

Clips or straps used for cable support should each be secured by two screws, except that clips for supporting one-cable, two-band strapping used for cable support. Metallic band strapping used for cable support should be steel and corrosion treated if not a corrosion-resistant material. The support for all cables should be such as to prevent undue sag, but in no case should exceed the distance between frames or 610 mm, whichever is less. Metallic band strapping should be applied so that the cables remain tight without damage to the cable.

Metal supports should be designed to secure cables without damage to the armor or the insulation. The supports should be arranged [so] that the cable would be supported for a length of at least 13 mm.

Cable retention devices should be installed not less than every 610 mm on vertical runs and not less than every 2.5 m on horizontal runs. At turns of horizontal runs, cable retention devices should be spaced not more than 610 mm apart. Nylon or plastic retaining devices may be used in horizontal runs where cables will not fall if the retention devices fail. When nylon or plastic cable retaining devices are employed on exterior cable runs, they should be of a type resistant to ultraviolet light (sunlight).

18.3.6 IEEE Std 45-2002, 25.6, Cables—radius of bends

Armored cables should not be bent to a radius of less than eight times the cable's diameter. Unarmored cables may not be bent to a radius of less than six times the cable's diameter.

18.3.7 IEEE Std 45-2002, 25.12, Propulsion cables (bold indicates change from IEEE Std 45-1998)

The effects of electromagnetic interference, impedance, system harmonics, short circuit bracing, and, for voltages above **2001 V**, corona should be taken into consideration when selecting cable constructions (single or multiple conductor) for electric propulsion systems.

Note that the voltage level has been changed from 3.3 kV (see IEEE Std 45-1998, 10.12) to 2001 V.

Propulsion system cables should be run as directly as possible and in accessible locations where they can be readily inspected. Propulsion cables rated 2001 V or greater should be terminated with a stress cone or other stress-terminating device, where necessary. A propulsion cable should not be spliced. Single-conductor propulsion cables should be adequately secured to prevent displacement by short-circuit conditions.

Propulsion system power cables interconnecting generators, main switchboards, main transformers, static power converters, and motors should be separated from ship service, control, and signal cables by at least 610 mm to reduce radiated electromagnetic interference.

If required for harmonic attenuation, propulsion cables should be armored or shielded and of the types identified in Clause 24 and Clause 25 of IEEE Std 45-2002 [see Section 19 of this Handbook for some discussion]. Cables should be sized to provide 105% of motor- and drive-rated current continuously and 125% of rated current for 10 min. Voltage drop should not exceed 5%.

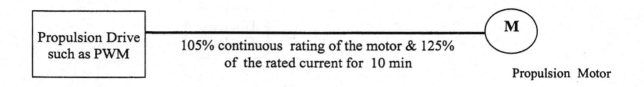

Figure 18-1: Propulsion Motor to Drive Feeder Rating (example only)

18.3.8 IEEE Std 45-2002, 31.14, Propulsion cables

Propulsion cable and cable installation should be as required by Clause 24 and Clause 25 of IEEE Std 45. Special considerations should be made to the harmonic losses and for attenuation of EMI when selecting cable, as recommended by the vendor of the propulsion converter. Propulsion cable should not have any splice or joint. For parallel-connected semiconductor devices, an equal current distribution should be ensured.

For additional details of variable frequency drive cable requirements refer to Section 18.10 of this Handbook.

18.4 IEEE Std 45-2002, 5.4.4, Individual and multiple motor circuits

Except as recommended in Clause 5.8.1 of IEEE Std 45-2002, conductors supplying an individual motor should have a continuous current-carrying capacity equal to 125% of the motor nameplate rating. Conductors supplying more than one motor should have a continuous current-carrying capacity equal to 125% of the largest motor plus the sum of the nameplate ratings of all other motors supplied, including 50% of the rating of spare switches on the distribution unit.

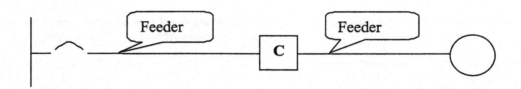

Figure 18-2: Branch Circuit Feeder Identification (example only)

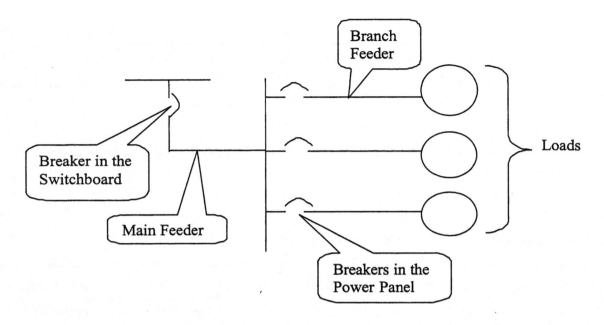

Figure 18-3: Main Feeder and Branch Circuit Feeder Identification (example only)

Conductors supplying a group of three or more workshop tool motors should have a continuous current-carrying capacity equal to 125% of the nameplate rating of the largest motor plus 50% of the sum of the nameplate ratings of all other motors of this group.

Conductors supplying two or more motors driving deck cargo winches or cargo elevators should have a continuous current-carrying capacity equal to 125% of the largest motor plus 50% of the sum of the nameplate ratings of all other motors.

Conductors supplying two or more motors driving cargo cranes should have a continuous current-carrying capacity equal to 125% of the largest motor plus 50% of the sum of the nameplate rating of all other motors. Where the conductor supplies two or more motors associated with a single crane, the current-carrying capacity should be equal to 125% of the largest motor plus 45% of the sum of the nameplate rating of the other motors.

Conductors between separate ship service switchboards having connected generators (ship service generator switchboards) should be sized on the basis of 75% of the switchboard having the greatest generating capacity. The drop in voltage from each generator to its switchboard should not exceed 1%, and the drop in voltage between switchboards should not exceed 2%.

Conductors between ships service switchboards and the emergency switchboard should be sized on the basis of the maximum operating load of the emergency switchboard or 115% of the emergency generator capacity, whichever is larger.

Conductors from storage batteries to the point of distribution should be sized for the maximum continuous charge or discharge rate, whichever is greater. Battery conductors for heavy-duty applications, such as diesel starting, should be sized for 125% of maximum-rated battery discharge.

Remarks:
1. For typical 460 V generator feeders and switchboard bus tie feeders refer to Figure 18-5
2. For medium voltage generator feeder details refer to Figure 18.6
3. For 460 V ship service generator feeder details refer to figure 18-7
4. For emergency generator feeder details refer to Figure 18-8
5. For emergency circuit feeders details refer to Figure 18-9

18.5 IEEE Std 45-2002, 5.8.1, Branch circuits (extract)

The branch circuit conductors should not be less than No. 15 AWG (1.5 mm^2 for IEC wiring systems; see 1.7 of IEEE Std 45-2002. Cable types and sizes … should be selected for compatibility with all local environmental conditions throughout the length and voltage drop limitations of this subclause. Each branch circuit should be provided with overcurrent protection in accordance with 5.9.5 of IEEE Std 45-2002 except as otherwise indicated. The maximum connected load should neither exceed the rated current-carrying capacity of the cable nor 80% of the overcurrent protective device setting or rating.

Note: Cable ampacity for Generator, feeder, and bus-tie cables per alternate compliance plan (ACP) for ABS and USCG (see ACP details in chapter 20 of this Handbook)

18.6 Shipboard cable insulation types and maximum ambient temperature

Table 18-3 is adapted from IEEE-1580, 5.3.1, Table 2 for maximum <u>conductor temperature</u> corresponding to the conductor insulation type. The table is also adapted from IEEE Std 45-2002, 24.9 Table 30 for cable insulation type and respective conductor maximum <u>ambient temperature</u> in centigrade.

It should be noted that allowable ambient temperature on board ship varies from 40°C to 55°C depending on specific location as well as regulatory body requirements. Sometimes ambient temperature requirements can further vary, depending on special application.

Cable Insulation type designation			Maximum conductor temperature (°C) (IEEE-1580)	Maximum ambient temperature (°C) (IEEE Std 45)
T	PVC	Polyvinylchloride	75	50
T/N	PVC/polyamide	Polyvinylchloride/nylon	90	60
E	EPR	Ethylene propylene rubber	90	60
X	XPLE	Cross-linked polyethylene	90	60
LSE	LSEPR	Low-smoke, halogen-free ethylene propylene rubber	90	60
LSX	LSXLPO	Low-smoke, halogen-free cross-linked polyolefin	90	60
S	Silicone	Silicone rubber	100	70
P	XLPO	Cross-linked polyolefin	100	70

Table 18-3: Cable Insulation – IEEE Std 45 & IEEE-1580

18.7 Cable designation type (typical service symbol or designation)

Shipboard cable designations are standardized with unique symbols, which directly correspond to the number of conductors constituting the cable. These typical cable symbols are listed for information. These abbreviations are often used to prepare shipboard working drawings such as one-line diagrams and power deck plans.

— "S"   Single conductor distribution
— "D"   Two conductor distribution
— "T"   Three conductor distribution
— "F"   Four conductor distribution
— "Q"   Five conductor distribution
— "C"   Control cable
— "TP"   Twisted pair
— "TT"   Twisted triad

18.8 Cable color code

The cable color code was in IEEE Std 45-1998, 8.31. This color code has been taken out from IEEE Std 45-2002 and can now be found in IEEE 1580-2001, Table 22. The color code is adapted from NEMA WC 57, Table E-1.

Conductor number	Base color	Tracer color	Tracer color	Conductor number	Base color	Tracer color	Tracer color
1	Black			45	White	Black	Blue
2	White			46	Red	White	Blue
3	Red			47	Green	Orange	Red
4	Green			48	Blue	Red	Orange
5	Orange			49	Blue	Red	Orange
6	Blue			50	Black	Orange	Red
7	White	Black		51	White	Lack	Orange
8	Red	Black		52	Red	Orange	Black
9	Green	Black		53	Green	Red	Blue
10	Orange	Black		54	Orange	Black	Blue
11	Blue	Black		55	Blue	Black	Orange
12	Black	White		56	Black	Orange	Green
13	Red	White		57	White	Orange	Green
14	Green	White		58	Red	Orange	Green
15	Blue	White		59	Green	Black	Blue
16	Black	Red		60	Orange	Green	Blue
17	White	Red		61	Blue	Green	Orange
18	Orange	Red		62	Black	Red	Blue
19	Blue	Red		63	White	Orange	Blue
20	Red	Green		64	Red	Black	Blue
21	Orange	Green		65	Green	Orange	Blue
22	Black	White	Red	66	Orange	White	Red
23	White	Black	Red	67	Blue	White	Red
24	Red	Black	White	68	Black	Green	Blue
25	Green	Black	White	69	White	Green	Blue
26	Orange	Black	White	70	Red	Green	Blue
27	Blue	Black	White	71	Green	White	Red
28	Black	Red	Green	72	Orange	Red	Black
29	White	Red	Green	73	Blue	Red	Black
30	Red	Black	Green	74	Black	Orange	Blue
31	Green	Black	Orange	75	Red	Orange	Blue

Conductor number	Base color	Tracer color	Tracer color	Conductor number	Base color	Tracer color	Tracer color
32	Orange	Black	Green	76	Green	Red	Black
33	Blue	White	Orange	77	Orange	White	Green
34	Black	White	Orange	78	Blue	White	Green
35	White	Red	Orange	79	Red	White	Orange
36	Orange	White	Blue	80	Green	White	Orange
37	White	Red	Blue	81	Blue	Black	Green
38	Black	White	Green	82	Orange	Green	White
39	White	Black	Green	83	Green	Red	
40	Red	White	Green	84	Black	Green	
41	Green	White	Blue	85	Green5	White	
42	Orange	Red	Green	86	Blue	Green	
43	Blue Red	Red	Green	87	Black	Orange	
44	Black	White	Blue	88	White	Orange	

Table 18-4: Cable Color Code–IEEE Std 45 & IEEE-1580 (extract)

18.9 Cable circuit designation

For commercial shipbuilding, IEEE Std 45-2002, Appendix B provides circuit designations, to be used in electrical engineering drawing preparation. Table B1 gives traditional system designation prefixes and Table B2 provides system designation. Table 18-5 below is an extract from the circuit designation table with cable designations and further explanation.

All electrical circuits, including those for power, lighting, interior communications, control, and electronics, are typically identified in the appropriate documentation and on-equipment labeling, such as nameplates, cable tags, wire markers, and so on, by a traditional system designation. These designations use a designation prefix, as given below.

Circuit designation	Prefix	Circuit number– example only	Explanation— example only
Ship service power (460 V)	P	P-412-T-250 (120 ft)	Power feeder, 460 V distribution, IEEE cable type T-250, three conductor , 120 ft run
Emergency power (460 V)	EP	3EP-407-T-106 (50 ft)	
Propulsion power (6600 V)	PP	4PP-6612-T-26(50 ft)	
Shore power (460 V)	SP	1SP-402-T-400 (125 ft)	
Lighting (120 V)	L	4L-114-T-212 (95 ft)	Lighting branch circuit, 120 V distribution, IEEE cable type T-212, three conductor , 95 ft run
Emergency lighting (120 V)	EL	3EL-118-T-52 (60 ft)	
Interior communication	C		
Control	K		
Announcing: General	C-IMC	C-1MC-5	
Electric plant control and	C-DE	C-DE-7	

Circuit designation	Prefix	Circuit number– example only	Explanation— example only
monitoring			
Emergency generator set control and indication	K-EG	K-EG-9	
Flooding alarm	C-FD	C-FD-32	
Radio antenna	R-RA	R-RA-5	
Radio satellite communication	R-RS		
Telephone: Automatic dial	C-J		
Telephone: Sound-powered (machinery control engineers)	C-2JV	C-2JV-15, TPS18-7 (55 ft)	Circuit 2JV, branch 15, 7 pairs of shielded twisted pairs. 18 AWG, 55 ft run
Telephone: Sound-powered (damage control)	C-JZ	C-JZ-9, TPS18-10 (70 ft)	
Sewage tank high-level alarm	C-5TD		
Shaft revolution indication	C-K	C-K-8, TPS18-14 (90 ft)	
Ship service generator local control and indication	K-SG	K-SG-21	
Smoke indicator	C-SM	C-SM-24	

Table 18-5: Circuit Designation –Commercial (example only)

18.10 Cable requirements for non-linear power equipment

The shipboard power cable requirement for non-linear power loads, such as variable frequency drive (VFD) or solid state UPS, is different from normal power equipment cable. The cable sizing for non-linear loads should consider system-related electric line noises such as harmonics, reflected wave, and induced voltages. Certain non-sensitive ship service loads may be able to withstand some non-linear noise effects, whereas some other sensitive loads may not be able to tolerate any system related non-linear noise.

18.10.1 Non-linearity related power system noise fundamentals – Electromagnetic Compatibility (EMC)

Electromagnetic Compatibility (EMC) is related to electromagnetic interference (EMI) and electromagnetic susceptibility (EMS) of electrical system. Figure 18-4 is a block diagram showing the relationship.

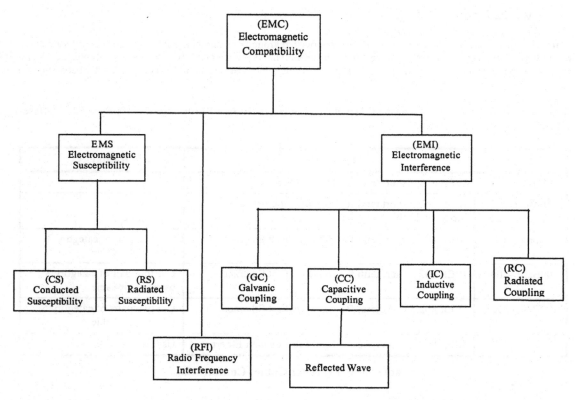

Figure 18-4: Non Linear Power System Electromagnetic Compatibility (EMC) Block Diagram for Cable

18.10.1.1 Normal Mode (NM) and Common Mode (CM) interference related to non-linear load for shipboard application

Normal mode (NM) interference voltage appears along the outgoing and returning conductors of a circuit. One of the causes of this interference is inductive coupling. Common mode (CM) interference is caused by interference voltage arising between the reference potential and the conductors in the circuit.

Non-linear power cable should be provided with cable shield and grounds. The cable shield and grounds must be terminated properly. The shipboard electric system designer should consult the equipment supplier to determine the non-linearity aspect of the equipment and then discuss with the cable manufacturer to determine cable size, cable shielding and armoring, etc requirements. Cable may require grounding conductor along the power conductor, to eliminate the possibility of the shield overloading. The cable voltage drop calculation should be performed, considering the non-linearity of the system, such as reflected voltage. The final cable procurement must be approved by regulatory bodies having the authority of certification.

18.11 Ethernet cable

Ethernet cable, used for data communications, comes in many different varieties. The most commonly used are described in the following sections.

18.11.1   Twisted pair cabling

Twisted pair cables are so named because pairs of wires are twisted around one another. Each pair consists of two insulated copper wires twisted together. The wire pairs are twisted because it helps reduce cross talk and noise susceptibility. High quality twisted pair cables have about 1 to 3 twists per inch. For best results, the twist rate should vary significantly between pairs in a cable.

Twisted pair cables are used with the following Ethernet physical layers: 10Base-T, 100Base-TX, 100Base-T2, 100Base-T4, and 1000Base-T.

Type	Description	Remarks
10Base-T	IEEE 802.3 short term for 10 Mb/s Ethernet	
10Base36	IEEE 802.3 short term for 10 Mb/s Ethernet on broadband cable	
100Base –T2	IEEE 802.3 short term for 100 Mb/s Fast Ethernet	Use two pair of category 3 twisted pair cable
100Base – T4	IEEE 802.3 short term for 100 Mb/s Fast Ethernet	Use four pair of category 3 twisted pair cable
100Base -TX	IEEE 802.3 short term for 100 Mb/s Fast Ethernet	Use two pairs of category 5 twisted pair cable
1000Base - CX	IEEE 802.3 short term for 1000 Mb/s Gigabit Ethernet	Use copper wire

Table 18-6: Ethernet Cable -Commercial

18.11.2   Unshielded twisted pair cabling (UTP)

Unshielded twisted pair (UTP) cabling is twisted pair cable that, obviously enough, contains no shielding. For networking applications, the term UTP generally refers to 100 ohm, Category 3, 4, & 5 cables. 10Base-T, 100Base-TX, and 100Base-T2 use only 2 of the twisted pairs, while 100Base-T4 and 1000Base-T require all 4 twisted pairs.

The following is a summary of the UTP cable categories:

—   **Category 1 (Cat 1)** – Unshielded twisted pair used for audio frequencies. Not suitable for use with Ethernet.

—   **Category 2 (Cat 2)** – Unshielded twisted pair used for transmission at frequencies up to 1.5 Mhz. Used in analog telephone application

—   **Category 3 (Cat 3)** – Unshielded twisted pair with 100 Ω impedance and electrical characteristics supporting transmission at frequencies up to 16 MHz. May be used with 10Base-T, 100Base-T4, and 100Base-T2.

—   **Category 4 (Cat 4)** – Unshielded twisted pair with 100 Ω impedance and electrical characteristics supporting transmission at frequencies up to 20 MHz. May be used with 10Base-T, 100Base-T4, and 100Base-T2.

—   **Category 5 (Cat 5)** – Unshielded twisted pair with 100 Ω impedance and electrical characteristics supporting transmission at frequencies up to 100 MHz. May be used with 10Base-T, 100Base-T4, 100Base-T2, and 100Base-TX. May support 1000Base-T, but cable should be tested to make sure it meets 100Base-T specifications.

—   **Category 5e (Cat 5e)** – also known as "Enhanced Cat 5", is a new standard that specifies transmission performance that exceeds Cat 5. Like Cat 5, it consists of unshielded twisted pair with 100 Ω impedance and electrical characteristics supporting transmission at frequencies up to 100 MHz. Targeted for 1000Base-T, but also supports 10Base-T, 100Base-T4, 100Base-T2, and 100BaseTX.

—   **Category 6** – Cat 6 is a proposed standard that aims to support transmission at frequencies up to 250 MHz over 100 Ω twisted pair.

—   **Category 7** – Cat 7 is a proposed standard that aims to support transmission at frequencies up to 600 MHz over 100 Ω twisted pair.

18.12 Coaxial cable

Coaxial cable is a type of communication transmission cable in which a solid center conductor is surrounded by an insulating spacer, which in turn is surrounded by a tubular outer conductor (usually a braid, foil or both). The entire assembly is then covered with an insulating and protective outer layer. Coaxial cables have a wide bandwidth and are capable of carrying many data, voice, and video conversations simultaneously.

COAX type	Description	Remarks
10Base 2	IEEE 802.3 short term for 10 Mb/s Ethernet over <u>thin</u> coaxial cable	
10Base 5	IEEE 802.3 short term for 10 Mb/s Ethernet over <u>thick</u> coaxial cable	

Table 18-7: Coaxial Cable – Commercial

18.13 Fiber optic cabling

Fiber optic cabling is a technology where electrical signals are converted into optical signals, transmitted through a thin glass fiber, and re-converted into electrical signals. It is used as transmission medium for the following Ethernet media systems: FOIRL, 10Base-FL, 10Base-FB, 10Base-FP, 100Base-FX, 1000Base-LX, and 1000Base-SX.

Fiber optic cabling is constructed of three concentric layers: The "core" is the central region of an optical fiber through which light is transmitted. The "cladding" is the material in the middle layer. It has a lower index of refraction than the core, which serves to confine the light to the core. An outer "protective layer", or "buffer", serves to protect the core and cladding from damage. The two primary types of fiber optic cabling are multi-mode fiber (MMF) and single-mode fiber (SMF).

Fiber type	Description	
10Base – F	IEEE 802.3 short term for 10 Mb/s Ethernet over Fiber Optics Cable	
100Base – FX	IEEE 802.3 short term for 100 Mb/s Fast Ethernet over Fiber Optics Cable	
1000Base – LX	IEEE 802.3 short term for 1000 Mb/s GigabitFast Ethernet over Fiber Optics Cable	Short Wavelength
1000Base – SX	IEEE 802.3 short term for 1000 Mb/s GigabitFast Ethernet over Fiber Optics Cable	Long Wavelength

Table 18-8: Fiber optic cable—Commercial

18.13.1 Multi-mode fiber (MMF)

MMF allows many "modes", or paths, of light to propagate down the fiber optic path. The relatively large core of a multi-mode fiber allows good coupling from inexpensive LED light sources, and the use of inexpensive couplers and connectors. Multi-mode fiber typically has a core diameter of 50 to 100 microns.

Two types of multi-mode fiber exist with a refractive index that may be "graded" or "stepped". With graded index fiber, the index of refraction of the core is lower toward the outside of the core and progressively increases toward the center of the core, thereby reducing modal dispersion of the signal. With stepped index fiber, the core is of uniform refractive index with a sharp decrease in the index of refraction at the core-cladding interface. Stepped index multi-mode fibers generally have lower bandwidths than graded index multi-mode fibers.

The most popular fiber for networking is the 62.5/125 micron multi-mode fiber. These numbers mean that the core diameter is 62.5 microns and the cladding is 125 microns. Other common sizes are 50/125 and 100/140.

The primary advantage of multi-mode fiber over twisted pair cabling is that it supports longer segment lengths. Multi-mode fiber can support segment lengths as long as 2000 m for 10 and 100 Mbps Ethernet, and 550 meters for 1 Gbps Ethernet.

18.13.2   Single-mode fiber (SMF)

SMF has a core diameter that is small (on the order of 10 microns), so that only a single mode of light is propagated. This eliminates the main limitation to bandwidth, modal dispersion. However, the small core of a single-mode fiber makes coupling light into the fiber more difficult, and thus expensive lasers must be used as light sources. The main limitation to the bandwidth of a single-mode fiber is material dispersion. Laser sources must also be used to attain high bandwidth, because LEDs emit a large range of frequencies, and thus material dispersion becomes significant.

Single-mode fiber is capable of supporting much longer segment lengths than multi-mode fiber. Segment lengths of 5000 m and beyond are supported at all Ethernet data rates through 1 Gbps. However, single-mode fiber has the disadvantage of being significantly more expensive to deploy than multi-mode fiber.

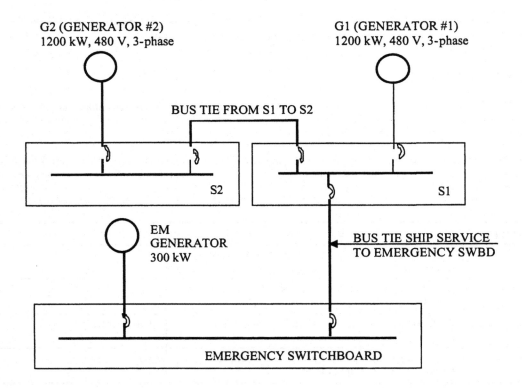

Figure 18-5: Electrical One-Line Diagram-Sizing Generator Feeder—Bus Tie Feeder – Commercial (example only)

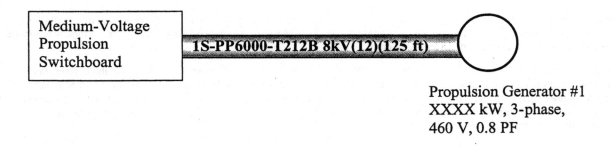

Figure 18-6: Medium-Voltage Generator to Switchboard Feeder – Commercial (example only)

In reference to Figure 18-6 above, the designation for the feeder 1S-PP6000-T212B-8kV(12)(125ft) is as follows:

1	Sequence number of power feeder. In this case, this sequence number is for 6600 V power distribution
S	Starboard side power feeder
PP	Propulsion Power Feeder
6000	6600 V power distribution
T-212	Three conductor 212MCM cable
B	Bronze armored cable
(12)	Total number of three conductor cable T-212
(125ft)	Total length of the feeder

Ship Service Generator #3
XXXX kW, 3-phase, 460 V, 0.8 PF

Figure 18-7: 480 V Ship Service Generator Feeder—Commercial Ship (example only)

In reference to Figure 18-7 above, the designation for the feeder 2S-P401-T250 (6)(200ft) is as follows:

2	Sequence number of power feeder. In this case, this sequence number is for 460 V power distribution .
S	Starboard side power feeder
P403	Power Feeder 403 sequence number for 400series feeders for 460 V distribution
T-250	Three-conductor 250MCM cable
(6)	Total number of three conductor cable T-250
(200ft)	Total length of the feeder

Figure 18-8: 460 V Emergency Generator Feeder – Commercial (example only)

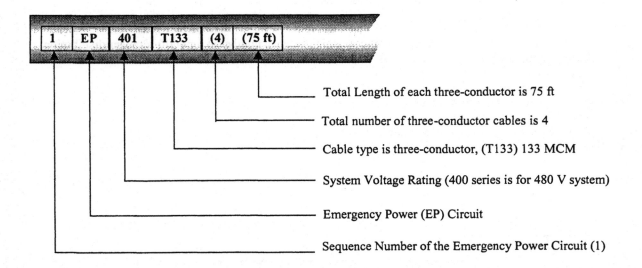

Total Length of each three-conductor is 75 ft

Total number of three-conductor cables is 4

Cable type is three-conductor, (T133) 133 MCM

System Voltage Rating (400 series is for 480 V system)

Emergency Power (EP) Circuit

Sequence Number of the Emergency Power Circuit (1)

Figure 18-9: 480 V Emergency Power System Feeder Designation—Commercial Ship (example only)

With reference to Figure 18-9 above, the cable designation for the feeder is as follows:

1	sequence number for emergency power circuit
EP	designation for emergency power circuit
401	460 V emergency power circuit related sequence number
T133	cable designation which stands for three conductor, 133MCM cable
(4)	total number of cables in the 1EP401 circuit
(75 ft)	F – 75 ft is the (4)T133 cable length of 1EP401 circuit

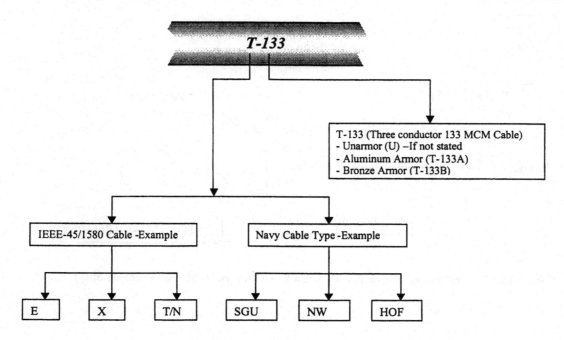

Figure 18-10: Typical Commercial and Navy Cable Specification

For further IEEE cable details, refer to IEEE Std 45-2000 and IEEE Std 1580-2001.

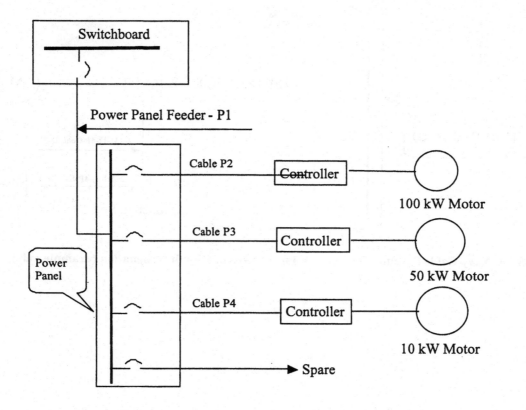

Figure 18-11: Power Panel Feeder and Load (example only)

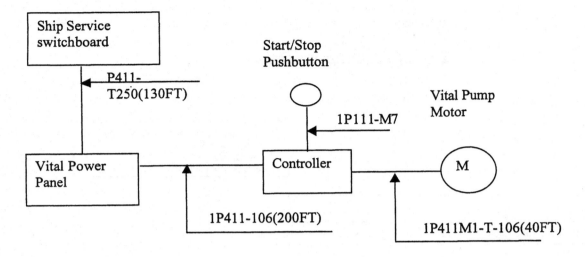

Figure 18-12: Vital Pump Motor Feeder from Switchboard to Motor (example only)

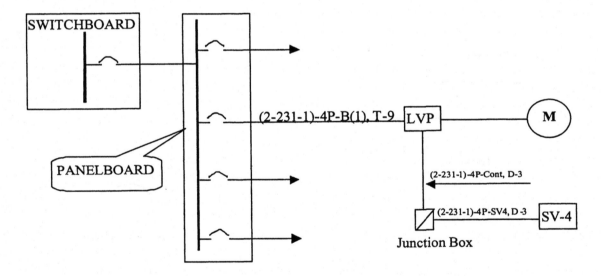

Figure 18-13: Navy Cable Specification—480 V Power System Circuit Designation (example only)

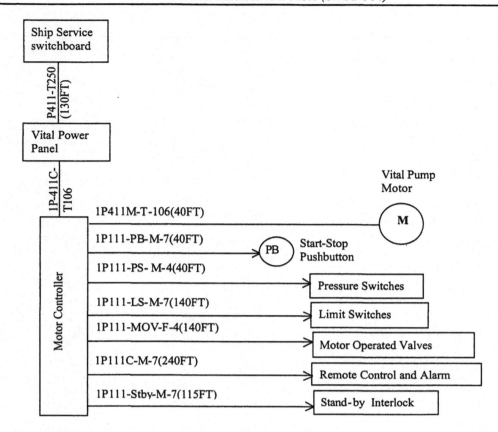

Figure 18-14: Vital Pump Motor and Controller Feeder (Switchboard to the Motor) (example only)

Cable construction requirements were in IEEE Std 45-1998, 8. However, in IEEE Std 45-2002 the cable construction section was removed, as it was decided that the IEEE Std 45 is a performance standard. Instead, the cable construction standard can be found in the new standard IEEE Std 1580-2001.

19 Shipboard Control, automation, and Power Management System (PMS)— (SWBS 202)

19.1 General

Shipboard propulsion engines, auxiliaries, electrical power generation and distribution systems are normally controlled from a central station. The central control station is called the engineering operator station (EOS), machinery centralized control station (MCCS) or machinery plant control and monitoring station (MPCMS). ABS classifies centralized machinery control in two categories, ACC and ACCU. The ACC category is for engine rooms, which are attended at all times, and ACCU is a classification for engine rooms, which are periodically unattended. The ACCU class ship must be controllable from the central control room as well as from the wheelhouse. The centralized control, monitoring, alarm, and safety shutdown are automatically performed by a computerized control and management system, which is called the power management system (PMS). See Figure 19-3 for a block diagram of the PMS.

The basic control requirement is that all machinery automatic controls must have manual override controls. The manual override can be further classified as manual control from the control room, or manual control from the equipment. The manual control at the local control station is usually provided with local control as well as a remote control feature to transfer control from the local station to a remote control station. The PMS system provides systematic hierarchical control functions, which provides a means for automatically maintaining load limits and power limits for propulsion engines and generators by continuously monitoring the load. This provision assures that required power is available to support the load demand.

For electric propulsion systems, multiple engines and generators are installed to drive propulsion motors as an integrated propulsion plant, the power management system automatically maintaining power demand under required operating conditions. This ensures optimal utilization of the power plant. The power management system monitors prime mover load setting, If the load requirement exceeds the rating of running generator(s), the standby engine will be started automatically. It will synchronize with the switchboard bus, close the generator circuit breaker and then automatically perform load sharing. If, for some reason, the standby generator does not come on line under power management, the appropriate alarm activates, and then non-critical loads are systematically removed so that the engines load remain within the permissible load limit. Alternatively (if permissible), the propulsion load can be automatically reduced to save the operating load level of the prime mover. In case a vital auxiliary fails, a stand-by auxiliary will come on line as a PMS permissive with appropriate alarms and indications at the console.

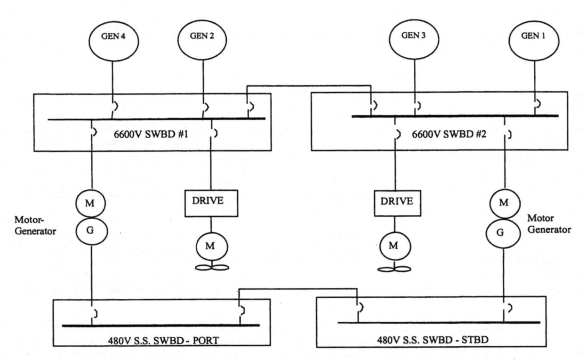

Figure 19-1: Electrical One-Line Diagram-Redundant Electrical Plant

Notes:
1. Each propulsion motor is of the two winding type. Each winding can run independently.
2. Ship Service power is supplied with motor generator set. In this arrangement, the ship service power is clean, meeting Mil 1399 power requirements.

The next three sections refer to Figure 19-1 above.

19.2 Power generation (Figure 19-1)

There are four 6600 V main generators, G1 through G4. Two generators are in the starboard engine room and two generators are in the port engine room. The port switchboard is dedicated for port generators and port side 6600 V distribution and the starboard switchboard is dedicated for starboard generators. Both switchboards are interconnected to facilitate parallel operation. Each 6600 V switchboard supplies power to the propulsion motor trough variable frequency drive. Each 6600 V switchboard supplies power to a dedicated 480 V ship service switchboard through a 6600 V/480 V motor-generator set. The control and power management system can automatically control, monitor, load share and maintain the most efficient power generation setup.

19.3 Power distribution (Figure 19-3)

The starboard 480 V switchboard supplies power to all starboard side auxiliaries and the port 480 V ship service switchboard supplies power to all portside ship service auxiliaries. Each 480 V ship service switchboard is connected to the emergency switchboard with a bus tie feeder. The control and power management system can automatically control, monitor, load share and maintain the most efficient running and stand-by operation set-up as programmed.

19.4 Controls

Each engine room control consists of the following:

— Two main engines
— Two main generators
— One 6600 V switchboard
— One 480 V distribution switchboard

— Power feeders and controllers for all auxiliaries

All equipment can be controlled from the following locations:

— **Manual control from the equipment location:** All equipment must be capable of manual operation from its location. This operation includes control, monitoring, alarm and safety shutdown. If all other remote operation fails, operator should be able to operate from this location. At this location there are additional features, such as local and remote manual switch. If the control functions operate properly, the switch is turned to the remote position, making remote control possible.
— **Manual control from the control room:** This operation is also a remote control feature of the equipment, but is otherwise similar to the manual operation from the equipment's physical location.
— **Automatic control from the control room (ABS ACC):** This feature is complete automatic control of equipment and system from the central control station. However, the control station is fully supervised.
— **Automatic control from the control room (ABS ACCU):** This feature is complete automatic control of equipment and system from the central control station. However, the control station is periodically unsupervised. In this mode of operation the ship is fully controlled from the wheelhouse.

19.5 Safety system

The safety system ensures safety of the equipment, and is a regulatory body requirement. Lube oil circulation is a prerequisite for running the engine: if the lube oil circulation fails, the system will initiate alarm. The system will also initiate shutdown. If lube oil circulation does not resume, the safety system will automatically shut down the engine.

19.6 Failure mode and effect analysis (FMEA) – general

It is a performance requirement to ensure operation of the propulsion system under all operating conditions, including emergency conditions. There are layers of redundant systems and auxiliaries, which remain in a stand-by operational mode in case of a failure of the running system. In case of one propulsion engine failure, the other engine should be capable of maintaining minimum propulsion load so that the ship is capable of maintaining minimum speed. The failure mode and effect analysis (FMEA) provides guidelines to provide step-by-step analysis of ship's systems. The following clauses of IEEE Std 45-2002 refer to FMEA.

19.6.1 IEEE Std 45-2002, 31.12, Failure mode and effect analysis (FMEA)

A failure analysis methodology is used during design to postulate every failure mode and the corresponding effect or consequences. Generally, the analysis is to begin by selecting the lowest level of interest (part, circuit, or module level). The various failure modes that can occur for each item at this level are identified and enumerated. The effect of each failure mode, taken singly and in turn, is to be interpreted as a failure mode for the next higher functional level. Successive interpretations will result in the identification of the effect at the highest function level, or the final consequence. A tabular format is normally used to record the results of each study. FMEA should be certified and acceptable to the authority having jurisdiction.

19.6.2 ABS-2002, 4-9-1 General Definition of FMEA

5.1.16 Failure Mode and Effect Analysis (FMEA): A failure analysis methodology used during design to postulate every failure mode and the corresponding effect or consequences. Generally, the analysis is to begin by selecting the lowest level of interest (part, circuit, or module level). The various failure modes that can occur for each item at this level are identified and enumerated. The effect for each failure mode, taken singly and in turn, is to be interpreted as a failure mode for the next higher functional level. Successive interpretations will result in the identification of the effect at the highest function level, or the final consequence. A tabular format is normally used to record the results of such a study.

19.6.3 ABS-2002, 4-9-1 FMEA for system design

9.11 Failure Mode and Effect Analysis (FMEA): Failure modes and effects analysis (FMEA) may be carried out during system design to investigate if any single failure in control systems would lead to undesirable consequences such as loss of propulsion, loss of propulsion control, etc. The analysis may be qualitative or quantitative.

19.7 Machinery Control

In general, shipboard propulsion machinery and auxiliary machinery control is provided from a central control station. However, due to redundancy requirements, there are hierarchical control requirements such as local control at the equipment, central control from the engine room, control from navigation bridge, and unmanned control. The recommendations of IEEE Std 45-2002 in this regard are summarized below.

19.7.1 IEEE Std 45-2002, 9.5.1, Machinery control—general

The centralized control system should support real-time monitoring and control of ship propulsion, electrical, auxiliary, and steering functions.

When provided for essential machinery functions, standby machines should incorporate automatic changeover features. The changeover function should be alarmed. Vital control, safety, and alarm systems should automatically transfer power sources to backup systems upon failure of the operating power source. The system should be designed to ensure the changeover operation does not further degrade the plant condition by inadvertently compounding the situation.

Effective means should be provided to allow propulsion units to be operated under all sailing conditions, including maneuvering. The speed, direction of thrust and, if applicable, the pitch of the propeller, should be fully controllable from the navigating bridge. This includes all thrust conditions, from that associated with maximum controllable open-water astern speed to that associated with the maximum controllable ahead speed.

Control functions from a console may be designed for either remote manual or automatic control. Vital systems that are automatically or remotely controlled should be provided with the following:

a) An effective primary control system.
b) A manual alternate control system. A method should exist that provides for alternative positive control of the vital equipment.
c) A safety control system. Methodologies should be incorporated into the control system either in hardware or software that preclude the unsafe operation of the equipment. Unsafe operation is considered to be operation that will cause severe and permanent damage of the equipment, or risk to operating personnel.
d) Instrumentation to monitor system parameters necessary for the safe and effective operation of the system. The instrumentation installed should be sufficient to provide a skilled worker with an adequate ability to assess the status, operational performance, and health of the equipment under control. Human factors and ergonometric considerations should be included in the instrumentation design.
e) An alarm system. A method should be provided to alert the operator to abnormal and dangerous operating conditions.

Provision should be made for independent manual control in the event of a loss of an element of, or the entire centralized control system. The provision should include methods and means of overriding the automatic controls and interlocks. The instrumentation and control provided should be capable of supporting the manual operation for indefinite periods of time.

19.7.2 IEEE Std 45-2002, 9.5.2, Control hierarchy

For ships with more than one control station, a decreasing authority should be assigned according to the following control station locations:

a) Local controls at the controlled equipment
b) Control station(s), in the machinery spaces, closest to the controlled equipment [local control station(s)]
c) Remote control station(s) outside of machinery spaces
d) Navigating bridge (or bridge wing) control station

The control station of higher authority should be designed to include a supervisory means for transferring control from a station of lower authority at all times, and to block any unauthorized request from any station of lower authority. A station of higher authority must be capable of overriding and operating independently of all stations of lower authority. The overriding action when executed should be alarmed at the remote location affected. Transfer of control from one station to another, except for the specific case of override by a station of higher authority, is to be possible only with acknowledgment by the receiving station.

The control function at a control station of higher authority should not depend on the proper functioning of the control station of the lower authority. The control functions from only one station should, except for emergency control actions, initiate a command signal to controlled equipment. Transfer of control between stations should be accomplished smoothly and control of the controlled equipment should be maintained during the transfer. Failure of a control function at a station of lower authority should automatically and smoothly transfer the control to the station of the next higher authority. This transfer should generate an alarm at both stations. Appropriate indications with discriminatory intelligence should be provided at each station, except locally at the controlled equipment, to identify the station in control and any loss of control.

19.7.3 IEEE Std 45-2002, 9.5.3, Control, start, stop, and shutdown conditions

The control system should include a means for emergency stopping the propulsion system from the navigation bridge. This feature should be independent of the navigation bridge control system.

For remote starting of the propulsion machinery, the control system should include interlocks that prevent the remote starting of propulsion machinery under conditions that may be hazardous to the machinery or operating personnel, such as when machinery is being maintained or inspected.

Automatic shutdown of machinery should be alarmed at all control stations. Restoration of normal operating conditions should be possible only after a manual reset.

Loss of control for any reason, including loss of power, logic failure, or power conditioning failure, should be alarmed at all stations. Upon failure of the control system, the system should be designed to continue to operate the plant in the condition last received, until local control is established.

19.7.4 IEEE Std 45-2002, 9.6, System design characteristics

Efforts should be taken to minimize the probability of a failure of any one component or device in the control circuit, causing unsafe operation of the machinery. There should be no single points of failure that will disable the control system. Where multiple auxiliaries such as normal and standby are installed for vital services, any single failure should be limited in effect to only one of the auxiliaries, and such failures should not render any reserve automatic or manual control, or both, inoperative.

All electric and electronic devices and components should be suitable for use in a marine environment, resistant to corrosion, treated for resistance to fungus growth, not affected by shipboard vibration or shock loading, and capable of providing their intended function at typical environmental temperature and humidity levels. All control equipment should be designed to perform satisfactorily without adverse susceptibility to electromagnetic fields, power switching spikes, or other electromagnetic noise that might typically be encountered in the operating environment, or induced on interconnecting cables (see 9.20 [not quoted in this Handbook]).

19.7.5 IEEE Std 45-2002, 9.7, Control system power supply

Feeders supplying power to the control console should be provided with overload and short-circuit protection. Where circuits are protected by fuses, control system protection should be subdivided and arranged so that failure of one set of fuses will not cause mal-operation or failure of other circuits or systems. Isolation of faults should be readily accomplished.

Power for monitoring, alarms, and vital controls should be supplied from an emergency source upon failure of the main power supply. Power conversion equipment, if required by the design, should be of the solid-state type. Common power conversion devices should be supplied in duplicate, arranged to operate in parallel, with either capable of supplying the full load of all units powered by the system they serve.

Designs in which the control system is isolated into separate sections controlling major equipment and systems that are operated independently of each other may use separate power conversion devices for each section, with one spare, readily interchangeable with any unit.

Individual protective devices should be provided in the input and output of each power conversion device.

19.7.6 IEEE Std 45-2002, 9.8, Continuity of power

The control system should be designed to include an uninterruptible power supply (UPS) that has sufficient capacity to maintain power to the control system for a period sufficient to bring the emergency power generator on line, or if necessary, to safely shut down the equipment. The quality and operation of the UPS supply should be such that the control's equipment experiences no disturbance or interruption in power. Nominal capacity should be for a period of not less than 15 min at full rated load with a voltage degradation not more than 10% of rated level.

19.7.7 IEEE Std 45-2002, 9.9, Communication systems (extract)

… Data communications between control stations and controlled equipment should be designed to incorporate industry standard protocol and interfaces. The system should include a built-in redundancy for the transmission of data. The data system should be separate from all other communication systems. The system should be designed to be relatively immune to negative effects induced by the operational environment. Within the data communication system, the fire alarm communication system should be a separate standalone system, not dependent on any other subsystem for its operation. …

19.7.8 IEEE Std 45-2002, 9.10, Alarms (extract)

Alarm devices should be provided that automatically sound and visually indicate the loss of power to the control system. Provision should be made for silencing audible alarms. The system should indicate any fault requiring attention and should contain the following features:

a) Audible and visual alarms in the machinery control room or at the propulsion machinery control position, which indicate each separate alarm function at a central position.
b) Alarm indication circuits for the engineers' public rooms and selectable alarm indications to each engineer's stateroom.
c) Supervisory alarm activation if the alarm system malfunctions.
d) Supervisory alarm activation in an attended location if the alarm has not been acknowledged in a limited time (not exceeding 5 min).
e) A continuously energized system, with automatic transfer to standby power upon failure of normal power.
f) Capability to indicate more than one alarm condition simultaneously, while retaining the first fault, and providing descriptive information on the nature of each fault.
g) Prevention of nuisance alarm activation due to ship's dynamic motion.
h) Ability to acknowledge an alarm condition without preventing the alarm indication of other abnormal conditions.
i) Ability to override an audible alarm condition from reoccurring continuously if the alarm condition could not be rectified immediately. …

19.8 ABS-2002, 4-9-1, Remote propulsion control and automation

The propulsion machinery Automatic Centralized Control (ACC) and Automatic Centralized Control, periodic Unmanned (ACCU) are sufficiently covered by ABS rules in ABS-2002 chapter 4-9-1. Some of those requirements are quoted below.

19.8.1 ABS-2002, 4-9-1, 3.1, ACC notation

Where, in lieu of manning the propulsion machinery space locally, it is intended to monitor it and to control and monitor the propulsion and auxiliary machinery from a continuously manned centralized control station, the provisions of Section 4-9-3 are to be followed. And upon verification of compliance, the class notation ACC will be assigned.

19.8.2 ABS-2002, 4-9-1, 3.3, ACCU notation

Where it is intended that propulsion machinery space be periodically unmanned and that propulsion machinery be controlled primarily from the navigating bridge, the provisions of Section 4-9-4 are to be complied with. And upon verification of compliance, the class notation ACCU will be assigned.

19.8.3 ABS-2002, 4-9-1, 5.1, General definitions

5.1.1 Alarm: Visual and audible signals indicating an abnormal condition of a monitored parameter.

5.1.2 Control: The process of conveying a command or order to enable the desired action to be effected.

5.1.3 Control System: An assembly of devices interconnected or otherwise coordinated to convey the command or order.

5.1.4 Automatic Control: A means of control that conveys predetermined orders without action by an operator.

5.1.5 Instrumentation: A system designed to measure and to display the state of a monitored parameter and which may include one of more of sensors, read-outs, displays, alarms and means of signal transmission.

5.1.6 Local Control: A device or array of devices located on or adjacent to a machine to enable it to be operated within sight of the operator.

5.1.7 Remote Control: A device or array of devices connected to a machine by mechanical, electrical, pneumatic, hydraulic or other means and by which the machine may be operated remote from, and not necessarily within sight of, the operator.

5.1.8 Remote Control Station: A location fitted with means of remote control and monitoring.

5.1.9 Monitoring System: A system designed to supervise the operational status of machinery or systems by means of instrumentation, which provides displays of operational parameters and alarms indicating abnormal operating conditions.

5.1.10 Safety System: An automatic control system designed to automatically lead machinery being controlled to a predetermined less critical condition in response to a fault which may endanger the machinery or the safety of personnel and which may develop too fast to allow manual intervention. To protect an operating machine in the event of a detected fault, the automatic control system may be designed to automatically:

— Slowdown the machine or to reduce its demand;
— Start a standby support service so that the machine may resume normal operation; or
— Shutdown the machine.

For the purposes of this Chapter, automatic shutdown, automatic slow down and automatic start of standby pump are all safety system functions. Where "safety system" is stated hereinafter, it means any or all three automatic control systems.

5.1.11 Fail-safe: A designed failure state which has the least critical consequence. A system or a machine is fail-safe when, upon the failure of a component or subsystem or its functions, the system or the machine automatically reverts to a designed state of least critical consequence.

5.1.12 Systems Independence: Systems are considered independent where they do not share components such that a single failure in any one component in a system will not render the other systems inoperative.

5.1.13 Propulsion Machinery: Propulsion machinery includes the propulsion prime mover, reduction gear, clutch, and controllable pitch propellers, as applicable.

5.1.14 Unmanned Propulsion Machinery Space: Propulsion machinery space which can be operated without continuous attendance by the crew locally in the machinery space and in the centralized control station.

5.1.15 Centralized Control Station: A propulsion control station fitted with instrumentation, control systems and actuators to enable propulsion and auxiliary machinery be controlled and monitored, and the state of propulsion machinery space be monitored, without the need of regular local attendance in the propulsion machinery space.

5.1.16 Failure Mode and Effect Analysis (FMEA): A failure analysis methodology used during design to postulate every failure mode and the corresponding effect or consequences. Generally, the analysis is to begin by selecting the lowest level of interest (part, circuit, or module level). The various failure modes that can occur for each item at this level are identified and enumerated. The effect for each failure mode, taken singly and in turn, is to be interpreted as a failure mode for the next higher functional level. Successive interpretations will result in the identification of the effect at the highest function level, or the final consequence. A tabular format is normally used to record the results of such a study.

5.1.17 Vital Auxiliary Pumps: Vital auxiliary pumps are that directly related to and necessary for maintaining the operation of propulsion machinery. For diesel propulsion engines, fuel oil pump, lubricating oil pump, cooling water pumps are examples of vital auxiliary pumps.

19.8.4 ABS-2002, 4-9-1, 5.3, Definition for Computerized System

5.3.1 Computer Based System: A system of one or more microprocessors, associated software, peripherals and interfaces.

5.3.2 Integrated System: A combination of computer based systems, which are interconnected in order to allow communication between computer systems; between computer systems and monitoring, control, and vessel management systems; and to allow centralized access to information and/or command/control. For example, an integrated system may consist of systems capable of performing passage execution (e.g. steering, speed control, traffic surveillance, voyage planning); machinery management and control (e.g. power management, machinery monitoring, fuel oil/lubrication oil transfer); cargo operations (e.g. cargo monitoring, inert gas generation, loading/discharging); etc.

5.3.3 Interface: A transfer point at which information is exchanged. Examples of interfaces include: input/output interface (for interconnection with sensors and actuators); communications interface (to enable serial communications/networking with other computers or peripherals).

5.3.4 Peripheral: A device performing an auxiliary function in the system, e.g. printer, data storage device.

19.8.5 ABS-2002, 4-9-1, 9, Conceptual requirements for system design

The following are conceptual requirements for control system design in general and are to be complied with except where specially exempted.

9.1 Fail-safe: Fail-safe concept is to be applied to the design of all remote control system, manual emergency control system and safety system. In consideration of its application, due regard is to be given to safety of individual machinery, the system of which the machinery forms a part, and the vessel as a whole.

9.3 System Independence: Systems performing different functions, e.g. monitoring systems, control systems, and safety systems, are to be, as much as practicable, independent of each other such that a single failure in one will not render the others inoperative. Specifically, the shutdown function of the safety system is to be independent of control and monitoring systems.

9.5 Local Control: In general, local manual controls are to be fitted to enable safe operation during commissioning and maintenance, and to allow for effective control in the event of an emergency or failure of remote control. The fitting of remote controls is not to compromise the level of safety and operability of the local controls

9.7 Monitoring Systems: Monitoring systems are to have the following detail features.

9.7.1 Independence of Visual and Audible Alarm Circuits: As much as practicable, a fault in the visual alarm circuits is not to affect the operation of the audible alarm circuits.

9.7.2 Audible Alarms: Audible alarms associated with machinery are to be distinct from other alarms such as fire-alarm, general alarm, gas detection alarm, etc., and are to be of sufficient loudness to attract the attention of duty personnel. For spaces of unusual high noise level, a beacon light or similar, installed in a conspicuous place may supplement the audible alarm in such spaces; however, red light beacons are only to be used for fire alarms.

9.7.3 Visual Alarms: Visual alarms are to be a flashing signal when first activated. The flashing display is to change to a steady display upon acknowledgment. The steady display is to remain activated, either individually or in the summarized fashion, until the fault condition is rectified. Other arrangements capable of attracting the operator's attention to an alarm condition in an effective manner will be considered.

9.7.4 Acknowledgment of Alarms: Newly activated alarms are to be acknowledged by manual means. This means is to mute the audible signal and change the flashing visual display to steady display. Other alarm conditions, occurring during the process of acknowledgment, are to be alarmed and displayed. The latter alarm is not to be suppressed by the acknowledgment of the former alarm. The acknowledgment of the alarm at an associated remote control station is not to mute and steady the same alarm signals at the centralized control station.

9.7.5 Temporarily Disconnecting Alarms: Alarm circuits may be temporarily disabled, for example for maintenance purposes, provided that such action is clearly indicated at the associated station in control and at the centralized control station, if fitted.

9.7.6 Built-in Alarm Testing: Audible alarms and visual alarm indicating lamps are to be provided with means of testing that can be operated without disrupting the normal operation of the monitoring systems.

9.7.7 Self-Monitoring: Monitoring system is to include a self-monitoring mechanism such that a fault, e.g. power failure, sensor failure, etc. may be detected and alarmed.

9.9 Safety Systems: In addition to complying with ABS-2002-4-9-1/9.1 through 4-9-1/9.7, safety systems are also to comply with the following:

i)   Means are to be provided to indicate the cause of the safety action.
ii) Alarms are to be given on the navigating bridge, at the centralized control station and at local manual control position, as applicable, upon the activation of safety system.
iii) Propulsion machinery shutdown by safety system is not to be designed to restart automatically, unless first actuated by a manual reset.
iv) Safety system for the protection of one machine unit is to be independent of that of the other units.

9.11 Failure Mode and Effect Analysis (FMEA): Failure modes and effects analysis (FMEA) may be carried out during system design to investigate if any single failure in control systems would lead to undesirable consequences such as loss of propulsion, loss of propulsion control, etc. The analysis may be qualitative or quantitative.

19.8.6 ABS-2002, 4-9-1, 13, Automatic safety shutdown

To avert rapid deterioration of propulsion and auxiliary machinery, the following automatic shutdowns are to be provided, regardless of the mode of control: manual, remote or automatic. These shutdowns are not to be fitted with manual override.

i)   For all diesel engines:
Over-speed

ii)   For all gas turbines (see 4-2-3/Table 1):
Failure of lubricating oil system
Failure of flame or ignition
High exhaust gas temperature
High compressor vacuum
Over-speed
Excessive vibration
Excessive axial displacement of rotors

iii)   For all steam turbines:
Failure of lubricating oil system
Over-speed
Back-pressure for auxiliary turbines

iv)   For all boilers:
Failure of flame
Failure of flame scanner
Low water level
Failure of forced draft pressure
Failure of control power

v)   For propulsion reduction gears:
Shutdown prime movers upon failure of reduction gear lubricating oil system.
For manned operation, where prime movers are diesel engines, shutdown is mandatory only for multiple high speed or medium speed diesel engines coupled to a reduction gear (see ABS-2002-4-6-5/5.3.4).

vi)   For generators:
For generators fitted with forced lubrication system only: shutdown prime movers upon failure of generator lubricating oil system (see ABS-2002-4-8-3/3.11.3).

vii)   For propulsion DC motor:
Overspeed (see ABS-2002- 4-6-7/5.17.7(b))

19.9 ABS-2002, 4-9-2, Remote propulsion control

5.11 Transfer Between Remote Control Stations: Remote control of the propulsion machinery is to be possible only from one location at a time. At each location there is to be an indicator showing which location is in control of the propulsion machinery. The following protocol is to be observed for transfer of control between stations:

i) The transfer of propulsion control between stations is to take effect only with acknowledgment by the receiving station. This, however, does not apply to transfer of control between the centralized control station and the local manual control.

ii) The transfer of propulsion control between the navigating bridge and the propulsion machinery space is to be possible only in the propulsion machinery space, i.e. at either the centralized control station or the local manual control position.

iii) The centralized control station is to be capable of assuming propulsion control at any time or blocking orders from other remote control stations. However, where special operating requirements of the vessel prevail, override control over the centralized control station will be considered.

iv) Propeller speed and direction of thrust are to be prevented from altering significantly when propulsion control is transferred from one control station to another.

5.13 Local Manual Control: Means are to be provided for local manual control so that satisfactory operation of the propulsion machinery can be exercised for lengthy periods in the event of the failure of the remote propulsion control system. For this purpose, indicators for propeller speed and direction of rotation (for fixed pitch propellers) or pitch position (for controllable pitch propellers) are to be provided at this local manual control station. The means of communication as required by 4-8-2/11.5 is to be fitted also at this manual control station. It is also to be possible to control auxiliary machinery, which are essential for propulsion and safety of the vessel, at or near the machinery concerned.

5.15 Communications Systems: For communication systems associated with propulsion control stations, the requirements in 4-8-2/11.5 are applicable.

9.1 Safety System—General: In all cases, automatic safety shutdowns in ABS-2002-4-9-1/13 are to be provided. Other safety system functions, such as automatic startup of standby pump or automatic slowdown, as appropriate, may be provided.

9.3 Safety System Alarms: Activation of safety system to automatic slowdown or automatic shutdown of propulsion machinery is each to be arranged with individual alarm at remote propulsion control station. Audible alarm may be silenced at the control station, however visual alarm is to remain activated until it is acknowledged in the machinery space.

9.5 Override of Safety System Functions: Automatic slowdowns and automatic shutdowns indicated in ABS-2002-4-9-4/Table 3A through 4-9-4/Table 6 may be provided with override except that specified in ABS-2002-4-9-1/13. Automatic slowdowns and automatic shutdowns where provided in excess of those indicated in 4-9-4/Table 3A through 4-9-4/Table 6 are to be provided with override. Overrides are to be as follows:

i) The activation of the override is to be alarmed and clearly identifiable at the remote propulsion control station and is to be so designed that it cannot be left activated.

ii) Overrides fitted on the navigating bridge are to be operable only when the propulsion control is from the navigating bridge.

iii) The override actuator is to be arranged to preclude inadvertent operation.

19.10 Automation Local Area Network (LAN) (Figure 19-2)

Figure 19-2 is a simplified block diagram of the main machinery centralized control local area network (LAN). The diagram shows a ring type local area network to establish required shipboard redundancy. In this redundancy configuration, any break in the LAN bus should have no disruption in the signal processing. The system consists of dual data collecting lines, dual centralized processing units, and dual video terminal units (VTU).

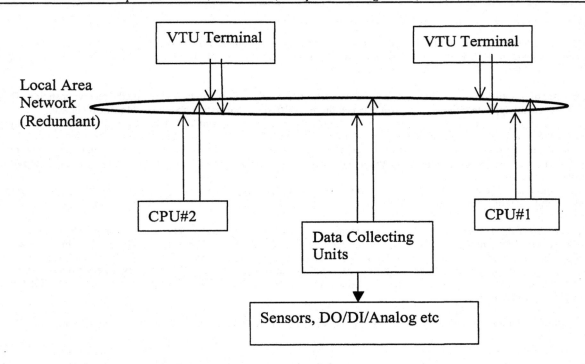

Figure 19-2: Automation Local Area Network

19.11 Central control redundancy distribution

The following three figures are block diagrams showing centralized control system block diagrams.

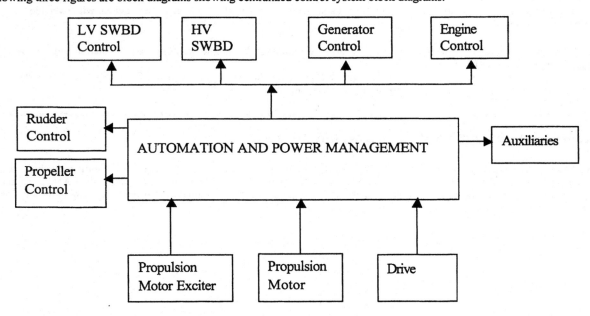

Figure 19-3: Automation and Power Management—Electric Propulsion System

Figure 19-3 shows an automation and power management block diagram connecting the engine controls, generator controls, HV switchboard controls, low voltage switchboard controls, rudder controls, propulsion motor controls, propulsion drive controls, and auxiliary controls. Figure 19-4 shows controls a block diagram for a machinery system with redundant LAN, two processors, and wheelhouse controls. Figure 19-5 shows a redundant machinery plant with dedicated, redundant remote terminal units (RTUs) for collecting field signals. This plant features completely redundant machinery configurations with local area network (LAN) and redundant control system processor units. The RTUs are installed in different locations for collecting signals and sensor data such as analog, digital for control, monitoring, and alarms of the machinery control system.

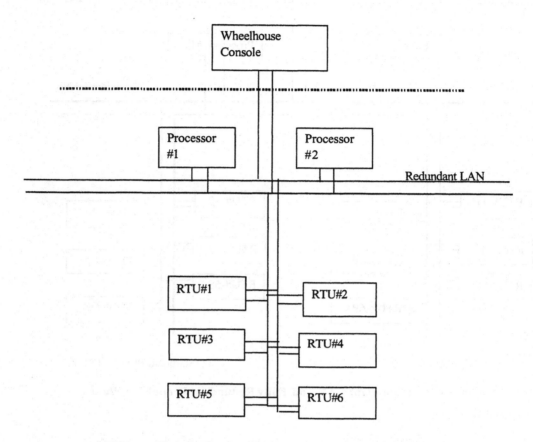

Figure 19-4: Redundant LAN for Single Drive System

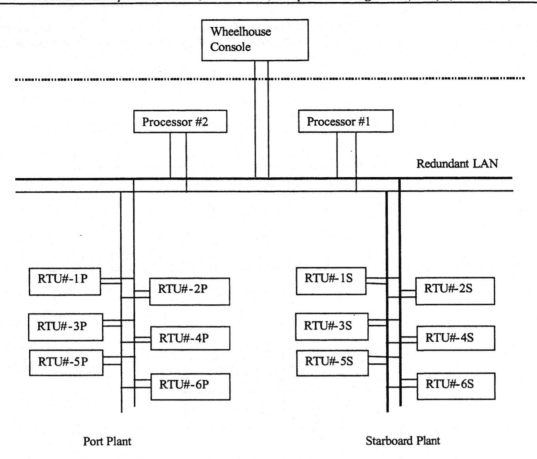

Figure 19-5: Redundant LAN and Fully Redundant Propulsion Plant

20 Alternate Compliance Program (ACP) – ABS and USCG (Ship Design)

20.1 General

This chapter covers a topic which is not part of IEEE Std 45-2002, but which is nevertheless of central importance in shipboard electrical system design. As a user of IEEE Std 45, the reader will doubtless also come into contact with the Alternate Compliance Program (ACP). This section covers the most relevant requirements of the ACP.

ABS and USCG jointly adapted the Alternate Compliance Program (ACP) for shipboard electrical installation requirements. The ACP is in compliance with "International Maritime Organization Safety of Life at Sea" IMO SOLAS requirements.

For details of ACP requirements, refer to USCG COMDTPB P16700.4A, NVIC 2-95, dated 1 August 1997, and ABS – USCG Alternate Compliance Program (ACP), dated 1 November 1999.

The ACP was prepared by using the ABS rules of 1998, but the ABS rules section numbering was completely changed to a different format in the ABS 2000 rule publication. In this section, we refer to ABS 1998 rules as well as ABS 2002 rules for easy understanding of ACP requirements. The user of this Handbook should be aware of the original ACP requirements and the new numbering system in relation to the ABS requirements and changes to the ABS rules to ensure that the interpretation of requirements is in line with the latest rules and regulations.

The large sections of ACP are discussed in the following sections of this chapter:

20.4	Generator: ABS-1998 rules related to ACP
	(ACP Cites and ABS-2002 rules related to the ACP)
20.5	Emergency Switchboard (ABS-1998 rules, ACP cites and ABS-2002)
20.6	Distribution (ABS-1998 rules, ACP cites and ABS-2002)
20.7	Circuit Protection (ABS-1998 rules, ACP cites and ABS-2002)
20.8	Lighting Circuits (ABS-1998 rules, ACP cites and ABS-2002)
20.9	Navigation Lights (ABS-1998 rules, ACP cites and ABS-2002)
20.10	Interior Communications (ABS-1998 rules, ACP cites and ABS-2002)

20.1.1 Overview of ACP

ACP is divided in six sections. The following is a portion of the ACP sections discussed here, which deals with electrical power generation, distribution, controls, interior communications, and external communications:

Oversight:

The USCG, in delegating surveys to ABS, still retains the ultimate responsibility that vessels meet regulatory requirements. Crucial to fulfilling this responsibility is active and viable oversight by the USCG of surveys conducted by ABS on behalf of the USCG. The foundation of this oversight is ABS's World Wide ISO 9001 Certified Quality System. As with any successful quality system, it is a smoothly functioning in-service process verification scheme. It provides a source of continuous and timely opinion related to the effectiveness of the processes in place to meet customer requirements. An added benefit is the information it provides to both clients and management to prove that controlled work is being accomplished. In this respect it is very important in facilitating the delegation to ABS of USCG vessel inspection. It provides a framework that will be used in the USCG oversight program for delegated responsibilities.

Oversight will consist of internal and external audits of ABS by the USCG. It will also consist of annual boarding of the vessels to conduct renewal inspections. The boardings will be similar to those done in Port State Inspections. A check sheet describing the considerations to expand the boardings is a part of the USCG Marine Safety Manual, Chapter 32.

In Section 20.2, the applicable ABS -1998 rules are quoted, followed by ACP requirement and ABS-2002 requirements to provide complete ACP requirements. ABS 2002 rules are also quoted to help understanding the transition from ABS-1998 requirements to ABS - 2002 rules and IEEE Std 45-2002 recommendations.

For ABS class ships, ABS requirements must be met. For ACP class ships, the ABS requirements *and* the ACP requirements must be met. IEEE Std 45 provides recommendations in addition to these regulatory body rules and regulations, which are supplemental recommendations for consideration of better operation, performance etc.

20.2 ACP requirements (Supplemental requirements to current ABS - 1998 rules)

20.2.1 US supplement, Section I, ABS-1998-4/5A3.5.2, Electrical Equipment – Emergency Sources of Power – Power Supply – Generator

A stop control for an emergency generator must only be in the space that has the emergency generator, except a remote mechanical reach rod is permitted for the fuel oil shutoff valve to an independent fuel oil tank located in the space.

20.2.2 US supplement- Section I, Cite: 4/5A3.9-ABS 1998 rules - Electrical Equipment – Emergency Sources of Power – Emergency Switchboard

Each bus-tie between a main switchboard and an emergency switchboard must be arranged to prevent parallel operation of the emergency power source with any other source of electric power, except for interlock systems for momentary transfer of loads.

If there is a reduction of potential of the normal source by 15 to 40 percent, the final emergency power source must start automatically without load. When the potential of the final emergency source reaches 85 to 95 percent of normal value, the emergency loads must transfer automatically to the final emergency power source. When the potential from the normal source has been restored, the emergency loads must be manually or automatically transferred to the normal source, and the final emergency power source must be manually or automatically stopped.

20.2.3 US supplement, Section I, Cite: 4/5A4, ABS – 1998 rules: Electrical Equipment – Distribution System

Each grounded system must have only one point of connection to ground regardless of the number of power sources operating in parallel in the system.

There must be ground detection for each ship's service power system, lighting system, and power or lighting distribution system that is isolated from the ship's service power and lighting system by transformers, motor generator sets, or other devices.

The ground indicators must be at the ship's service generator distribution switchboard for the normal power, normal lighting, and emergency lighting systems; and be at the propulsion switchboard for propulsion systems.

20.2.4 US supplement- Section I, Cite: 4/5A5 –ABS rules - Electrical Equipment – Circuit Protection System

Each fuse must meet the general provisions of Article 240 of the National Electrical Code or IEC 92-202, as appropriate, and have an interrupting rating sufficient to interrupt the maximum asymmetrical RMS short-circuit current at the point of application. Each fuse must provide for ready access to test the condition of the fuse.

20.2.5 US supplement - Section I, Cite: 4/5C4.11.1-ABS rule - Electrical Equipment –

Bus Bars – Design

A busway must not need mechanical cooling to operate within its rating. Each busway must meet Article 364 of the National Electric Code.

20.2.6 US supplement-Section I, Cite: 4/5C4.11.4 – ABS rule - Electrical Equipment – Switchboards – Internal Wiring

Instrument and control wiring for switchboards is to be rated for 9⊘C or higher; 14AWG or larger or must be ribbon cable or similar conductor size cable recommended for use in low power instrumentation, monitoring, or control circuits by the equipment manufacturer; and be flame retardant meeting ANSI/UL 1581 test VW-1 or IEC 332-1.

20.2.7 US supplement, Section I, Cite: 4/5C4.15.4 – ABS rule - Electrical Equipment – Switchboards – Equipment and Instrumentation

Each AC switchboard must have a voltage regulator functional cutout switch for transferring from automatic to manual control mode and a manual control rheostat for exciter field.

20.2.8 US supplement, Section I, Cite: 4/5C7-ABS rules-Electrical Equipment – Cables and Wires

General:

Generator, feeder, and bus-tie cables must be selected on the basis of a computed load of not less than the demand load given in the following table:

Demand loads (ACP-Cable Ampacity Table 19-1)

Generator cable	115% of continuous generator rating or 115 percent of the overload for a machine with a 2-hour or greater overload rating.
Switchboard bus-tie, except ship's service to emergency switchboard bus-tie	75% of generator capacity of the larger switchboard.
Emergency switchboard bus-tie	115% of continuous rating of emergency generator.
Motor feeders	Article 430, National Electrical Code.
Galley equipment feeder	100% of either the first 50 kW or one-half the connected load, whichever is the larger, plus 65% of the remaining connected load, plus 50% of the rating of the spare switches or circuit breakers on the distribution panel.
Lighting feeder	100% of the connected load plus the average active circuit load for the spare switches or circuit breakers on the distribution panel.
Grounded neutral of a dual voltage feeder	100% of the capacity of the ungrounded conductors when grounded neutral is not protected by a circuit breaker over-current trip, or not less than 50% of the capacity of the ungrounded conductors when the grounded neutral is protected by a circuit breaker over-current trip or over-current alarm.

Cables

Marine shipboard cables must meet all construction and identification requirements of:

IEC 92-3, 1982.

IEEE Std-45, 1983 Edition.

Cables complying with MIL Standards MIL-C-24640A, 1995 and MIL-C-24643A, 1994.

Direct current electrical cable, for industrial applications only, may be applied in accordance with IADC-DCCS-1/1991.

Flame retardant materials may not be used as a substitute for fire stops or in lieu of cables required to meet the fire resistance requirements of IEEE Std 45 or IEC 92-3.

Internal Wiring

Wire must be of the copper stranded type and in an enclosure. Wire, other than in switchboards, must meet Sections 19.6.4 and 19.8 of IEEE Std. 45, MIL-W-76D, MIL-W-16878F, UL 44, UL83 or equivalent standard.

20.2.9 US supplement Section I, Cite: 4/5C7.5-ABS rule - Electrical Equipment – Cables and Wires – Portable and Flexing Electric Cables

Each flexible electric cord and cable must meet UL 62, NEMA WC 3, NEMA WC 8, Section 19.6.1 of IEEE Std. 45, or Article 400 of the NEC.

A flexible cord must be used only as allowed under Sections 400-7 and 400-8 of the National Electrical Code; and in accordance with Table 400-4 of the National Electrical Code.

A flexible cord must not carry more current than allowed under Table 400-5 of the National Electrical Code, NEMA WC 3 or NEMA WC 8.

Each flexible cord must be 0.82 mm 2 (No. 18 AWG) or larger.

Each flexible cord and cable must be without splices or taps except for a cord or cable 3.3 mm 2 (No. 12 AWG) or larger.

20.2.10 US supplement, Section I, Cite: 4/5E1-ABS rule - Electrical Equipment – Specialized Vessels and Services – Oil Carriers

Each vessel that carries combustible liquid cargo with a closed-cup flashpoint of 60° C (140° F) or higher must have only intrinsically safe electric systems in cargo tanks and intrinsically safe apparatus or explosion-proof electrical lighting, communication, general alarm, and fire-extinguishing medium release alarm systems in any cargo handling room.

An enclosed hazardous space that has explosion-proof lighting fixtures must have at least two lighting branch circuits and be arranged so that there is light for relamping any deenergized lighting circuit.

No tank vessel that carries petroleum products grades A through D contracted for on or after June 15, 1977 may have an air compressor or an air compressor intake installed in: a cargo handling room; an enclosed space containing cargo piping; a space in which cargo hose is stowed; a space adjacent to a cargo tank or cargo tank hold; a space within three meters of a cargo tank opening, an outlet for cargo gas or vapor, a cargo pipe flange, a cargo valve, or an entrance or ventilation opening to a cargo handling room; the cargo deck space, except for tank barges (the cargo deck space is the open deck over the cargo area extending 3 m (10 ft) forward and aft of the cargo area and up to 2.4 m (8 ft) above the deck); an enclosed space having an opening into any of the above locations; or a location similar to those described above in which cargo vapors or gases may be present.

20.2.11   US supplement, Section I, Cite:4/5E5.3-ABS rule - Electrical Equipment – Specialized Vessels and Services – Chemical Carriers

Each vessel that carries liquid sulphur cargo or inorganic acid cargo must meet the requirements of 4/5E1 of the Rules and Cite 4/5E1 above for a vessel carrying oil with a flashpoint not exceeding 60 degrees C, except that a vessel carrying carbon disulfide must have only intrinsically safe electric equipment in cargo tanks, cargo handling rooms, enclosed spaces, cargo hose storage spaces, spaces containing cargo piping and hazardous locations in the weather.

20.3   US supplement, Section I, Cite: 4/8.9.6-ABS rule - Steering Gear – Controls

The main steering gear is to be provided with full follow up control in the pilothouse. Follow up control means closed-loop (feedback) control that relates the position of the helm to a specific rudder angle by transmitting the helm-angle order to the power actuating system and, by means of feedback, automatically stopping the rudder when the angle selected by the helm is reached.

20.3.1   US supplement, Section I, Cite: 4/8.9.7-ABS rule - Steering Gear – Instrumentation and Alarms

This requirement applies to each vessel of 1600 gross tons and over that has power driven main or auxiliary steering gear.

The steering failure alarm system must be independent of each steering gear control system, except for the input received from the steering wheel shaft.

The steering failure alarm system must have audible and visible alarms in the pilothouse when the actual position of the rudder differs by more than 5° from the rudder position ordered by the follow-up control systems for more than:

(a)   30 seconds for ordered rudder position changes of 70 degrees,
(b)   6.5 seconds for ordered rudder position changes of 5 degrees, and
(c)   The time period calculated by the following formula for ordered rudder position changes between 5 degrees and 70 degrees:

   $t = (R/2.76) + 4.64$ Where:

   t = maximum time delay in seconds

   R = ordered rudder change in degrees

Each steering failure alarm system must be supplied.

20.3.2   US supplement, Section I, Cite: 4/11.3.5b-ABS rule - Safety Systems – Characteristics

Safety systems must not operate as a result of failure of the normal electric power source unless it is determined to be the failsafe state.

20.3.3 US supplement, Section I, Cite: 4/11.5.1-ABS rule - Automatic or Remote Propulsion Control and Monitoring Systems – General

> Sensors for the primary speed, pitch or direction of rotation control in closed loop propulsion control systems must be independent and physically separate from required safety, alarm or instrumentation sensors.

20.3.4 US supplement, Section I, Cite: 4/11.7.3-ABS rule - Shipboard Automatic or Remote Control and Monitoring Systems – Automatic or Remote Propulsion Control and Monitoring Systems for Propulsion – Machinery Spaces Covered Under Class Symbols – Station in Navigating Bridge

> The Main Control Station must include control of the main machinery space fire pumps.

20.3.5 US supplement, Section I, Cite: 4/11.7.6 & 4/11.9.5 - ABS rule - Continuity of Power

> The standby electric power is to be available in no more than 30 seconds.

20.3.6 Comments on US supplement - Section I

The section numbering system and format of the ABS rules have been changed, and therefore ACP references to the ABS section numbering are not the same. We have made an attempt to relate the previous numbering system to ABS-2002 rules numbering system. The user of the IEEE Std 45-2002 should be aware of the original ACP requirements and the new numbering system in relation to the ABS requirements to ensure that the interpretation of requirements is in line with the latest rules and regulations.

20.4 Generator: ABS-1998 rules related to ACP (ACP Cites and ABS-2002 rules related to the ACP)

20.4.1 ABS-1998, 4/5A3.5.2: Generator

> Where the emergency source of electrical power is a generator, it is to be:
>
> a)   driven by a prime mover with an independent supply of fuel, having a flashpoint (closed cup test) of not less than 43 degrees C (110 degrees F), and
>
> b)   1) started automatically upon failure of the main source of electrical power supply and connected automatically to the emergency switchboard;
>
> — then, those services referred to in 4/5A3.7 are to be connected automatically to the emergency generator as quickly as is safe and practicable subject to a maximum of 45 seconds, or
>
> 2) provided with a transitional source of emergency electrical power as specified in 4/5A3.7 unless an emergency generator is provided capable both of supplying the services referred to in 4/5A3.7 of being automatically started and supplying the required load as quickly as is safe and practicable subject to a maximum of 45 seconds, and
>
> c)   An adequate fuel capacity for the emergency generation prime mover is to be provided.

20.4.2 ACP Supplement I of Section I -Generator

ABS Cite: 4/5A3.5.2   Electrical Equipment – Emergency Sources of Power – Power Supply – Generator

A stop control for an emergency generator must only be in the space that has the emergency generator, except a remote mechanical reach rod is permitted for the fuel oil shutoff valve to an independent fuel oil tank located in the space.

20.4.3 ABS-2002, 4-8-2, 5.9.1, Generator 4-8-2/5.9.1

Where the emergency source of electrical power is a generator, it is to be:

i) driven by a prime mover with an independent supply of fuel, having a flash point (closed cup test) of not less than 43°C (110°F);

ii) started automatically upon failure of the main source of electrical power supply; and

iii) automatically connected to the emergency switchboard supplying those services referred to in 4-8-2/5.5 in not more than 45 seconds.

V) Where the emergency generator as specified above is not provided with automatic starting, a transitional source of emergency electrical power as specified in 4-8-2/5.11 is to be fitted.

20.5 Emergency switchboards

The requirements below are called out by ABS rule 1998 related to the ACP, ACP cites and ABS-2002 rules related to the ACP.

20.5.1 ABS-1998, 4/5A3.9, Emergency Switchboard

ABS-98-4/5A3.9.1 General

The emergency switchboard is to be installed as near as is practicable to the emergency source of electrical power.

ABS-98-4/5A3.9.2 Emergency Switchboard for Generator

Where the emergency source of electrical power is a generator, the emergency switchboard is to be located in the same space unless the operation of the emergency switchboard would thereby be impaired.

ABS-98-4/5A3.9.3 Accumulator Battery

No accumulator battery fitted in accordance with 4/5A3.5.3 or 4/5A3.7 is to be installed in the same space as the emergency switchboard. An indicator is to be mounted on the main switchboard or in the machinery control room to indicate when these batteries are being discharged.

ABS-98-4/5A3.9.4 Inter-connector Feeder Between Emergency and Main Switchboards

The emergency switchboard is to be supplied during normal operation from the main switchboard by an inter-connector feeder, which is to be protected at the main switchboard against overload and short circuit. The inter-connector feeder is to be disconnected automatically at the emergency switchboard upon failure of the main source of electrical power. Where the system is arranged for feedback operation, the inter-connector feeder is also to be protected at the emergency switchboard against short circuit. In addition, the circuit protection device at the emergency switchboard on the inter-connector feeder is to trip to prevent overloading of the emergency generator.

ABS-98-4/5A3.9.5 Disconnection of Non-emergency Circuits

For ready availability of the emergency source of electrical power, arrangements are to be made where necessary to disconnect automatically non-emergency circuits from the emergency switchboard so that electrical power is to be available automatically to the emergency circuits.

20.5.2 ACP US supplement, Section I

ABS-1998-Cite: 4/5A3.9 Electrical Equipment – Emergency Sources of Power – Emergency Switchboard

---

Each bus-tie between a main switchboard and an emergency switchboard must be arranged to prevent parallel operation of the emergency power source with any other source of electric power, except for interlock systems for momentary transfer of loads.

If there is a reduction of potential of the normal source by 15 to 40 percent, the final emergency power source must start automatically without load. When the potential of the final emergency source reaches 85 to 95 percent of normal value, the emergency loads must transfer automatically to the final emergency power source. When the potential from the normal source has been restored, the emergency loads must be manually or automatically transferred to the normal source, and the final emergency power source must be manually or automatically stopped.

20.5.3 ABS-2002, 4-8-2, 5.13, Emergency Switchboard

ABS-2002-4-8-2-5.13.1 Location of Emergency Switchboard

The emergency switchboard is to be installed as near as is practicable to the emergency source of electrical power.

Where the emergency source of electrical power is a generator, the emergency switchboard is to be located in the same space unless the operation of the emergency switchboard would thereby be impaired.

No accumulator battery fitted in accordance with 4-8-2/5.9.2 or 4-8-2/5.11 is to be installed in the same space as the emergency switchboard. An indicator is to be mounted on the main switchboard or in the machinery control room to indicate when these batteries are being discharged.

ABS-2002-4-8-2-5.13.2 Inter-connector Feeder between Emergency and Main Switchboards

The emergency switchboard is to be supplied during normal operation from the main switchboard by an inter-connector feeder which is to be protected at the main switchboard against overload and short circuit and which is to be disconnected automatically at the emergency switchboard upon failure of the main source of electrical power.

ABS-2002-4-8-2-5.13.3 Feedback Operation

Where the emergency switchboard is arranged for feedback operation, the inter-connector feeder is also to be protected at the emergency switchboard at least against short circuit, which is to be coordinated with the emergency generator circuit breaker. In addition, this inter-connector feeder protective device is to trip to prevent overloading of the emergency generator, which might be caused by the feedback operation.

ABS-2002-4-8-2-5.13.4 Non-emergency Services and Circuits

The emergency generator may be used, exceptionally and for short periods, for services such as routine testing (to check its proper operation), and dead ship (blackout) start-up provided that measures are taken to safeguard the independent emergency operation as required in 4-8-2/5.9.1.

For ready availability of the emergency source of electrical power, arrangements are to be made where necessary to disconnect automatically non-emergency circuits from the emergency generator to ensure that electrical power is available automatically to the emergency circuits upon main power failure.

ABS-2002-4-8-2-5.13.5 Arrangements for Periodic Testing

Provision is to be made for the periodic testing of the complete emergency system and is to include the testing of automatic starting system.

20.6 Distribution systems

The following requirements on distribution systems are called out by ABS rule 1998 related to the ACP, ACP cites and ABS-2002 rules related to the ACP.

20.6.1 ABS-1998, 4/5A4, Distribution System

ABS-98-4/5A4.1.1 General

Current-carrying parts with potential to earth are to be protected against accidental contact. For recognized standard distribution systems, see 4/5.7. Separate feeders are to be provided for essential and emergency services.

ABS-98-4/5A4.1.2 Method of Distribution

The output of the ship's service generators may be supplied to the current consumers by way of either branch system, meshed network system, or ring main system. The cables of a ring-main or other looped circuit (e.g. interconnecting section boards in a continuous circuit) are to be formed of conductors having sufficient current carrying and short-circuit capacity for any possible load and supply configuration.

ABS-98-4/5A4.1.3 Through-feed Arrangements

The size of feeder conductors is to be uniform for the total length, but may be reduced beyond any intermediate section board and distribution board, provided that the reduced size section of the feeder is protected by an overload device.

ABS-98-4/5A4.1.4 Motor Control Center

Feeder cables from the main switchboard or any section board to the motor control centers are to have a continuous current-carrying capacity not less than 100% of the sum of the nameplate ratings of all the motors supplied. The over-current protective device is to be in accordance with 4/5A5.1.3.

ABS-98-4/5A4.1.5 Motor Branch Circuit

A separate circuit is to be provided for each fixed motor having a full-load current rating of 6 amperes or more and the conductors are to have a carrying capacity of not less than 100% of the motor full-load current rating. No branch circuit is to have conductors less than 1.5 mm2 wire. Circuit-disconnecting devices are to be provided for each motor branch circuit and to be in accordance with 4/5B2.13.2 and 4/5C4.17.2.

20.6.2 ACP US supplement of Section I - Distribution System

Cite: 4/5A4   Electrical Equipment – Distribution System

Each grounded system must have only one point of connection to ground regardless of the number of power sources operating in parallel in the system.

There must be ground detection for each ship's service power system, lighting system, and power or lighting distribution system that is isolated from the ship's service power and lighting system by transformers, motor generator sets, or other devices.

The ground indicators must be at the ship's service generator distribution switchboard for the normal power, normal lighting, and emergency lighting systems; and be at the propulsion switchboard for propulsion systems.

20.6.3 ABS-2002, 4-8-2, 7, Distribution System

ABS-2002-4-8-2-7.1 General

The following are recognized as standard systems of distribution. Distribution systems other than these will be considered.

— Two-wire direct current
— Two-wire single-phase alternating current
— Three-wire three-phase alternating current
— Four-wire three-phase alternating current

ABS-2002-4-8-2-7.3 Hull Return Systems

ABS-2002-4-8-2-7.3.1 General

A hull return system is not to be used, with the exception as stated below:

impressed current cathodic protection systems; limited locally earthed system, provided that any possible resulting current does not flow through any hazardous locations; insulation level monitoring devices, provided the circulation current does not exceed 30 mA under all possible conditions.

ABS-2002-4-8-2-7.3.2 Final Sub-circuits and Earth Wires

Where the hull return system is used, all final sub-circuits, i.e., all circuits fitted after the last protective device, are to consist of two insulated wires, the hull return being achieved by connecting to the hull one of the bus bars of the distribution board from which they originate. The earth wires are to be in accessible locations to permit their ready examination and to enable their disconnection for testing of insulation.

ABS-2002-4-8-2-7.5 Earthed AC Distribution System

ABS-2002-4-8-2-7.5.1 General Earthing Arrangement

For earthed distribution systems, regardless of the number of power sources, the neutral of each power source, including that of the emergency generator where applicable, is to be connected in parallel and earthed at a single point. Reference should be made to manufacturer-specified allowable circulating currents for neutral-earthed generators.

ABS-2002-4-8-2-7.5.2 System Earthing Conductor

System earthing conductors are to be independent of conductors used for earthing of non-current carrying parts of electrical equipment. See 4-8-4/23.3 for installation details and earth conductor sizing. Four-wire three-phase AC systems having an earthed neutral are not to have protective devices fitted in the neutral conductors. Multipole switches or circuit breakers, which simultaneously open all conductors, including neutral, are allowed. In multiple generator installations, each generator's neutral connection to earth is to be provided with a disconnecting link for maintenance purpose.

ABS-2002-4-8-2-7.7 Cable Sizing

This Paragraph applies to cables conforming to IEC Publication 60092-353 or IEC Publication 60092-3. Cables conforming to other standards are to be sized in accordance with corresponding provisions of that standard. For marine cable standards acceptable to the Bureau, see 4-8-3/9.1.

ABS-2002-4-8-2-7.7.1 Cable's Current Carrying Capacity

ABS-2002-4-8-2-7.7.1(a) General. Cable conductor size is to be selected based on the current to be carried such that the conductor temperature, under normal operating conditions including any overload condition that may be expected, does not exceed the maximum rated temperature of the cable insulation material. The selected cable type is to have a maximum rated temperature at least 10°C (18°F) higher than the maximum ambient temperature likely to exist at the location where the cable is installed.

ABS-2002-4-8-2-7.7.1(b) Current carrying capacities. The maximum current carrying capacities of cables are to be obtained from 4-8-3/Table 6. These values are applicable, without correction factors for cables installed either in single- or double-layer in cable tray, or in a bunch in cable trays, cable conduits or cable pipes where the number of cables in the bunch does not exceed six. The ambient temperature is to be 45°C (113°F) or less.

ABS-2002-4-8-2-7.7.1(c) Current carrying capacity correction. Where more than six cables which may be expected to operate simultaneously are laid close together in a bunch in such a way that there is an absence of free air circulation around them a reduction factor is to be applied to the current carrying capacity of the cables; this reduction factor is to be 0.85 for seven to twelve cables in one bunch. The correction factor for cable bunches of more than twelve cables each is to be specially considered in each case based on cable type and service duty.

20.7 Circuit protection systems

The following requirements on circuit protection systems are dictated by ABS rule 1998 related to the ACP, ACP cites and ABS-2002 rules related to the ACP.

20.7.1 ABS-1998, 4/5A5, Circuit Protection System

ABS-1998-4/5/A5.1.1 General

Electrical installations are to be protected against accidental overload and short circuit, except:

a) as permitted by 4/5A6.3
b) where it is impracticable to do so, such as engine starting battery circuit, and
c) where by design, the installation is incapable of developing overload, in which case it may be protected against short circuit only.

The protection is to be by automatic protective devices for:

i) continued supply to remaining essential circuits in the event of a fault, and
ii) minimizing the possibility of damage to the system and fire.

Three-phase, three-wire alternating current circuits are to be protected by a triple-pole circuit breaker with three overload trips or by a triple-pole switch with a fuse in each phase. All branch circuits are to be protected at distribution boards only and any reduction in conductor sizes is to be protected. Dual-voltage systems having a earthed neutral are not to have fuses in the neutral conductor, but a circuit breaker which simultaneously opens all conductors may be installed when desired. In no case is the dual-voltage system to extend beyond the last distribution board.

ABS-1998-4/5A5.1.2 Protection Against Short-circuit

a) Protective Devices

Protection against short-circuit is to be provided for each non-earthed conductor by means of circuit breakers or fuses.

b) Rated Short-circuit Breaking Capacity

The rated short-circuit breaking capacity of every protective device is not to be less than the maximum available fault current at that point. For alternating current (AC), the rated short-circuit breaking capacity is not to be less than the root mean square (rms) value of the AC component of the prospective short-circuit current at the point of application. The circuit breaker is to be able to break any current having an AC component not exceeding its rated breaking capacity, whatever the inherent direct current (DC) component may be a the beginning of the interruption.

c) Rated Short-circuit Making capacity

The rated short-circuit making capacity of every switching device is to be adequate for maximum peak value of the prospective short -circuit current at the point of installation. The circuit breaker is to be able to make the current corresponding to its making capacity without opening within a time corresponding to the maximum time delay required.

20.7.2 ACP Supplement I of Section I

Cite: ABS-1998-4/5A5   Electrical Equipment – Circuit Protection System

---

Each fuse must meet the general provisions of Article 240 of the National Electrical Code or IEC 92-202, as appropriate, and have an interrupting rating sufficient to interrupt the maximum asymmetrical RMS short-circuit current at the point of application. Each fuse must provide for ready access to test the condition of the fuse.

20.7.3 ABS-2002, 4-8-2, 9.1, General

Each electrical system is to be protected against overload and short circuit by automatic protective devices, so that in the event of an overload or a short circuit the device will operate to isolate it from the systems: to maintain continuity of power supply to remaining essential circuits; and to minimize the possibility of fire hazards and damage to the electrical system.

These automatic protective devices are to protect each non-earthed phase conductors (e.g. multipole circuit breakers or fuses in each phase). In addition, where possibility exists for generators to be overloaded, load-shedding arrangements are to be provided to safeguard continuity of supply to essential services

The following are exceptions:

— Where it is impracticable to do so, such as engine starting battery circuits.
— Where, by design, the installation is incapable of developing overload, in which case, it may be protected against short circuit only. Steering circuits; see 4-8-2/9.17.5.

ABS-2002, 4-8-2- 9.3 Protection Against Short-circuit

ABS-2002, 4-8-2- 9.3.1 General

Protection against short-circuit is to be provided for each non-earthed conductor (multipole protection) by means of circuit breakers, fuses or other protective devices.

ABS-2002, 4-8-2- 9.3.2 Short Circuit Data

In order to establish that protective devices throughout the electrical system (e.g. on the main and emergency switchboards and sub-distribution panels) have sufficient short circuit breaking and making capacities, short circuit data as per 4-8-1/5.1.3 are to be submitted.

ABS-2002, 4-8-2- 9.3.3 Rated Breaking Capacity

The rated breaking capacity of every protective device is not to be less than the maximum prospective short circuit current value at the point of installation. For alternating current (AC), the rated breaking capacity is not to be less than the root mean square (rms) value of the prospective short-circuit current at the point of installation. The circuit breaker is to be capable of breaking any current having an AC component not exceeding its rated breaking capacity, whatever the inherent direct current (DC) component may be at the beginning of the interruption.

ABS-2002, 4-8-2- 9.3.4 Rated Making Capacity

The rated making capacity of every circuit breaker which may be closed on short circuit is to be adequate for the maximum peak value of the prospective short-circuit current at the point of installation. The circuit breaker is to be capable of closing onto a current corresponding to its making capacity without opening within a time corresponding to the maximum time delay required.

ABS-2002, 4-8-2- 9.3.5 Backup Fuse Arrangements

Circuit breakers having breaking and/or making capacities less than the prospective short circuit current at the point of application will be permitted provided that such circuit breakers are backed up by fuses which have sufficient short circuit capacity for that application. Current-limiting fuses for short circuit protection may be without limitation on current rating, see 4-8-2/9.5.

ABS-2002, 4-8-2- 9.3.6 Cascade Protection

Cascade protection will be permitted where the combination of circuit protective devices has sufficient short circuit capacity at the point of application. Where used in circuits of essential services, such services are to be duplicated and provided with means of automatic transfer.

20.8 Lighting circuits

The following requirements for lighting circuits are laid out by ABS rule 1998 related to the ACP, ACP cites and ABS-2002 rules related to the ACP.

20.8.1 ABS-1998- 4/5A7.1.3, Lighting Circuits

a. Machinery Space and Accommodation Space In spaces such as:

— public spaces;
— main machinery spaces;
— galleys;
— corridors;
— stairways leading to boat-decks;

b. xx there are to be more than one final sub-circuit for lighting, one of which may be supplied from the emergency switchboard, in such a way that failure of any one circuit does not leave these spaces in darkness.

c. Cargo Spaces

Fixed lighting circuits in cargo spaces are to be controlled by multipole-linked switches situated outside the cargo spaces. Means are to be provided on the multipole linked switches to indicate the live status of circuits.

20.8.2 ACP US supplement of Section I

Cite: ABS-1998- 4/5A7.1.3   Electrical Equipment – Lighting Circuits

The construction of each lighting fixture not in a hazardous location must meet UL standard 595 until May 3, 1999; UL 1570, UL 1571 or UL 1572, as applicable, including the Marine Supplement; or IEC 92-306. Each fixture located in the weather or in a ro-ro space, cargo hold or machinery space is to be provided with a suitable lens and guard.

Each lighting fixture must be firmly secured. Each pendent-type fixture must be supported by a short length of rigid conduit.

No fixture may be used as a connection box for other than the branch circuit supplying the fixture.

Non-emergency and decorative interior lighting fixtures in environmentally protected, non-hazardous locations need only meet the applicable UL type fixture standards in UL 1570 through UL 1574 (and either the general section of the Marine Supplement or the general section of UL 595), UL 595 or IEC 92-306. These fixtures must have vibration clamps on fluorescent tubes longer than 102 cm (40 inches), secure mounting of the glassware and rigid mounting.

20.8.3 ABS-2002, 4-8-2, 9.21, Protection for Branch Lighting Circuits

Branch lighting circuits are to be protected against overload and short circuit. In general, overload protective devices are to be rated or set at not more than 30 A. The connected load is not to exceed the lesser of the rated current carrying capacity of the conductor or 80% of the overload protective device rating or setting.

20.9 Navigation light systems: ABS rule 1998 related to the ACP, ACP citations, and ABS-2002 rules related to ACP

20.9.1 ABS-1998-4/5A7.3, Navigation Light System

ABS-1998-4/5A7.3.1 Feeders- Navigation Light System

The masthead, side and stern lights are to be separately connected to a distribution board reserved for navigation lights, placed in an accessible position on bridge, and connected directly or through transformers to the main or emergency switchboard. These lights are to be fitted with duplicate lamps or other dual light sources and are to be controlled by an indicator.

20.9.2 ACP Supplement I of Section I- Navigation Light System

Cite: 4/5A7.3   Electrical Equipment – Navigation Light System

---

Each navigation light must meet the following:

(a)   Meet the technical details of the applicable navigation rules.

(b)   Be certified by an independent laboratory to the requirements of UL 1104 or an equivalent standard.

(c)   Be labeled with a label stating the following:

(1) "MEETS _____" (Insert the identification name or number of the standard under paragraph (b) above to which the light was tested.)

(2) "TESTED BY _____" (Insert the name or registered certification mark of the independent laboratory that tested the fixture to the standard under paragraph (b) above.)

(3) Manufacturer's name.

(4) Model number.

(5) Visibility of the light in nautical miles.

(6) Date on which the fixture was Type Tested.

(7) Identification of the bulb used in the compliance test

20.9.3 ABS-2002: 4-8-2, 11.3, Navigation Light System

ABS-2002-4-8-2-11.3.1 Feeder

Navigation lights (mast head, side and stern lights) are to be fed by its own exclusive distribution board located on the navigating bridge. The distribution board is to be supplied from the main as well as from the emergency source of power (see 4-8-2/5.5.3). A means to transfer the power source is to be fitted on the navigating bridge.

ABS-2002-4-8-2-11.3.2 Branch Circuit

Each navigation light is to have its own branch circuit and each branch circuit is to be fitted with a protective device.

ABS-2002-4-8-2-11.3.3 Duplicate Lamp

Each navigation light is to be fitted with duplicate lamps.

ABS-2002-4-8-2-11.3.4 Control and Indication Panel

A control and indication panel for the navigation lights is to be provided on the navigating bridge. The panel is to be fitted with the following functions:

— A means to disconnect each navigation light.
— An indicator for each navigation light.
— Automatic visual and audible warning in the event of failure a navigation light. If a visual signal device is connected in series with the navigation light, the failure of this device is not to cause the extinction of the navigation light. The audible device is to be connected to a separate power supply so that audible alarm may still be activated in the event of power or circuit failure to the navigation lights.

20.10 Interior communication systems

The following requirements are called out by ABS rule 1998 related to the ACP, ACP cites and ABS-2002 rules related to the ACP.

20.10.1 ABS-1998, 4/5A8, Interior Communication Systems

ABS-1998-4/5A8.5 Voice Communications

ABS-1998-4/5A8.5.1 Propulsion and Steering Control Stations

A common talking means of voice communication and calling is to be provided between the navigating bridge, main propulsion control station, and the steering gear compartment so that the simultaneous talking among these spaces is possible at all times and the calling to these spaces is always possible even if the line is busy.

ABS-1998-4/5A8.5.2 Elevator

Where an elevator is installed, a telephone is to be permanently installed in all cars and connected to a continuously manned area. The telephone may be sound powered, battery operated or electrically powered from the emergency source of power.

20.10.2 ACP Supplement of Section I – Interior Communication Systems

Cite:ABS-1998-4/5A8   Electrical Equipment – Interior Communication Systems

Each vessel must have an effective means of voice communications not dependent upon the main or emergency source of power for communications between the navigating bridge and the bow or forward lookout station, unless direct voice communication is possible.

Audible signaling devices in the weather must be external to the enclosure.

If a communications station is in the weather and on the same circuit as other required stations, there must be a cutout switch on the navigating bridge that can isolate this station from the rest of the stations, unless the system possesses other effective means of station isolation during a fault condition.

For sound-powered voice communication systems, each calling circuit must be independent of each talking circuit and any fault on the calling circuit must not affect the talking circuit.

Circuits must be insulated from ground. Cables should not be run through areas of high fire risk, such as machinery rooms and galleys, unless it is technically impractical to route them otherwise or they are required to serve circuits in the high fire risk area. In high ambient noise installations, accommodations to facilitate conversation must be provided.

Jack boxes or portable headsets are prohibited on any circuit that includes any required station. However, a hard-wired headset, with a push-to-talk feature, and cutout switch if in the weather is acceptable.

20.10.3   ABS-2002, 4-8-2, 11.5: Interior Communication Systems

ABS-2002, 4-8-2-11.5.1 General

Means of communication are to be provided between the navigating bridge and the following interior locations:

i)   Radio room, if separated from the navigating bridge.

ii)   Centralized propulsion machinery control station, if fitted.

iii)   Propulsion machinery local control position.

iv)   Engineers' accommodation, where propulsion machinery space is to be periodically unattended.

v)   Steering gear compartment.

vi)   Any other positions where the speed and direction of thrust of the propellers may be controlled, if fitted.

ABS-2002, 4-8-211.5.2 Engine Order Telegraph

An engine order telegraph system, which provides visual indication of the orders and responses both in the machinery space (the centralized control station, if fitted, otherwise propulsion machinery local control position) and on the navigating bridge is to be provided. A means of communication is to be provided between the centralized propulsion machinery control station, if fitted, and the propulsion machinery local control position. This can be a common talking means of voice communication and calling or an engine order telegraph repeater at the propulsion machinery local control position

ABS-2002, 4-8-2-11.5.3 Voice Communication

Means of voice communication are to be provided as follows. A common system capable of serving all the following will be acceptable.

i)   A common talking means of voice communication and calling is to be provided among the navigating bridge, centralized control station if fitted (otherwise the propulsion machinery local control position) and any other position where the speed and direction of thrust of the propellers may be controlled. Simultaneous talking among these positions is to be possible at all times and the calling to these positions is to be always possible even if the line is busy.

ii)   A means of voice communication is to be provided between the navigating bridge and the steering gear compartment.

iii)   For vessels intended to be operated with unattended propulsion machinery space, the engineers' accommodation is to be included in the communication system in i).

ABS-2002, 4-8-2-11.5.4 Public Address System

A public address system is to be provided to supplement the general emergency alarm system in 4-8-2/11.7.1, unless other suitable means of communication is provided. The system is to comply with the following requirements:

i)   The system is to have loudspeakers to broadcast messages to muster stations and to all spaces where crew are normally present.

ii)   The system is to be designed for broadcasting from the navigating bridge and at least one other emergency alarm control station situated in at least one other location for use when the navigating bridge is rendered inaccessible due to the emergency (see 4-8-2/11.7.1ii). The broadcasting stations are to be provided with an override function so that emergency messages can be broadcast even if any loudspeaker has been switched off, its volume has been turned down, or the public address system is used for other purposes.

iii)   With the ship underway, the minimum sound pressure level for broadcasting messages in interior spaces is to be 75dB(A) and at least 20dB(A) above the corresponding speech interference level, which is to be maintained without action from addressees.

iv)   The system is to be protected against unauthorized use.

ABS-2002, 4-8-2-11.5.5 Power Supply

The above communication systems are to be supplied with power (not applicable to sound powered telephones) from the emergency switchboard. The final power supply branch circuits to these systems are to be independent of other electrical systems.

21 Steering and maneuvering system—shipboard (SWBS 561)

21.1 General

The ship steering gear and maneuvering system is considered one of the most critical systems on board ship. This section is dedicated to the electro-hydraulically operated steering gear systems onboard ship. The hydraulic system is run by electric motors running at 480 V, three-phase, 60 Hz. The steering gear system must be designed to meet regulatory body requirements for 100% operational availability with required redundancy. The power source and various protections are as follows (Refer to Figure 21-1 through Figure 21-6 for example – these figures do not depict specific applications):

— There are two independent steering gear motors for one hydraulic unit. The electric power to one motor is supplied from the ship service switchboard and the other is supplied from the emergency switchboard (Refer to Figure 21-1 through Figure 21-5)
— The control power for each steering gear motor should be from the same steering motor feeder circuit.
— The steering gear controller must be LVR type.
— The steering gear controller must be in the steering gear room, and must be fully operational and controlled locally from the steering gear room
— The steering gear control, monitoring, and alarms are as shown in Figure 21-6.
— In case of main power loss to the steering gear, power must be totally restored within 45 seconds. This can be accomplished by the automatic bus transfer feature of the emergency switchboard. In case of a blackout of ship service power, there must be an automatic start-up of the emergency generator and transfer to emergency power.
— The alternate power must be capable of continuously operating for half an hour for steering the ship from 15° degree to 15° degree to either side in not more than 60 seconds, at maximum design draft loading, and half of the maximum design or 7 knots, whichever is the greater.
— The motor control system should be capable of accelerating the motor under torque requiring a current of 150% motor rating.

21.2 Steering and maneuvering system – comparison of various standards

The steering gear motors should be supplied from two different circuits, and, where practical, from two different switchboards or switchboard bus sections. These circuits should be widely separated so as to minimize failure of both feeders due to collision, fire, or other casualty. Both feeders may be connected to the ship service switchboard or, where required by the regulatory agencies, one feeder may be connected to the ship service switchboard and one to the emergency switchboard. Each circuit should have a continuous current-carrying capacity of not less than 125% of the rating of the motor or motors simultaneously operated. The following table is a comparison of requirements in IEEE Std 45, ABS, USCG, and IEC standards.

IEEE Std 45-2002		ABS-2002		USCG CFR-46-Chapter 1		IEC
32.5 .2	Each steering gear feeder circuit should have a current-carrying capacity of 125% of the full-load current rating of the electric steering gear motor or power unit plus 100% of the normal current of one steering control system, including any associated motor.	4-3-4, 11.7.1	**Steering Gears with Intermittent Working Duty:** Electric motors of, and converters associated with, electro-hydraulic steering gears with intermittent working duty are to be at least of 25% non-periodic duty rating (corresponding to S6 of IEC Publication 60034-1) as per 4-8-3/3.3.3 and 4-8-3/Table 4. Electric motors of electro-mechanical steering gears are, however, to be at least of 40% non-periodic duty rating (corresponding to S3 of IEC Publication 60034-1).	111.70. 1 (a) (1) 58.25-10(b1) 58.25-10(2)	Each Steering gear motor circuit and protection must meet part 58-25, subpart of USCG CFR – 46 Chapter 1. Adequate strength and capable of steering the vessel at max speed. Capable of moving the rudder from 35 degrees either side to 35 degrees oh the other side with the vessel at its deepest draft and running at maximum ahead service speed and from 35 degrees on either side to 30 degrees on the other in not more than 28 seconds under the same condition.	IEC Publication 60034-1
32.5 .2	The over-current protection for each steering gear circuit at the main and emergency switchboards should be an instantaneous circuit breaker set to trip at not less than 200% of locked rotor current of one steering gear motor plus other loads that may be on this feeder.	4.3.4, 11.7.2	**Steering Gears with Continuous Working Duty:** Electric motors of, and converters associated with, steering gears with continuous working duty are to be of continuous rating (corresponding to S1 of IEC Publication 60034-1) as per 4-8-3/3.3.4 and 4-8-3/Table 4.	58.25-10(c.2)	Capable of moving the rudder from 15 degree on either side to 15 degree on the other in not more than 60 seconds with the vessel at its deepest loadline draft and running at one-half maximum service speed or 7 knots, whichever is greater.	IEC Publication 60034-1)

Table 21-1: Steering gear requirement comparison among IEEE Std 45-2002, ABS-2002, USCG CFR, and IEC

21.3 IEEE Std 45-2002 recommendations for steering control systems

21.3.1 IEEE Std 45-2002, 32.6.1, Steering control systems—general

Electric control may be of the self-synchronous "follow-up" type or "non-follow-up" type.

The "follow-up" type uses a remotely controlled servomotor that, by means of signal feedback, gives a definite rudder position for each steering wheel position. A follow-up control system may be provided that produces at the steering gear machinery a motion and positioning that is in synchronism with the position and motion of the steering wheel in the remote location.

The "non-follow-up" type electric control consists of a master switch with spring return to the OFF position, which gives right or left rudder motion. Rudder motion is continuous in the direction indicated, until the limits of travel of the rudder are reached or the master switch returned to the OFF position. The rudder remains in the last ordered position until the master switch is moved from the OFF position.

Any of these electrically powered systems may function in conjunction with automatic steering systems. See Clause 30 [not quoted in this Handbook].

21.3.2 IEEE Std 45-2002, 32.6.2, Steering control system installation

Each steering power unit should have at least one steering control system capable of being operated from the navigating bridge. Additional control stations may be provided as required elsewhere on the ship. Each steering control system on vessels of 300 gross tons and above should be arranged so that each steering gear power unit can be controlled in the steering gear room. A selector switch should be provided in the navigating bridge for delegating the control to any one of these stations. All circuits from each steering station should be entirely disconnected by the selector switch, except those circuits to the station in use. A rudder angle indicator should be provided at each such station (see 26.3 [not quoted in this Handbook]).

The steering control system for a steering power unit should be separated as widely as practicable from each other steering control system and each steering power unit that it does not control.

Each navigating bridge steering control system should have a switch in the navigating bridge that is arranged in such a way that one action of the switch's handle automatically puts into operation a complete steering control system and associated steering power units. If there is more than one steering control system, this switch should be

a) Operated by one handle
b) Arranged so that each one individually or both steering control systems and all associated steering power units can be energized from the navigating bridge
c) Arranged so that the handle passes through an "off" position when transferring from one steering control mode to another
d) Arranged so that the switches for each system are in separate enclosures or separated by fire-resistant barriers

Each steering control system should receive its power from the feeder circuit for its steering power unit in the steering gear room and have a switch that is in the steering gear room and disconnects the steering control system from its power source. Each motor controller for a steering gear should be in the steering gear room and have low-voltage release. A means should be provided to start and stop each steering gear motor in the steering gear room.

21.3.3 IEEE Std 45-2002, 31.13.2.3, Steering system (podded electric propulsion drive) (extract; italics in original)

The steering gear system should be of electro-hydraulic or electrical type. If the pod is used as a steering device, class and regulatory requirements for steering systems shall apply for the pod steering system and the term *rudder angle* should be interpreted as *pod azimuth angle.*

The podded propulsion unit should be provided with a dual redundant steering system. If more than one azimuthing pod is provided, each shall have a fully independent steering system. The steering system should be capable of moving, stopping, and holding the pod unit at any desired angle within design limits. ...

21.3.4 IEEE Std 45-2002, 32.5.1, Steering gear—general

When the main and auxiliary steering gears are electrically powered and controlled or when an arrangement of two or more identical power units are utilized, the vessel should have two separate steering systems, each consisting of a power unit, steering control system, steering gear feeder, and associated cable and ancillary equipment. The two systems should be separate and independent on a port and starboard basis.

Each steering gear motor controller should include the following apparatus:

   a) Power input disconnect switch or non-automatic circuit breaker
   b) Power available indicator light
   c) Steering motor START/STOP pushbutton
   d) Motor running indicator light
   e) Steering control system power supply transformer
   f) Steering control power supply transformer output circuit breaker having only an instantaneous trip
   g) Control power available indicator light
   h) Steering control power transfer switch, LOCAL/NAVIGATING BRIDGE
   i) Steering motor overload alarm relay
   j) Input power failure alarm relay
   k) Control power failure alarm relay
   l) Phase failure trip relay (for three-phase fused power sources)

21.3.5 IEEE Std 45-2002, 32.5.2, Feeder circuits

Vessels with one or more electric-driven steering power units should have at least two feeder circuits. One of these feeder circuits should be supplied from the main switchboard. On vessels where the rudder stock is required to be over 230 mm in diameter in way of the tiller (excluding strengthening for navigation in ice) and an emergency power source is required, the other feeder circuit should be supplied from the emergency switchboard or an alternative power supply. Where an alternative power supply is provided, it should be available automatically within 45 s of loss of power supply from the main switchboard, be located in the steering gear compartment, and be used for no other purpose. The alternative power supply should have capacity sufficient for one-half hour of continuous operation of the rudder from 15° on one side to 15° on the other side in not more than 60 s with the ship at its deepest sea-going draft while running at one-half of its maximum ahead service speed or 7 kn, whichever is the greater.

Vessels that have a steering gear with two electric motor-driven power units should be arranged so that one power unit is supplied by one feeder and the other power unit is supplied by the other feeder. Each steering gear feeder circuit should be separated as widely as practicable from the other and should have a disconnect switch in the steering gear room. Each feeder circuit should have a current-carrying capacity of 125% of the full-load current rating of the electric steering gear motor or power unit plus 100% of the normal current of one steering control system, including any associated motors.

The over-current protection for each steering gear circuit at the main and emergency switchboards should be an instantaneous circuit breaker set to trip at not less than 200% of the locked rotor current of one steering gear motor plus other loads that may be on this feeder for AC installations. For DC installations, the instantaneous trip should be set not less than 300% and not more than 375% of the rating of the steering gear motor. No other overload device or fuse that will open the power circuit should be provided in the motor or control circuits.

The opening of the main or emergency switchboard steering gear circuit breaker should operate audible and visual alarms located at the main propulsion control station and the navigating bridge.

Each steering gear motor circuit should be equipped with an over-current relay to operate audible and visual alarms located at the main propulsion control station and the navigating bridge. No other functions should be performed by this relay.

The steering gear motor circuit should be capable of accelerating the motor under a torque requiring a current of 150% of motor rating. If the inherent design of the steering gear does not prevent overhauling of the rudder, a magnetic brake should be installed.

A pilot light for each steering gear motor to indicate motor running should be provided at the main propulsion control station and the navigating bridge. This pilot light should be fused.

21.4 ABS-2002 steering motor requirements

21.4.1 ABS-2002, 4-3-4, 11.7, Motor Rating

11.7.1 Steering Gears with Intermittent Working Duty

Electric motors of, and converters associated with, electro-hydraulic steering gears with intermittent working duty are to be at least of 25% non-periodic duty rating (corresponding to S6 of IEC Publication 60034-1) as per 4-8-3/3.3.3 and 4-8-3/Table 4. Electric motors of electro-mechanical steering gears are, however, to be at least of 40% non-periodic duty rating (corresponding to S3 of IEC Publication 60034-1).

11.7.2 Steering Gears with Continuous Working Duty

Electric motors of, and converters associated with, steering gears with continuous working duty are to be of continuous rating (corresponding to S1 of IEC Publication 60034-1) as per 4-8-3/3.3.4 and 4-8-3/Table 4.

21.4.2 ABS-2002, 4-8-3, 3.3, Definitions

3.3.1 Periodic Duty Rating: The periodic duty rating of a rotating machine is the rated kW load at which the machine can operate repeatedly, for specified period (N) at the rated load followed by a specified period (R) of rest and de-energized state, without exceeding the temperature rise given in 4-8-3/Table 4; where N+R = 10 min, and cyclic duty factor is given by N/(N+R) %.

3.3.2 Short Time Rating: The short time rating of a rotating electrical machine is the rated kW load at which the machine can operate for a specified time period without exceeding the temperature rise given in 4-8-3/Table 4. A rest and de-energized period sufficient to re-establish the machine temperature to within 2 °C (3.6 °F) of the coolant prior to the next operation is to be allowed. At the beginning of the measurement the temperature of the machine is to be within 50 °C (90 °F) of the coolant.

3.3.3 Non-periodic Duty Rating: The non-periodic duty rating of a rotating electrical machine is the kW loads which the machine can operate continuously, for a specific period of time, or intermittently under the designed variations of the load and speed within the permissible operating range, respectively; and the temperature rise, measured when the machine has been run until it reaches a steady temperature condition, is not to exceed those given in 4-8-3/Table 4.

3.3.4 Continuous Rating: The continuous rating of a rotating electrical machine is the rated kW load at which the machine can continuously operate without exceeding the steady state temperature rise given in 4-8-3/Table 4.

21.5 Examples and figures

The following figures show several common steering configurations (example only). Figures 21-1 through 21-6 show sample detailed design of shipboard steering gear systems. For additional propulsion and steering system redundant configurations, refer to sections 2 and 14.

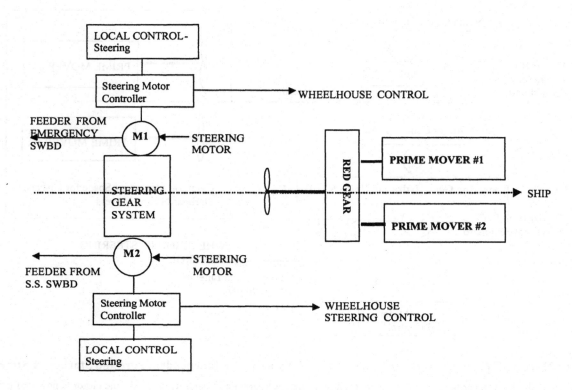

Figure 21-1: ABS Propulsion Redundancy—"R1" Notation for Propulsion and Steering

Figure 21-1 shows one steering system with redundant hydraulic pump motors. The motor M1 power feeder is from emergency switchboard and the motor M2 power feeder is from ship service switchboard. This configuration is in compliance with ABS-R1 redundancy notation.

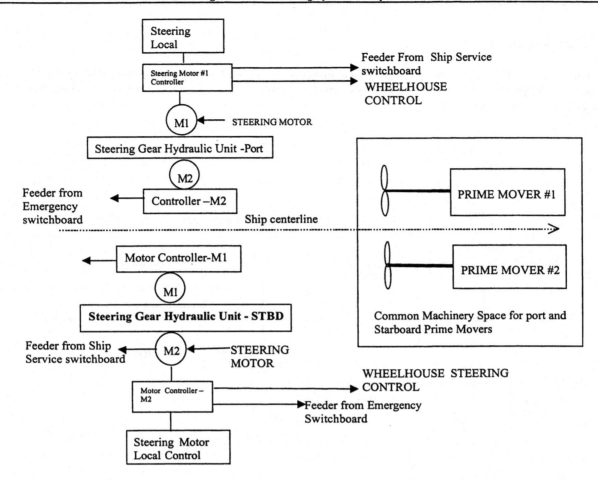

Figure 21-2: ABS Propulsion Redundancy—"R2-S" Notation for Independent Port and Starboard Steering

Figure 21-2 is for duplicate steering gear units. Each unit has redundant hydraulic motors. In each side one motor is powered from the ship service switchboard and other motor is powered from the emergency switchboard. Both steering gear units are physically located in the same space. This configuration is in compliance with ABS R2-S notation.

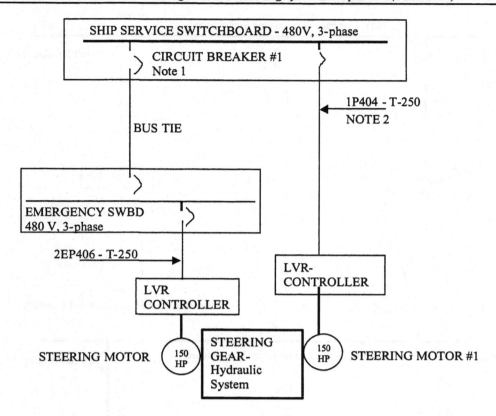

Figure 21-3: Single Steering Gear with Redundant Hydraulic Pumps

Figure 21-3: This configuration is the same as Figure 21-1. However, it is shown with detail ship service and emergency switchboard configuration.

Notes:
1. Steering gear feeder circuit breaker with no overload protection device. The breaker is "instantaneous trip" only.
2. The feeder ampacity shall be 125% of the steering gear motor ampacity plus 100% of the other loads, such as the control transformer.

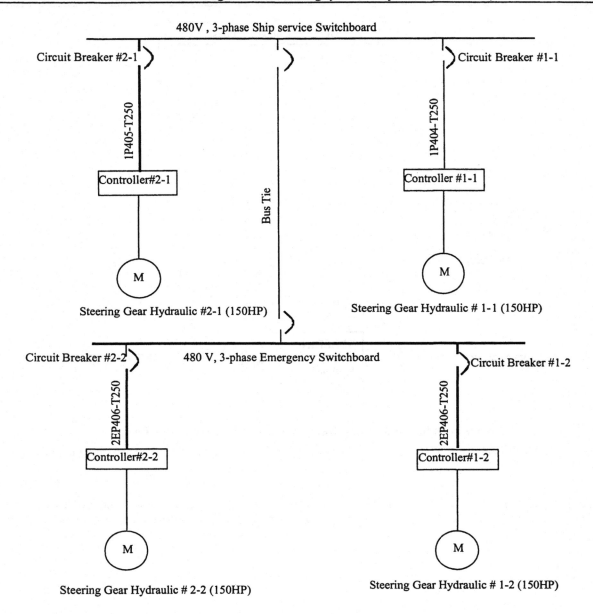

Figure 21-4: Electrical One-Line Diagram—Fully Redundant Steering Gear System
for ABS "R2-S" or ABS "R2-S+" classification

Figures 21-4 and 21-5: The configuration show in these figures is the same as Figure 21-2. However, it is shown with detailed ship service and emergency switchboard configurations. If the steering gear units are physically separated by a watertight bulkhead, this configuration qualities for ABS R2-S+ notation. For additional details, refer to ABS rules.

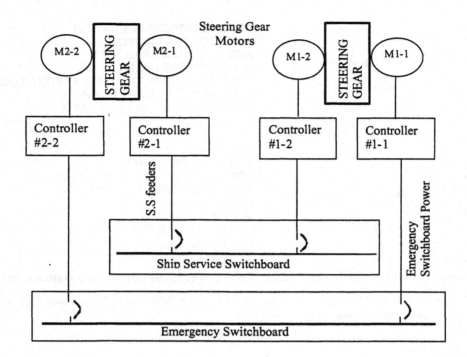

Figure 21-5: Fully Redundant Steering Gear Units for ABS "R2-S" or ABS "R2-S+" Classification

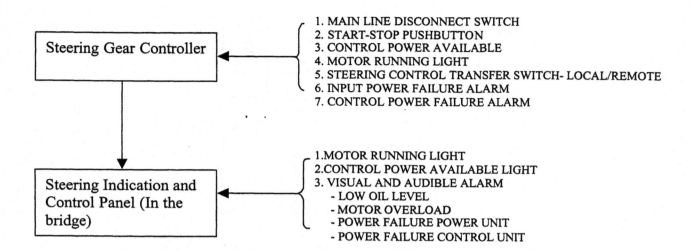

Figure 21-6: Steering Gear Controller with Control Functions at Local and Remote Stations—Block Diagram

Figure 21-6: This configuration is to show typical steering gear motor controller control functions. This is applicable to all controller configurations, individual or redundant.

22 Shipboard interior communication systems—(SWBS 432-433)

22.1 General

Shipboard interior communication systems (IC Systems)--such as sound powered telephone system, Dial Telephone system, Wireless communications, and voice communications--are required, like all other shipboard electrical systems, to be in accordance with regulatory body requirements. In IEEE Std 45-2002 IC system recommendations are minimal, as the Navy general specifications SWBS 432 and SWBS 433 both cover interior communications in a fair degree of detail. Table 22-1 provides a cross-reference of different systems between the Handbook, IEEE Std 45, and Navy system interior communications requirements.

Handbook section and title		IEEE Std 45-2002 section and title		Navy specifications SWBS number and title	
22.2	Sound-powered telephone system	26.7	Sound Powered Telephone System	SWBS 432	Sound-powered Telephone System
22.3	Automatic dial telephone system	26.7.3	Dial Telephone System	SWBS 433	Automatic dial telephone system
22.4	Wireless hands-fee communications (WIFCOM)	26.7			
22.5	Integrated voice communication system (IVCS)	26.6	Voice communication systems	SWBS 432	Wireless Communications
		26.7.2	Public Address/ General Announcing/talk back systems		

Table 22-1: Cross-References of Shipboard Interior Communication—
Handbook, IEEE Std 45 and Navy Military Specifications

22.2 Sound-powered telephone system (SWBS 432)

The sound-powered telephone system consists of individual sound-powered telephone circuits, each of which is required to operate without any external source of energy. The individual systems are as follows (example only):

— C-1JV – Maneuvering and Docking
— C-2JV – Engineer's Circuit
— C-3JV – Fueling Engineer's
— C-2JZ – Damage Control

22.3 Automatic dial telephone system (SWBS 432)

A marine type automatic dial telephone system is suitable for shipboard installation and service, complete with all auxiliary equipment and panels. The power supply to the automatic dial telephone system is required to be backed up with a battery power source. The system shall have multi-line shore connections, and is an integral part of the ship's integrated voice communication system (IVACS). It provides communication between living, mess, and recreation spaces; offices, operational areas, quarterdeck stations, and machinery spaces. All dial telephones have access capability to access the Public Address System.

22.4 Wireless hands-free communications (WIFCOM) (SWBS 432)

Wireless, hands-free communications provide support of machinery control, maintenance, fueling, mooring and line handling, damage control, physical security, and mission operations. Wireless communication systems interface and have access to integrated voice communications system (IVCS), including the Public Address System.

The system should meet the following system performance requirements: (example only)

- The Wirefree Communications System shall be implemented using commercial technology at a given frequency range.
- The system shall provide a number of radios for normal mobile interior communications topside and throughout the ship in all spaces except voids and tanks.
- Normal WIFCOM communications in support of operations other than damage control performs with multi-channel low power portable radios operating from 20 to 100 mW in the semi-duplex trunked mode.
- Emergency WIFCOM communications, in support of damage control, provides multi-channel portable radios operating at 20 to 100 m W in the simplex (point-to-point) mode should the trunking subsystem fail.
- Control head with microphone, vehicle adapter and power supply with an Uninterruptable Power Supply (UPS). The UPS shall accept and condition single phase, 115 Vac ship power and provide a minimum four-hour back-up power supply. DCC shall also be provided with one 20 to 100 m W multi-channel ruggadized portable radio, a single unit battery charger, a belt carrying accessory and a remote microphone.

22.5 Integrated voice communication system (IVCS) (SWBS 433)

The shipboard integrated voice communication system (IVCS) provides support for the transmission of orders and information and the required audible and visual general emergency alarm signals throughout the ship, simultaneously by means of microphones and loudspeakers connected through a central amplifier. IVCS also provides public address, amplified voice intercom/talk-back, loud hailer, fog alarm, and general emergency alarm sub-systems. The IVCS interfaces to the automatic dial telephone system.

Loudspeakers should be strategically located to provide intelligibility and audibility. A sufficient number of loudspeakers should be installed on weather decks and interior areas to provide an even distribution of sound, and to eliminate uncomfortable sound levels near loudspeakers and poor intelligibility in areas between loudspeakers. Loudspeakers shall not face each other. Loudspeakers are provided with volume controls that are accessible for adjustment. The quantity of loudspeakers shall be sufficient to provide satisfactory sound coverage, as required by regulatory body rules.

22.6 IEEE Std 45-2002 recommendations for interior communications systems

The recommendations of IEEE Std 45-2002, beyond Navy requirements, are fairly sparse in this area. They are summarized below.

22.6.1 IEEE Std 45-2002, 26.1, Interior communications systems—general (extract)

Each required interior communication system (voice, data, control, or alarm) should be designed and installed to provide clear, distortion-free, and uninterrupted operation under all operating conditions. Refer to 9.9 [see 20.6.8 in this Handbook] for requirements for communication equipment integrated into control and alarm systems. Interior communication system power sources should be in accordance with emergency power and lighting systems. Communications data may be conveyed by electrical (copper) or fiber-optic conductors.

All components should function continuously without adverse effect when supplied, with power having the characteristics described in Table 4 [see 5.4.1 in this Handbook].

All enclosures for use in essential circuits should be watertight. Other enclosures installed in exposed or wet locations, machinery spaces, galleys, and similar locations should also be watertight. Enclosures for nonessential interior communication circuits (e.g., entertainment systems) when installed in dry locations may be drip-proof. Each terminal should be identified, and each individual wire should be marked at the terminal. When more than one voltage level is found in an enclosure, conductors of differing voltages should be kept separated as much as practicable. All voltages within an enclosure should be indicated, and recommended creepage distances observed. Each enclosure should be of a corrosion-resistant material, and nonmetallic enclosures should be flame- and impact-resistant.

All indicators and controls should be installed to provide for operational accessibility, serviceability, visibility, and mechanical protection. Each installation should not hinder the operation or the accessibility of other equipment. Vibration mounting of equipment should be considered where necessary due to abnormal conditions at the installed location. …

22.6.2 IEEE Std 45-2002, 26.7.2, Public address/general announcing/talkback systems

> Each vessel required to have a general emergency alarm shall also have an amplifier-type public address announcing system to supplement the alarm.

22.6.3 IEEE Std 45-2002, 26.7.2.1, Public address/general announcing/talkback systems—general requirements (extract)

a) Each vessel shall have an amplifier-type announcing system that will supplement the general emergency alarm. This system shall provide for the transmission of orders and information throughout the vessel via loudspeakers connected through a redundant amplifier system. This system shall be capable of interfacing with the general emergency alarm and fire detecting and alarm systems. See 26.5 [noted quoted in its entirety in this Handbook; but see 23.1.7.4 and 23.1.7.5 of this Handbook for some discussion] and 28.4 [not quoted in this Handbook] for details of required interface. The public address system shall be protected against unauthorized use

b) Input to the public address system shall be by both the master control station and the dial telephone system. The master control station shall operate the system as follows:

    1) Public address announcements, including zone paging and all call paging, shall be possible from the master control station.

    2) The master control station shall have provisions to initiate a minimum of four different distinct alarms tones.

    3) The system shall allow for the ability to prioritize announcements, emergency announcements, and alarms.

    4) System status shall be monitored and alarms shall be generated at the master control station when the system detects speaker run or amplifier trouble. The alarm shall be both visual and audible. Audible alarms shall have the ability to be silenced, but shall recur hourly in order to ensure system stability. Fault indication may only be permanently silenced by remedy.

    5) The master control station shall include individual back-illuminated pushbuttons for each paging, talkback, alarm, and alarm monitoring function. These pushbuttons shall be singular in function and not require the user to operate more than one button to activate a single function.

    6) If a dial telephone system is provided on the vessel, it may be desirable to integrate the master control station with the PBX such that it provides full function of a dial telephone extension, including the ability to make station-to-station calls as well as access the trunk lines.

c) There shall be a means to silence all other audio distribution systems during an announcement or alarm.

d) The system shall be arranged to allow broadcasting separately to, or to any combination of, various areas on the vessel. If the amplifier system is used for the general emergency alarm required by 26.5, the operation of a general emergency alarm shall activate all speakers in the system, except that a separate crew alarm may be used as allowed by 26.5.7.2 [IEEE Std 45-2002 Clause]....

22.7 IEEE Std 45-2002, 26.7.3, Dial Telephone Systems

> Dial telephone systems and equipment shall be suitable for marine use. Telephones may be bulkhead, panel (surface or flush mount), or desk type. Telephones exposed to weather or hazardous locations shall be suitable for the location.
>
> The power source for the central power and switching unit (PBX switch) shall be in accordance with the recommendations in this clause. The PBX switch shall, as a minimum, be drip-proof or in a drip-proof enclosure, and be readily accessible.
>
> The dial telephone system shall be capable of two-way voice communications, transmitting and receiving data (e.g., fax and modem). It should be able to interface with commercial landline systems, public address systems, commercial satellite communications, and cellular telephone systems as required.

23 Shipboard exterior communications (SWBS 440)

23.1 General

The title of IEEE Std 45-2002 Section 27 is "External communication and navigation systems". However, this section mostly deals with external communications. Navy general specification SWBS 440 covers external communications and SWBS 426 addresses the navigation system. This chapter discusses external communications only; chapter 24 covers the navigation system.

Radio equipment in generally is in compliance with FCC. However, the maritime external communications system must comply with the requirements of Internal Maritime Organization (IMO) Safety of Life at sea (SOLAS). Excerpts from SOLAS are provided for understanding those requirements. SOLAS regulation Chapter IV, Part C, Regulations 6 through 17 provide detailed requirements of radio communication systems. Some of those SOLAS regulations are quoted below.

23.2 Global Maritime Distress Safety System (GMDSS)

The radio equipment provided shall comply with FCC guidelines, and shall be global maritime distress safety system (GMDSS) compliant for Sea Areas A-1, A-2, and A-3, with shore-based maintenance.

For ocean-going ships, a complete SOLAS-compliant GMDSS radio communication system is required to be installed. The GMDSS is a radio console assembly providing complete operation and supervision of radio equipment. The radio transmitters and receivers are designed and electrically interconnected to provide complete self-contained units with convenient operator front panel-mounted control features.

1. GMDSS terms:
    a.   Digital Selective Calling (DSC) – A technique of using digital codes which enables the radio station to establish contact with , and transfer information to another station or group of stations.

    b.   INMARSAT – The organization established by the Convention on the International Maritime Satellite Organization

    c.   Sea area A1- The area within the radiotelephone coverage of at least one VHF coast station in which continuous DSC alerting is available.

    d.   Sea area A2- The area excluding the sea area A1, within the radiotelephone coverage of at least one MF coast station in which continuous DSC alerting is available.

    e.   Sea area A3- The area excluding the sea area A1 and A2, within the radiotelephone coverage of an INMARSAT geostationary satellite in which continuous alerting is available.

    f.   Sea area A4 – The area outside sea area A1, A2, and A3

Refer to SOLAS Chapter IV for Radio communications details.

23.2.1 SOLAS radio equipment requirement

a. Radio equipment – General

1.   VHF radio installation capable of transmitting and receiving:
    1.1 DSC on the frequency 156.525 MHz (channel 70). It shall be possible to initiate the transmission of distress alerts on channel 70 from the position from which the ship is normally navigated, and
    1.2 Radiotelephony on the frequency 156.300 MHz (channel 6), 156.650 MHz (channel 13), and 156.800 MHz (channel 16)
    2.1 A radio installation capable of maintaining continuous DSC watch on VHF channel 70 which may be separate from, or combined with that required by above.
    3.1 A radar transponder capable of operating in the 9 GHz band
    4.1 A receiver capable of receiving international NAVTEX service broadcasts if the ship is engaged on voyages in any area in which an international NAVTEX service is approved.

    5.1 A radio facility for reception of maritime safety information by the INMARSAT enhanced group calling system if the ship is engaged on voyages in any area of INMARSAT coverage but in which an international NAVTEX service is not provided. However, ship engaged exclusively on voyage in area where an HF direct-printing telegraphy maritime safety information service is provided and fitted with equipment capable of receiving such service, may be exempt from this equipment.

6.1   Subject to the provisions of regulation 8.3 a satellite emergency position indicating radio beacon (satellite EPIRB) which shall be:

6.2   Capable of transmitting a distress alert either through the polar orbiting satellite service operating in the 406 MHz band or, if the ship is engaged only on voyages within INMARSAT coverage, through the INMARSAT geo-stationery satellite service operating in the 1.6 GHz band;

(For further details refer to SOLAS regulation)

—   Chapter IV-Regulation 8 is for radio equipment for Sea area A1
—   Chapter IV-Regulation 9 is for radio equipment for Sea area A1 and A2
—   Chapter IV-Regulation 10 is for radio equipment for Sea area A1, A2, and A3
—   Chapter IV-Regulation 11 is for radio equipment for Sea area A1, A2, A3, and A4

24 Shipboard navigation system (SWBS-426)

24.1 General

Navy general specification SWBS 426 describes shipboard navigation systems. Table 24-1 provides cross-reference between the Handbook, IEEE Std 45, and Navy General specification shipboard navigation system.

Handbook section and title		IEEE Std 45-2002 section and title		Navy military specification section and title	
25.1	Satellite communications terminal (as required)	27.6	Antenna		
25.2	Navigation system	30	Gyro compass systems		
25.3	Integrated navigation system (SWBS 426)	30.1	General	426	Gyrocompass System
25.4	ABS navigational bridge and equipment/systems	30.2	Installations and location	426	Doppler Speed Log

**Table 24-1: Shipboard Navigation System Cross-References—
Handbook, IEEE Std 45-2002, ABS and Navy Military Specifications**

24.2 Integrated navigation system (SWBS 426)

Vessel navigation, control and exterior communication should be provided in a functionally integrated bridge configuration in the wheelhouse. The Wheelhouse should be arranged to provide maximum visibility in all directions, including aft. All wheelhouse navigational equipment shall be IMO-SOLAS compliant as required.

All integrated navigation system components, as well as navigation, communication and ship control and monitoring equipment, should be arranged in a navigation console. The console should be set back from the Wheelhouse windows. In general, controls and instrumentation related to propulsion and maneuvering control (Ship Control Console) should be incorporated in the center of the console, along with the navigation conning display position of the integrated navigation system. Electronic chart main displays, electronic chart slave display, radars and other console mounted equipment should be installed to port and starboard of the center console section. The console should be designed to allow for full use, day or night, without impact vessel operation.

A complete, integrated navigation system should perform the following functions:

— Development and storage of voyage plans.
— Display of electronic charts and voyage plans.
— Real-time comparison of actual track vs. programmed voyage plan.
— Display of navigation sensor data.
— Automatic position plot and data logging.

Conning information will be displayed at the following locations:

— Integrated Bridge Console
— Bridge Wings
— Navigation Station

The system includes two independent command stations for monitoring and control, each capable of displaying electronic charts and navigation and command information on full color, high resolution, trackball or keyboard-controlled monitors. The command stations include real -time status, advisory and control screens for monitoring of all integrated navigation system elements. Output signals from the following navigation equipment are connected to the integrated navigation system via the navigation local area network:

— Two digital gyro compasses, with inter-switchable units for deviation alarm and functions.
— One fully adaptive autopilot.
— One course and P/S rudder angle recorder.

—   Two rudder angle indicators.
—   One echo sounder.
—   One speed log.
—   One automated weather station with two transmitters.
—   Three ARPA/radar (interswitchable).
—   Two Differential Global Positioning System receivers.

24.3 ABS – navigational bridge and equipment/systems

Refer to the ABS Publication " Guide for Bridge Design and Navigational Equipment/Systems", dated January 2000, for details. Below is an extract of ABS navigational bridge and equipment/systems, for general information.

24.3.1 ABS-2000, A1, Application

A1.1   The requirements of this Guide are applicable to vessels possessing valid SOLAS certificates, and having the bridge so designed and equipped as to enhance the safety and efficiency of navigation. When a vessel is designed, built and surveyed in accordance with the requirements of this Guide, and when found satisfactory, a classification notation as specified in A3 will be granted. Application of the requirements of this Guide is optional.

A1.2   The composition and qualifications of the crew remains the responsibility of the flag Administrations.

24.3.2 ABS-2000-A3, Optional notations

A3.1, Notation NBL (Navigational Bridge Layout)

Where requested by the Owner, a vessel having its bridge found to comply with the requirements in Parts A and B of this Guide, as applicable, and which has been constructed and installed under survey by the Surveyor, will be assigned the notation NBL.

A3.2, Notation NBLES (Navigational Bridge Layout and Equipment/systems)

Where requested by the Owner, a vessel which is found to comply with the requirements specified in Parts A through C of this Guide and which has been constructed and installed under survey by the Surveyor, will be assigned the notation NBLES.

A3.3, Notation NIBS (Navigational Integrated Bridge System)

Where requested by the Owner, a vessel which is fitted with an integrated bridge system (IBS) for navigational purpose, is found to comply with the requirements specified in Parts A through D of this Guide, and which has been constructed and installed under survey by the Surveyor, will be assigned the notation NIBS.

24.3.3 ABS-2000-A5, Operational assumptions

The requirements contained in this Guide are based on the following assumptions:

A5.1   Plans for emergencies and the conditions under which the vessel is intended to operate are clearly defined in an operational manual acceptable to the flag Administration. The manual should clearly state the bridge crew composition required under any particular set of circumstances.

A5.2   The requirements of the International Conventions on Standards of Training, Certification and Watch-keeping for Seafarers (STCW) and other applicable statutory regulations are complied with.

24.3.4 ABS-2000-A7, Regulations

For the purpose of this Guide, the International Regulations for Preventing Collisions at Sea, and all other relevant Regulations relating to radio and safety of navigation required by Chapters IV and V of 1974 SOLAS, as amended, are to be complied with. Valid statutory certificates issued by the pertinent flag Administration are to be provided onboard the vessel and made available to the Surveyor upon request.

24.3.5 ABS-2000-A9, Flag administration and national authorities

Vessel owners or other interested parties are urged to consult the flag Administration and relevant National Authorities concerning required manning levels on the bridge and any additional requirements which may be imposed by them.

25 Shipboard miscellaneous electrical alarm, safety, and warning systems (SWBS 436)

25.1 General

Navy general specifications SWBS 436 deals with miscellaneous shipboard electrical alarm, safety and warning systems. The systems in Section 436 corresponding to the commercial specifications such IEEE Std 45 are discussed in this section, with appropriate cross-references.

IEEE Std 45-2002 Clause and title		Navy General specification section and title	
Clause	**Title**	**Section**	**Title**
28	Fire detection, alarm, and sprinkler systems	436	Fire and Smoke Alarm
28.3	Automatic fire alarm systems	436	High temperature fire alarm – Circuit "F"
28.4	Fire detection and fire alarm systems for periodically unattended machinery space		
28.7	Automatic sprinkler, fire detection, and fire alarm systems	436	Sprinkling Alarm System – Circuit "FH"
		436	Flooding Alarm – Circuit "FD"
26.4	Refrigerated and cold storage alarm systems		
		436	Steering Emergency Signal Alarm
26.5	General emergency alarm system	436	General Alarm System
		436	CO_2 Release Alarm – Circuit "1FR"
		436	Halon Release Alarm-Circuit "2FR"
26.6	Alarm system for lubricating oil, refrigeration, and other fluid systems		
		436	Watertight Door Alarm

Table 25-1: Cross-References of Shipboard Miscellaneous Electrical Alarm, Safety, and Warning Systems - IEEE Std 45-2002, and Navy General Specifications

25.2 IEEE Std 45-2002, 28, Fire detection, alarm, and sprinkler systems

Fire detection, alarm, and sprinkler systems are required by regulatory agencies for various vessels types and applications, including periodically unattended machinery spaces. Specific system and equipment requirements can be found in NFPA Standards, SOLAS, classification society rules, and regulatory agency regulations. System/equipment type approval is often required. The latest edition of these documents should be consulted prior to system design.

Automatic fire detection systems are recommended for all vessels, with detection in all accommodation and machinery spaces, including passageways, stairwells, work spaces, store rooms, and pump rooms. Cargo holds and vehicle ro/ro spaces should be fitted with smoke/fire detection in accordance with current regulations.

25.3 IEEE Std 45-2002, 28.1, Fire detection, alarm, and sprinkler systems—general

Each component of an installed shipboard fire detection and fire alarm system should be constructed for the marine environment, including electromagnetic interference (see 1.5.1) [Clause IEEE Std 45-2002] and 9.20 [Clause IEEE Std 45-2002]. The system, and its components, should have been tested by an independent laboratory to nationally or internationally recognized testing protocols. See NFPA 72-1999 for more information about fire detection and alarm systems.

All detectors should be capable of being operationally tested and restored to normal operation without replacing parts. Detectors installed in damp or wet locations, such as ro/ro spaces or machinery spaces, should be specifically designed and tested for such use.

Every fire detection and fire alarm system should be provided with two sources of power, one of which should be from an emergency source that switches automatically to provide power upon failure of the main power source. The emergency source can be a storage battery.

Detecting systems should be designed so that the indication of a fire on any circuit will not interfere with the operation of an alarm on any other circuit. The system should be designated so that failure of a detection circuit will not interfere with the operation of any other detection circuit.

System cables should be so arranged to avoid galleys, machinery spaces, and other enclosed spaces having a high fire risk, except to connect fire detection equipment in those spaces.

Fire alarm systems should not be used for the transmission of other than fire alarm signals.

26 Shipboard UPS (Uninterruptible Power Supply)

26.1 General

It is a requirement that certain critical loads must be provided with normal ship service power as well as an uninterruptible power supply (UPS) to ensure operation under blackout conditions. The centralized machinery control system is required to be supported for 30 minutes by a UPS system. The emergency generator starter consists of multiple starting systems, where one of the starting systems is a battery and battery charger-supported UPS system. The emergency generator usually takes over the emergency loads when the ship service power source is dead. During a ship service power source blackout, the critical loads are usually powered by UPS until the emergency power or ship service power is available.

There are different types of UPS available. The choice is in general between an integrated UPS consisting of small secondary batteries/chargers, and a large UPS consisting of batteries, switching, AC/DC and DC/AC inverter/rectifiers as required to continue to supply power to their own selected loads without interruption and without any action on the part of the operator. In the case of the large UPS, the inverter/rectifier supplies power for normal operation. The batteries should be capable of supplying power to the loads for a predetermined time, in accordance with the size of UPS system. The battery capacity is chosen to support the predetermined time requirement (measured in amp-hours). When the normal power is restored, the inverter/rectifier/charger resumes supplying the power and starts recharging the batteries.

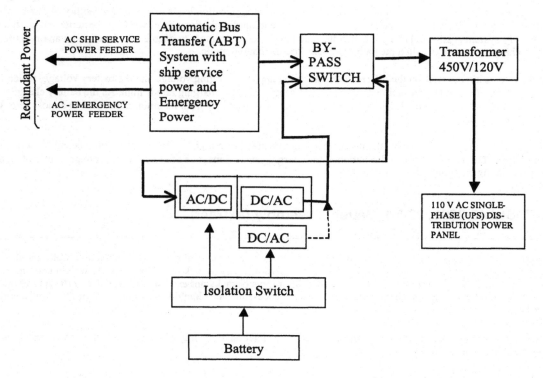

Figure 26-1: 110 V AC UPS Distribution System with Redundant Ship's Power Supply with ABT, Charger Unit, Batteries, and Inverter for Shipboard Automation

26.2 Battery charging unit (inverter/converter/charger units) (typical system; example only)

The battery charger unit consists of transformers and solid-state circuitry. The battery is permanently connected across the consumer DC output, and is maintained on a slight float charge by arranging a small excess of voltage output. The charging system consist of a main line switch, which is normally closed position for normal use, and a center zero battery ammeter indicates whether the charging unit is in charging state or discharge state. If the rectifier or its AC supply should fail, the battery, being permanently connected, starts to feed the load without interruption and without any switching operation.

Battery chargers have facilities for automatic boost, trickle charging and are sized to supply the total demand load while charging the discharged battery. In case of failure of the automatic circuit, manual charging facilities are also provided. The automatic sequence consists of the following steps:

1. Charge the battery by automatically adjusting the voltage
2. Final charge of the battery at a higher voltage in a period controlled by a timer
3. Switch automatically from battery boast-charge to floating-charge as soon as the battery has been fully recharged
4. Trickle charge at constant voltage and current

The rectifier has constant voltage control and a current limiting feature, with three voltage levels: floating charging, high rate charging , and boost charging. The switch from floating charging to high rate charging level is generally automatic; however manual switching is also available in case of failure in the automatic charging system. The boost charging feature is available, and can be used to rapidly charge the batteries, such as for first-time charging of newly installed batteries or when the recharging of the depleted batteries must be accomplished in the shortest possible time. High rate charge is used for fast recovery of the battery's full capacity after a discharge. Upon restoration of the mains, and for low battery voltage, the rectifier starts the automatic and regulated boost charge by means of a static adjustable relay. If one of the systems fails, the appropriate isolating switches, mounted inside each unit, will allow the load of the failing system to be supplied by the other operating unit.

26.3 Storage batteries

The storage batteries are connected as a secondary power source to vital application loads to ensure availability of power when the main distribution power is down. The batteries operate in parallel with the main distribution power while the batteries are continuously charged. The battery system is adjusted at a per-cell voltage and charged accordingly to enable it to function on an on-demand basis at full charge. This charging system is called the battery system floating charge mode.

The rate of battery charging depends on the difference between the applied charging voltage and the battery voltage, which falls as battery discharges. Therefore, at the beginning of a battery charging cycle, the charging rate is high. As the battery charging proceeds the emf rises, and voltage level is maintained. The battery charging rate will taper off from its initial high value to almost zero, which is called constant voltage charging.

If the initial rate of the battery charge is low, the initial charging rate may be so high that the charger could become overloaded or the battery itself damaged. Therefore, automatic limitation in the charging system must be provided, in order to prevent system overloading by maintaining regulating charging rate.

26.3.1 IEEE Std 45-2002, 22.5.1, Battery assembly (extract)

Cells should be assembled in trays or racks of suitable corrosion-resistant material and rigid construction. Cells should be equipped with handles for convenient lifting. The number of cells in a tray or rack will depend on the weight and on the space available for installation. It is recommended that the weight of trays or racks not exceed approximately 110 kg [250 lb]. Battery trays or racks should be arranged so that the trays or racks are accessible and should have a minimum of 250 mm [10 in] of headroom. ...

Each cell/battery tray or rack should have a nameplate securely attached to the tray or rack or molded onto the tray case. The nameplate should contain the following:

a) Battery manufacturer's name or trademark
b) Battery type designation
c) Ampere hour rating at some specific rate of discharge (rating and rate of discharge should correspond to the specific application of the tray)
d) Specific gravity of electrolyte when charged (for lead-acid batteries)

Inter-cell connections and terminals for connections between trays or racks and for external wiring should be suitable for the maximum current produced by the cells/batteries. For diesel engine cranking batteries with high discharge rates, copper inserts in posts or other special provisions may be required.

26.3.2 IEEE Std 45-2002, 22.6.1, Battery size categories

Battery installations are classified based on the power output of the battery charger. These categories are large, moderate sized, and small.

26.3.3 IEEE Std 45-2002, 22.6.1.1, Large batteries

A large battery installation is one connected to a charging device with an output of more than 2 kW computed from the highest possible charging current and the rated voltage of the battery installation.

26.3.4 IEEE Std 45-2002, 22.6.1.2, Moderate-sized batteries

A moderate-sized battery installation is one connected to a charging device with a power output of 0.2 kW up to and including 2 kW computed from the highest possible charging current and the rated voltage of the battery installation.

26.3.5 IEEE Std 45-2002, 22.6.1.3, Small batteries

A small battery installation is one connected to a charging device with a power output of less than 0.2 kW computed from the highest possible charging current and the rated voltage of the battery installation.

26.3.6 IEEE Std 45-2002, 22.6.2.1, Battery installation—general (extract)

Batteries should be located where they are not exposed to excessive heat, extreme cold, spray, steam, or other conditions that would impair performance or accelerate deterioration. Batteries for emergency service, including emergency diesel engine cranking, should be located where they are protected as far as practicable from damage due to collision, fire, or other casualty.

Alkaline batteries and lead-acid batteries should not be installed in the same compartment or enclosure. In addition, on every vessel where both alkaline and lead-acid storage batteries are installed, a separate set of necessary maintenance tools and equipment for each battery type should be provided.

Sealed-gelled electrolyte batteries may be installed in locations containing standard marine or industrial electrical equipment if protected from falling objects and mechanical damage, provided all ventilation requirements are met.

Where more than one, normally operating, charging device is installed for any battery or group of batteries in one location, the total power output should be used to determine battery installation requirements. ...

26.3.7 IEEE Std 45-2002, 22.6.3, Arrangement of batteries

Batteries should be arranged to permit ready access to each cell or tray of cells from the top and at least one side for inspection, testing, watering (if required), and cleaning. Shelves or racks should be not more than 760 mm [30 in] deep and 1067 mm high.

For cells that required [the] addition of electrolyte (i.e., wet cells), there should be at least 300 mm [12 in], with 380 to 460 mm [15 to 18 in] recommended, clear space above the levels of the filling openings. Trays, when used, should be readily removable for repair or replacement.

When batteries are arranged in two or more tiers, each shelf should have at least 51 mm [2 in] of space front and back for air circulation.

26.3.8 IEEE Std 45-2002, 22.6.4, Battery trays and racks

Battery trays and racks should be securely chocked with wood strips or equivalent to prevent movement.

Each battery rack should be secured to prevent movement. Each tray or rack should be fitted with nonabsorbent insulating supports, not less than 20 mm [0.75 in] high on the bottom, and with similar spacer blocks at the sides, or with equivalent provision to ensure 20 mm [0.75 in] of space around each tray for air circulation.

Each battery tray or rack should be accessible with at least 250 mm [10 in] of headroom.

26.3.9 IEEE Std 45-2002, 22.6.5, Battery storage lining (extract)

Each battery room or locker should have a watertight lining for the storage of batteries as follows:

a) Storing batteries on shelves: Install lining to a minimum height of 76 mm [3 in] with lining thickness and material as follows:
   1) For vented lead-acid type batteries: 1.6 mm [1/16 in] minimum lead or other material that is corrosion-resistant to the battery electrolyte.
   2) For alkaline type batteries: 0.8 mm [1/32 in] minimum steel or other material that is corrosion resistant to the battery electrolyte.
b) Storing batteries on racks: The battery racks must be made of a material that is corrosion-resistant to the battery electrolyte and have containment space under the rack, as required in 22.6.3 [see 26.3.8 of this Handbook].

Alternatively, a battery room may be fitted with a watertight lead (steel for alkaline batteries) pan over the entire deck, carried up not less than 150 mm [6 in] on all sides.

Battery boxes should have a watertight lining to a height of 76 mm [3 in] with the same thickness requirements as shelf-lining requirements as given in item a) of this sub-clause. ...

The interior of all battery compartments, including shelves, racks, and other structural parts therein, should be made of corrosion-resistant materials or painted with corrosion-resistant paint.

26.3.10   IEEE Std 45-2002, 22.7.1, Ventilation—general

All rooms, lockers, and boxes for storage batteries should be arranged and ventilated to avoid accumulation of flammable gas. Particular attention should be given to the fact that the gas involved is lighter than air, and will tend to accumulate in any pockets at the top of the space.

26.3.11   IEEE Std 45-2002, 22.7.2, Battery rooms (extract)

All battery rooms should be adequately ventilated. Natural ventilation may be employed where the number of air changes is small and if ducts can be run directly from the top of the room to the open air above, with no part of the cut more than 45° from vertical. If natural ventilation is impracticable, mechanical exhaust ventilation should be provided. ...

If mechanical ventilation or cooling system is to be provided, the following recommendations should be adhered to:

a) Battery room ventilation system or cooling system return air should be separate from ventilation systems for other spaces. ...

b) Ventilation exhausts shall be at the top of the room and ventilation supplies at the bottom of the room.

c) Each blower should have a non-sparking fan.

d) Fans should be capable of completely changing the air in a maximum of 20 min.

e) Electric fan motors should be outside the duct and compartment.

f) When required for large battery rooms, electric fan motors shall be suitable for the area classification in which they are installed.

g) Electric fan motors should be at least 3 m [10 ft] from the exhaust end of the duct.

h) The ventilation or cooling system should be interlocked with the battery charger such that the battery cannot be charged without operating the ventilation system.

i) Interior surfaces of ducts and fans should be painted with corrosion-resistant paint.

j) Adequate openings for air inlet should be provided near the floor.

26.3.12   IEEE Std 45-2002, 22.7.3, Battery lockers

Battery lockers should be ventilated, if practicable, similarly to battery rooms by a duct led from the top of the locker to the open air or to an exhaust ventilation duct. However, in machinery spaces and similar well-ventilated compartments, the duct may terminate not less than 910 mm [3 ft] above the top of the locker.

Louvers or the equivalent should be provided near the bottom of the locker for entrance of air.

26.3.13   IEEE Std 45-2002, 22.7.4, Battery boxes

Deck boxes should be provided with a duct from the top of the box to at least 1.2 m [4 ft] above the box ending in a gooseneck, mushroom head, or equivalent to prevent the entrance of water. Holes for air entrance should be provided on at least two opposite sides of the box.

The entire deck box, including openings for ventilation, should be sufficiently weather tight to prevent entrance of spray or run.

Boxes for small batteries should have openings near the top to allow the escape of gas.

26.3.14   IEEE Std 45-2002, 22.7.7, Moderate-sized and small battery installation ventilation

Battery rooms or battery lockers for moderate-sized or small battery installations should have louvers near the bottom of the room or locker for the intake of ventilation air.

The ventilation rate for moderate-sized and small battery installations should meet the same requirements as for large battery installations.

26.3.15   IEEE Std 45-2002, 22.9, Battery rating

The capacity of any battery should have minimum output sufficient for its application and duty. In determining battery capacity, consideration should be given to time and rate of discharge.

The capacity of batteries for emergency lighting and power should be as stated in Clause 6.3 [not quoted in this Handbook]. Where the voltage of the emergency lighting system is the same as the voltage of the general lighting system, battery voltage, at the rated rate of discharge, should be a maximum of 105% of generator voltage when fully charged, and a minimum of 87.5% of generator voltage at the end of rated discharge.

Batteries for diesel engine cranking should have a maximum output sufficient to ensure breakaway torque at the lowest expected temperature. The battery should have a capacity capable of providing a minimum of 1.5 min of cranking at a speed sufficient to ensure engine starting and have sufficient capacity to provide a minimum of six consecutive engine starts without recharging.

Appendix A   Abbreviations and symbols

A.1 Abbreviations

ABS	American Bureau of Shipping
ABT	Automatic Bus Transfer
ACCU	Automatic Centralized Control Unmanned
AFT	After
ALM	Alarm
AMM	Ammeter
AMP	Ampere
ANSI	American National Standards Institute
ANT	Antenna
AT	Acceptance Trial
AUX	Auxiliary
AVR	Automatic Voltage Regulator
AWG	American Wire Gage
BHD	Bulkhead
BL	Baseline
BT	Builder's Trial
BW	Bandwidth
C2	Command and Control
C4ISR	Command, Control, Communications, Computers, Intelligence, Surveillance, and Reconnaissance
CANDI	Commercial and Non-Developmental Items
CL	Centerline
CLG	Cooling
COMPR	Compressor
CONT	Controller
COTS	Commercial Off The Shelf
CPU	Central Processing Unit
CSL	Console
CT	Current Transformer
Db	Decibel
DBL	Double
DCS	Damage Control system
DET	Detail
Dia	Diagram
DIM	Dimension
DK	Deck
DNV	Det Norske Veritas (Norway)
DOD	Department of Defense
DOT	Department of Transportation
DPDT	Double Pole Double Through
DS	Disconnect Switch
DSP	Digital Signal Processing
DWN	Down
E3	Electromagnetic Environmental effect
EMI	Electromagnetic Interference
EMP	Electromagnetic Pulse
ENG	Engine
EOS	Engineering Operating Station

EOT	Engine Order Telegraph
EQUIP	Equipment
EXH	Exhaust
F.O.	Fuel Oil
FCC	Federal Communications Commission
FDN	Foundation
FIFO	First in first out
FLA	Full Load Ampere
FMEA	Failure Mode Effect Analysis
FR	Frame
FWD	Forward
GEN	Generator
GFCI	Ground Fault Circuit Interrupter
GFE	Government Furnished Equipment
GFI	Government Furnished Information
GL	GL – Germanischer Lloyd (Germany Classification Society)
GMDSS	Global Maritime Distress Signal System
GND	Ground
GPS	Global positioning System
HD	Head -Pressure
HERO	Hazards of Electromagnetic Radiation to Ordnance
HP	Horse Power
HVAC	Heating, Ventilation, Air Conditioning
IBNS	Integrated Bridge Navigation System
IEC	International Electro-technical Committee
ILS	Integrated Logistic Support
IMO	International Maritime Organization
IP	Ingress Protection
IPDE	Integrated product Data Environment
IPT	Integrated Product Team
ISA	Instrument Society of America
ISO	International Standard Organization
IT	Information Technology
L.O.	Lube Oil
LAN	Local Area Network
LC	Load Center
LCI	Load Commutating Inverter
LF	Load Factor
LIFO	Last in First out
LKR	Locker
LTG	Lighting
LVL	Level
LVP	Low-Voltage Protection
LVR	Low-Voltage Release
LVRE	Low-Voltage Release Effect
MBT	Manual Bus Transfer
MCC	Motor Control Center
MG	Motor Generator
MMI	Man Machine Interface
MODU	Mobile Offshore Drilling Unit
MTBF	Mean Time Between Failure
NEC	National Electrical Code (NFPA 70)
NEMA	National Electrical Manufacturer's Association
NIBS	Navigational Integrated Bridge System
NSN	National Stock Number

NTS	Not to scale
NVIC	Navigation and Vessel Inspection Circular
OL	Overload
OSHA	Occupational Safety and Health Administration
P	Port side
PA	Public Address
PB	Pushbutton
PBX	Private Branch Exchange
PF	Power Factor
PLC	Programmable Logic Controller
PNL	Panel
PP	Power Panel
PWM	Pulse Width Modulation
QA	Quality assurance
RADAR	Radio Direction and Ranging
RCM	Reliability Centered Maintenance
RFI	Radio Frequency Interference
RM	Room
RMS	Root Means Square
ROM	Rough Order Of Magnitude
RPM	Revolution Per Minute
SCI	Serial Communication Interface
SCR	Silicon Control rectifier
SF	Service Factor
SG	Specific Gravity
SOLAS	Safety of Life at Sea
SONAR	Sound Navigation And Ranging
SPDT	Single Pole Double Through
Spkr	Speaker
SS	Ship Service
STA	Station
STBD	Starboard
STL	Steel
SUP	Supply
Sw	Switch
SWBD	Switchboard
SWBS	Ship Work Breakdown Structure
SYM	Symbol
SYS	System
TCP/IP	Transmission Control Protocol/Internet Protocol
TEL	Telephone
THD	Total Harmonic Distortion
TK	Tank
UL	Upper Level
UL	Underwriter's Laboratory
UPS	Uninterruptible Power Supply
VERT	Vertical
WT	Watertight
WIFCOM	Wireless Hand Free Communications
WW	Wireway
XFMR	Transformer

NOTES:

There are abbreviations listed above in support of Military shipbuilding, such as **C4ISR** - Command, Control, Communications, Computers, Intelligence, Surveillance, and Reconnaissance. This is for combat systems, system of system integration.

COTS - Commercial Off The Shelf. COTS refers to non-mil spec equipment for military ship installation. In general marinized commercial off the shelf equipment is in compliance with American Bureau of Shipping (ABS) rules and IEEE Std 45 recommendations.

A.2 Symbols

A	Ampere
Ac	Alternating Current
Dc	Direct Current
Emf	Electromotive force
F	Frequency
Ft	Feet
G	Acceleration of free fall (Acceleration due to gravity)
Gal	Gallon
GHz	gigahertz
Hp	Horse Power
Hz	Hertz
Kg	Kilogram
kPa	KiloPascal
Kva	kilovoltampere
kW	Kilowatt
kWh	Kilowatthour
L	Liter
M	Meter
mA	Milliamp
MHz	Megahertz
Min	Minute
MKS	Meter-Kilogram-Second
Mm	millimeter
MW	Megawatt
N	Newton
N/m	Newton-meter
Pa	Pascal
Rpm	revolution per minute
S	Second
SI	International System of Unit
V	Voltage
Var	Reactive Power
W	Watt

Appendix B   Glossary

Most of the definitions come from IEEE Std 45, USCG and ABS standards. Where relevant, we have included a designation of USCG or ABS. Where not, it is an IEEE definition or a common-sense definition.

In some cases (fail-safe for example) the definitions are so important that both have been included.

ACC: Automatic Centralized Control-This notation is assigned by ABS to a vessel having the means to control and monitor the propulsion-machinery space from a continuously manned centralized control and monitoring station installed within or adjacent to, the propulsion machinery space.

ACCU: Automatic Centralized Control Unmanned -This notation is assigned by ABS to a vessel having the means to control and monitor the propulsion-machinery space from the navigation bridge and from centralized control and monitoring station installed within or adjacent to, the propulsion machinery space.

alarm: Visual and audible signals indicating an abnormal condition of a monitored parameter. (ABS)

alarm: Alarm means an audible and visual indication of a hazardous or potentially hazardous condition that requires attention. (USCG)

ambient temperature: Ambient temperature is the temperature of surrounding media such as air, fluid where equipment is operated, or positioned.

ASD: Adjustable speed drive. ASD can be cycloconverter type, Load commutating inverter type, pulse width modulation type, etc.

automated: Automated means the use of automatic or remote control, instrumentation or alarms. (USCG) (4-9-1/5.1.4)

automatic control: means of control that conveys predetermined orders without action by an operator. (ABS)

automatic control: Automatic control means self-regulating in attaining or carrying out an operator-specified equipment response or sequences. (USCG)

auxiliary services system: All support systems (e.g. fuel oil system, lubricating system, cooling water system, compressed air system, and hydraulic system, etc) which are required to run propulsion machinery and propulsors. (ABS definition)

azimuth thruster: Rotatable mounting thruster device where the thrust can be directed to any desirable direction.

bandwidth: Generally, frequency range of system input over which the system will respond satisfactory to a command.

braking: Braking provides a means of stopping and can be accomplished by Dynamic Braking, Regenerative Braking, DC Injection Braking, and Positive action brake.

breakdown torque: The maximum torque which can be developed with rated parameters, such as rated voltage applied at rated frequency.

clearing time: The total time between the beginning of the over-current and the final opening of the circuit at rated voltage by an over-current protective device. (ABS 4-9-1/5.1.15)

centralized control station: A propulsion control station fitted with instrumentation, control systems and actuators to enable propulsion and auxiliary machinery be controlled and monitored, and the state of propulsion machinery space be monitored, without the need of regular local attendance in the propulsion machinery space. (ABS)

constant horsepower range: A range of motor operation where motor speed is greater than base rating of the motor, in the case of AC motor operation usually above 60 Hz where the voltage remains constant as the frequency is increased.

constant torque range: A speed range in which the motor is capable of delivering a constant torque, subject to motor thermal characteristics. This essentially is when the inverter/motor combination is operating at constant volts/Hz.

constant volts/hertz (V/Hz): This relationship exist in AC drives where the output voltage is varied directly proportional to frequency. This type of operation is required to allow the motor produce constant rated torque as speed is varied.

continuous rated machine: The continuous rating of a rotating electrical machine is the rated kW load at which the machine can continuously operate without exceeding the steady state temperature rise. (ABS) (4-9-1/5.1.2)

control: The process of conveying a command or order to enable the desired action to be effected. (ABS) (-1/5.1.3)

control system: An assembly of devices interconnected or otherwise coordinated to convey the command or order.

controllable pitch propeller: Propeller pitch is changeable, mostly hydraulic.

converter: The process of changing AC to DC, AC to DC to AC.

current limiting: An electronic method of limiting the maximum current available to the motor. This is adjusted so that the motor's maximum current can be controlled. It can also be preset as a protective device to protect both the motor and control from extended overloads.

cycloconverter: Type A direct frequency converter without intermediate DC link which can convert power from one fixed frequency to a lower variable frequency

decibel: Decibel is a ratio (one-tenth of a bel), expressed as dB. It is a means of comparing the relative strength of two values, such as current, power. It is not an absolute value, as it is a comparison unit. As an example: dB=10 log P2/P1, P is power.

deviation: Difference between an instantaneous value of a controlled variable and the desired value of the controlled variable corresponding to the set point.

DOL starter: Direct on line starting system

duty cycle: The relationship between the operating and rest times or repeatable operation at different loads.

engineering control center (ECC): ECC means a centralized engineering control, monitoring, and communications location. (USCG)

efficiency: Ratio of mechanical output to electrical input indicated by percent.

enclosure: Enclosure refers to the housing in which the equipment in mounted. (ABS 4-9-1/5.1.11)

fail-safe: A designed failure state which has the least critical consequence. A system or a machine is fail-safe when, upon the failure of a component or subsystem or its functions, the system or the machine automatically reverts to a designed state of least critical consequence. (ABS)

fail safe: Fail safe means that upon failure or malfunction of a component, sub-system the output automatically reverts to a predetermined design state of least critical consequence. (USCG)

failure mode and effect analysis (FMEA): A failure analysis methodology used during design to postulate every failure mode and the corresponding effect or consequences. Generally, the analysis is to begin by selecting the lowest level of interest (part, circuit, or module level). The various failure modes that can occur for each item at this level are identified and enumerated. The effect for each failure mode, taken singly and in turn, is to be interpreted as a failure mode for the next higher functional level. Successive interpretations will result in the identification of the effect at the highest function level, or the final consequence. A tabular format is normally used to record the results of such a study. (ABS)

fixed pitch propeller: The propeller blades are fixed. There is no possibility of changing propeller pitch.

full-load torque: The full load torque of a motor is the torque necessary to produce rated horsepower at full-load frequency.

fuse: An over-current protective device with fusible link that opens the circuit on an over-current condition.

GFE: Government furnished equipment.

GMDSS: Global Maritime Distress Signal System.

GPS: Global Positioning System.

induction motor: An alternating current motor in which the primary winding on one member (usually stator) is connected to the power source. A secondary winding on the rotor carries the induced current. (ABS 4-9-1/5.1.5)

instrumentation: A system designed to measure and to display the state of a monitored parameter and which may include one of more of sensors, read-outs, displays, alarms and means of signal transmission. (ABS)

IPDE: Integrated Product Data Environment which features the capability to concurrently develop, update,, and reuse data in electronic form.

IPT: Integrated product team composed of representatives from appropriate disciplines working together to build successful program, identify and resolve issues, and make sound and timely recommendations to facilitate decision making.

isochronous: Constant speed and frequency irrespective of load.

LCI: Type of adjustable speed current source drive called load commutating inverter.

local control: A device or array of devices located on or adjacent to a machine to enable it to be operated within sight of the operator. (ABS)

local control: Local control means operator control from a location where the equipment and its output can be directly manipulated or observed, e.g. at the switchboard, motor controller, propulsion engine, or other equipment. (USCG)

locked rotor torque: The minimum torque of a motor which will develop at rest for all angular positions of the rotor, with rated voltage applied at rated frequency

manual control: Manual Control means operation by direct or power assisted operator intervention. (USCG) (4-9-1/5.1.9)

monitoring system: A system designed to supervise the operational status of machinery or systems by means of instrumentation, which provides displays of operational parameters and alarms indicating abnormal operating conditions. (ABS)

NIBS: Navigational Integrated Bridge System.

non-periodic duty rated machine: The non-periodic duty rating of a rotating electrical machine is the kW loads which the machine can operate continuously, for a specific period of time, or intermittently under the designed variations of the load and speed within the permissible operating range, respectively; and the temperature rise, measured when the machine has been run until it reaches a steady temperature condition, is not to exceed those given in 4-8-3/Table 4. (ABS)

over-current: A condition which exists on an electrical circuit when the normal full load current is exceeded. The overcurrent conditions are overloads and short circuits.

periodic duty rating machine: The periodic duty rating of a rotating machine is the rated kW load at which the machine can operate repeatedly, for specified period (N) at the rated load followed by a specified period (R) of rest and de-energized state, without exceeding the temperature rise given in 4-8-3/Table 4; where N+R = 10 min, and cyclic duty factor is given by N/(N+R) %. (ABS) (4-8)

podded propulsion: Propulsion electric motor in installed in watertight enclosure, where the motor is mounted on the same motor shaft.

propulsion machine: A device (e.g. diesel engine, turbine, electric motor etc.) which develops mechanical energy to drive a propulsor. (ABS)

propulsion machinery space: Any space containing machinery or equipment forming part of the propulsion system. (ABS)

propulsion system: A system designed to provide thrust to a vessel consisting of : one or more propulsion machines; one or more propulsors; all necessary auxiliaries, and associated control, alarm and safety system. (ABS)

propulsor: A device (e.g. propeller, waterjet) which imparts force to a column of water in order to propel a vessel, together with any equipment necessary to transmit the power from the propulsion machinery to the device (e.g. shafting, gearing etc.) (ABS)

PWM: PWM is a DC inverter system with the main power being rectified to produce DC and a self commutated inverter to invert DC to AC to a variable frequency.

remote control: A device or array of devices connected to a machine by mechanical, electrical, pneumatic, hydraulic or other means and by which the machine may be operated remote from, and not necessarily within sight of, the operator. (ABS)

remote control: Remote control means non-local automatic or manual control. (USCG) (ABS 4-9-1/5.1.8)

remote control station: A location fitted with means of remote control and monitoring. (ABS)

R1: This is ABS optional notation assigned to a vessel fitted with multiple propulsion machines but only a single propulsor and steering.

R2: This is ABS optional notation assigned to a vessel fitted with multiple propulsion machines and also multiple propulsors and steering system.

R1-S: This is ABS optional notation assigned to a vessel fitted with only single propulsor but having the propulsion machinery arranged in separate spaces such that a fire or flood in one space will not effect the propulsion machinery in the other space.

R2-S: This is an optional notation assigned to a vessel fitted with multiple machines and propulsors, and associated steering systems arrangement in separate spaces such that a fire or flood in one space will not effect the propulsion machine(s) and propulsor(s), and associated steering systems in the other space.

R1, R2, R1-S & R2-S (Plus)(+): This notation signifies that the redundant propulsion is capable of maintaining position under adverse weather conditions to avoid uncontrolled drift.

RMS current: The RMS is root-mean-square. The RMS current is the root-mean-square value of any periodic current.

safety system: An automatic control system designed to automatically lead machinery being controlled to a predetermined less critical condition in response to a fault which may endanger the machinery or the safety of personnel and which may develop too fast to allow manual intervention. To protect an operating machine in the event of a detected fault, the automatic control system may be designed to automatically (ABS)

— slowdown the machine or to reduce its demand;
— start a standby support service so that the machine may resume normal operation; or
— shutdown the machine.

For the purposes of this Chapter, automatic shutdown, automatic slow down and automatic start of standby pump are all safety system functions. Where "safety system" is stated hereinafter, it means any or all three automatic control systems.

short time duty rating machine: The short time rating of a rotating electrical machine is the rated kW load at which the machine can operate for a specified time period without exceeding the temperature rise given in 4-8-3/Table 4. A rest and de-energized period sufficient to re-establish the machine temperature to within 20°C (3.60°F) of the coolant prior to the next operation is to be allowed. At the beginning of the measurement the temperature of the machine is to be within 50°C (90°F) of the coolant. (ABS)

slip: The difference between rotating magnetic field speed (synchronous speed) and rotor speed of AC induction motor which usually expressed as a percentage of synchronous speed.

steering system: A system designed to control the direction of a vessel, including the rudder, steering gear etc. (ABS) (4-9-1/5.1.12)

systems independence: Systems are considered independent where they do not share components such that a single failure in any one component in a system will not render the other systems inoperative. (ABS)

THD: Total harmonic distortion. It is accumulated distortion of fundamentals in a variable frequency drive

TOC: Total ownership cost comprised of costs to research, develop, acquire, own, operate, maintain, and dispose of any and all assets comprising the ship.

tunnel thruster: Produces fixed directional thrust. (ABS 4-9-1/5.1.14)

Thyristor: Is a component which conducts current in only one direction. Unlike diode, the thyristor needs a firing pulse for it to start conducting current, after which it continues to conduct as long as there is current through it.

unmanned propulsion machinery space: Propulsion machinery space which can be operated without continuous attendance by the crew locally in the machinery space and in the centralized control station. (ABS 4-9-1/5.1.17)

vital auxiliary pumps: Vital auxiliary pumps are that directly related to and necessary for maintaining the operation of propulsion machinery. For diesel propulsion engines, fuel oil pump, lubricating oil pump, cooling water pumps are examples of vital auxiliary pumps. (ABS)

vital system or equipment: Vital system or equipment is essential to the safety of the vessel, its passengers and crew. (USCG)

Appendix C   Units and conversion tables

General

The use of International System of Units (SI) was adopted in 1960. The worldwide use of SI has opened the global marketplace; however, it has not been implemented as much as should have been in the United States, particularly in the shipbuilding industry. In the shipbuilding industry, the measurement units are often addressed as English Unit (US unit), SI Unit, hard Metric Unit, soft Metric Unit. American Bureau of Shipping (ABS) rules use US Unit, SI Unit, and MKS (Meter-Kilogram-System) Unit. The IEEE Std 45-2002 working group took measures to provide US units as well as SI units. The measurement conversion tables below are for measurements and conversions from one to another unit tailored to the shipbuilding application.[1]

SI units

Length is in meter (m), area is in square meter (m^2), time is in second (s), temperature is in Kelvin, flow rate is in volume, cubic meters per second (m^3/s), electric current is in ampere (A), mass is in kilogram (kg), weight is in force, kilogram-meter per second (kg-m/s), acceleration is in kilogrammeter per second square (Newton) (kgm/s^2), force is in Newton, pressure is in Pascal (Pa), and moment of force is in Newton-meter (Nm).

MKS units

MKS unit of measurement is meter-kilogram-second. SI unit was derived from MKS system. In MKS system length is in meter (m), area is in square meter (m^2), volume is in cubic meter (m^3), time in second (s), mass is in kilogram (kg), weight (force) is in kilogram-meter per second square ($kg-m/s^2$), density is in kilogram per centimeter square (kg/cm^2), acceleration is in kilogram-meter per second square (kgm/s^2), force(pressure) is in kilogram-force per square centimeter (kgf/cm^2).

Sample unit conversions as used by ABS

Sample unit conversions as used by ABS are given below. The SI unit has been given first and then within parenthesis, the MKS and US units.

Unit	= SI (MKS, US):
Force	= N (kgf, lbf)
Area	= mm^2 (mm^2, in^2)
Pressure	= N/m^2 (kgf/mm^2, lbf/in^2)
Torque	= N-mm (kgf-mm, lbf-in^2)
Volume	= mm^3 (mm^3, in^3)
Length	= mm (mm, in)
Speed	= m/s (m/s, ft/s)

Adapted from IEEE/ASTM SI-10-1997 Standards of Units (SI): Modern Metric System.

Frequently used unit of conversion in SI unit, MKS unit, and US unit

1 psi	= 0.0703 kgm/cm^2 = 0.068 atmosphere
1 (atm)	= 14.67 lbs/in^2 (in US unit) = 101.325 Pa (in SI unit) = 101.325 N/m (in MKS unit)
1Newton	= 1 kgm/s^2
1 Pascal	= Newton per meter square (N/m^2)
1 knot	= 1 nautical mile per hour = 1852/3600 (m/s) = 0.514 m/s
Energy (Joules-J)	= Newton-meter (N-m) = kg-m2/s^2
Acceleration due to gravity	= 9.807 m/s^2 = 32.2 ft/s^2
Pressure = Force/area	= 1 N/m2 = Pa (Pascal

In SI unit the pressure is Pascal (Pa). Pascal is a pressure of one Newton per square meter.

One atmosphere (atm) = 14.696 lbs/in^2 (psi)
Pressure head in feet = 2.31 psi

One standard atmosphere is the pressure exerted by a column of mercury exactly 76 cm high of density 13.505 g/cm2, in a place where the acceleration due to gravity is 980.7 cm/s^2.

1 atm   = (0.76 m) (1.3595 x 104) (9.807 m/s^2)
          = 101.325 N/m^2
          = 101.325 Pa

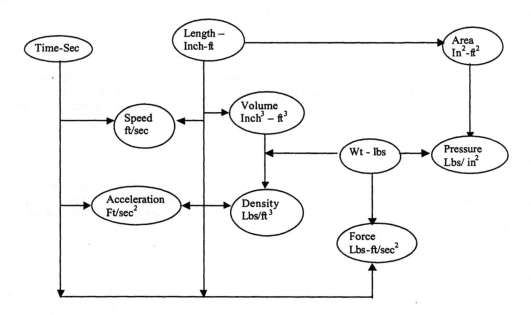

Figure C-1: Units of Measurement—US Units

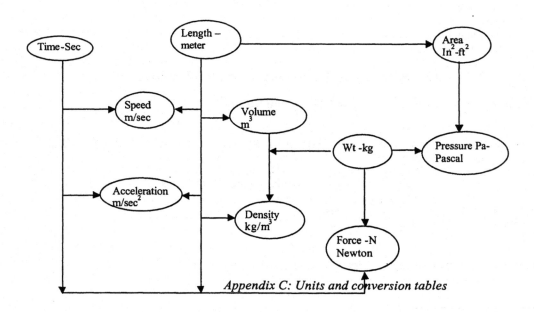

Appendix C: Units and conversion tables

Figure C-2: Units of Measurement—SI Units

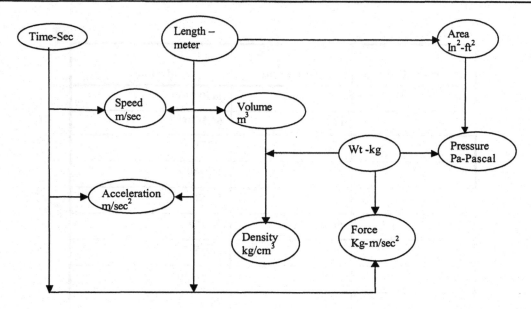

Figure C-3: Unit of Measurement—MKS Units

From	Unit	To US unit	Multiplier
1 mm	SI	in	0.03937
1 mm	SI	ft	0.00328
1 cm	SI	mm	10
1 m	SI	ft	3.2808
1 m	SI	cm	100
1 in	US	mm	25.4
1 ft	US	cm	30.48
1 mil	US	in	0.001
1 knot	US	mile	
1 knot	US	km	
1 mile	US	km	1.609
1 km	SI	mile	0.6215
1 Degree	Angle	radian	0.0175
1 footcandle		lux (lx)	10.76
1 foot pound (ft-lbs)		joules(J)	0.042
1 mile	US	ft	5280
1 mile	US	meter(m)	1609.34
1 nautical mile(Knot)		meter(m)	1852

Table C-1: Conversion Table—SI-MKS Units

From SI unit	To US unit	Multiplier
mil^2	in^2	1×10^{-6}
mil^2	mm^2	64.5×10^{-6}
mil^2	cirmil	1.273
mm^2	in^2	155×10^{-5}
in^2	cm^2	6.452
ft^2	m^2	0.0929
Cm^2	in^2	0.155
Cirmil	mm^2	5.067×10^{-4}
Cirmil	in^2	785×10^{-9}
m^2	in^2	1550
m^2	ft^2	10.76

Table C-2: Conversion Table—Units of Area

	(Pound per cubic foot) Lb/ft 3 - density	(Gram per cubic centimeter) Gm/cm^3 – density	Specific gravity
Aluminum	161	2.67	2.67
Copper	555.4	8.9	8.9
fresh water	62.4	1.0	1.0
Gold	1205	19.3	19.3
Ice	57.2	0.917	0.917
Iron	491.3	7.87	7.87
Lead	686.7	11.25	11.25
Mercury	848.7	13.6	13.6
Nickel	549.4	8.9	8.9
Platinum	1342	21.5	21.5
sea water	64	1.025	1.025
Silver	655.5	10.5	10.5
Tin	448	7.29	7.29
Tungsten	1161	18.4	18.4
Zinc	448.6	7.19	7.19

Table C-3: Conversion Table—Units of Density and Specific Gravity
Note: Density = Specific Gravity (sp. Gr) = 1 gram per cubic centimeter

Appendix D   Formulae

Algebraic formulae

a)   Complex Algebra – Rectangular Form

$z_1 = x_1 + j\,y_1$ and $z_2 = x_2 + jy_2$ Where $(j)^2 = -1$

Addition: $z_1 + z_2 = (x_1 + x_2) + j(y_1 + y_2)$

Subtraction : $z_1 - z_2 = (x_1 - x_2) + j(y_1 - y_2)$

Multiplication: $z_1\,z_2 = (x_1 + j\,y_1)\,(x_2 + jy_2) = (x_1x_2 - y_1y_2) + j\,(x_1\,y_1 + x_2y_1)$

Division : $z_{1/}\,z_2 = (x_1\,x_2 + y_1\,y_2)\,/\,((x_2)^2 + (y_2)^2) + j\,(x_2\,y_1 - x_1y_2)\,/\,((x_2)^2 + (y_2)^2)$

(NOTE—For imaginary number equation $a + jb = c + jd$ is correct when $a = c$ and $b = d$)

830 Complex algebra – polar form

$z_1 = x_1 + j\,y_1$ and $z_2 = x_2 + jy_2$ Where $(j)^2 = -1$

$x = [z]\,\text{Cos}\,\varphi$ and $y = [z]\,\text{Sin}\,\varphi$ Where $[z]$ is absolute value

$$[z] = \tan^{-1}\left(\frac{y}{x}\right)$$

In Polar form $z = [z]\,e^{j}\,\varphi = [z]$ angle $\varphi = ([z]\,\text{Cos}\,\varphi + [z]\,\text{Sin}\,\varphi)$

831 Rectangular to polar conversion

$$8 + j\,6 = \sqrt{8^2 + 6^2}\;\tan^{-1}\frac{8}{6} = 10\angle 36.9°$$

Polar to rectangular form:

10 angle 36.9 = 10Cos 36.9 + j 10 Sin 36.9 = 8 + j6

Electrical engineering formulae

832 Inductive reactance $X_L = 2\,\pi\,f\,L = \omega L = 6.283\,f\,L$

$$\int = \frac{X_L}{2\pi L}$$

Frequency (f) in Hz, Inductance (L) in henry and X_L inductance

Example 1:

Calculate inductive reactance of a circuit with 100mHenry coil at a frequency of 5 KHz

Calculation: $X_L = 2 \times \pi \times 5\text{ KHz} \times 100\text{ mH} = 2 \times 5 \times 10^3 \times 100 \times 100 \times 10^{-3}$

$X_L = 10^5\,(\Omega)$

833 Capacitive Reactance

$X_C = \dfrac{1}{2\pi f C}$ Where frequency (f) in Hz, Capacitance I in farads

Example 2:

Calculate Capacitive reactance of a circuit of 90 µf at 50 Hz.

$$\text{Calculations: } X_C = \frac{1}{2 \times 3.14 \times 50 \times 90 \times 10^{-6}}$$

834 Circuit resonance

Impedance is at resonance when $X_L = X_C$

$$2\pi f L = \frac{1}{2\pi f C}$$

Resonance frequency $f_\circ = \dfrac{1}{2\pi\sqrt{LC}}$

835 Power calculations:

836 Apparent Power $P = V \times I$

2) True Power

Power in single-phase reactive circuit

$P = V \times I \times pf \Rightarrow pf$ is power factor $= \text{Cos } \varphi$

$P = I^2 \times R = V \times I \times \text{Cos } \varphi$

3) Power in three-phase reactive circuit

$P = V \times I \times pf \times 1.732$

Appendix E   Tables

Tables E1 through E5 are provided for reference as follows:

  837   Table E1: List of code letters per NEC. The motor kVA/hp is identified by NEC Table 430-7(b) with code letters.

  838   Table E2: List of permissible temperature rise in motors, by classification society and standard

- Table E3: IEC 440 V, 60 Hz and 50 Hz three-phase motor technical data--kilowatt-voltage-ampere
- Table E4: NEMA AC 60 Hz motor technical data
- Table E5: IEC 380 V, 60 Hz and 50 Hz three-phase motor technical data--kilowatt-voltage-ampere (Full load amperes)

Code letter	kVA/hp with locked rotor	Code letter	kVA/hp with locked rotor
A	0-3.14	L	9.0 – 9.99
B	3.15 – 3.54	M	10.0 – 11.19
C	3.55 – 3.99	N	11.2 – 12.49
D	4.0 – 4.49	P	12.5 – 13.99
E	4.5 – 4.99	R	14.0 – 15.99
F	5.0 – 5.59	S	16.0 – 17.00
G	5.6 – 6.29	T	18.0 – 19.99
H	6.3 – 7.09	U	20.0 – 22.39
J	7.1 – 7.99	V	22.4 and UP
K	8.0 – 8.99		

Table E-1: NEMA Code Letters Indicating Motor with Locked Rotor [2]

Classification society and standards		Ambient temperature in Celsius (°C)	Permissible Temperature rise in Celsius for stator winding insulation class (°C)	
			B	F
ABS	American Bureau of Shipping	45/50	70	95
BV	Bureau of Varitas	45/50	70	90
CCS	China Classification Society	45/50	70	90
CR	China Corp. Register of Ship	45	70	90
DNV	Det Norske Veritas	45	75	95
GL	Germanischer Lloyd	45	70	90
IEC Pub. 34-1	International Electrotechnical Committee	40	80	105
IEC 92-301	International Electrotechnical Committee	45/50	70	90
KR	Korean Register of Shipping	45/50	75	95
LR	Lloyd's Register of Shipping	45	70	90
RINA	Registro Italiano Navale	45/50	70	90

Table E-2: Motor Permissible Temperature Rise by Classification Society and Standard

[2] Source: NEC Table 430-7(b) – Locked-Rotor Indicating Code Letters

kW Output	IEC Frame	440 V 60 Hz Amp	440 V pf	440 V Eff	440 V 50 Hz Amp
0.26	63	0.75	0.76	0.63	0.712555
0.37	71	0.95	0.8	0.685	0.885973
0.5	71	1.2	0.78	0.728	1.15543
0.7	80	1.6	0.8	0.75	1.530898
0.9	80	2	0.79	0.785	1.904343
1.3	90	2.7	0.84	0.784	2.590284
1.9	90	3.8	0.84	0.8	3.710084
2.6	100	5	0.85	0.828	4.847564
3.5	100	6.6	0.85	0.834	6.47862
4.6	112	8.5	0.84	0.865	8.307338
6.4	132	11.4	0.86	0.865	11.28924
8.6	132	15.4	0.85	0.89	14.91726
13	160	23.5	0.85	0.88	22.80558
18	160	30.5	0.85	0.91	30.53596
22	180	37.5	0.85	0.91	37.32173
26.5	180	44	0.85	0.91	44.95572
35	200	59	0.83	0.926	59.75557
44	225	74	0.84	0.93	73.90773
55	225	92	0.85	0.94	90.32653
65	250	107	0.86	0.94	105.5083
88	280	144	0.86	0.95	141.3384
105	280	167	0.87	0.95	166.7039
125	315	200	0.87	0.95	198.4571
150	315	238	0.87	0.95	238.1485
185	315	290	0.87	0.956	291.8731
230	355	362	0.87	0.96	361.3573
285	355	450	0.87	0.96	447.7688
345	355	535	0.87	0.96	542.0359
400	355	615	0.89	0.96	614.325
450	355	690	0.89	0.96	691.1156

kW	IEC	440 V 60 Hz	440 V	440 V	440 V 50 Hz
Output	Frame	Amp	pf	Eff	Amp
500	400	750	0.9	0.96	759.374
550	400	825	0.9	0.966	830.1231

Note that the IEC motor rating is in kilowatt.

Table E-3: IEC 440 V, 60 Hz, and 50 Hz Three-Phase Motor Technical Data – Kilowatt-Voltage-Ampere

NEMA Size	115	115	115	230	230	230	460	460	460
Output-HP	Volts	Volts	Volts	Volts	Volts	Volts	Volts	Volts	Volts
	PF	EFF	AMP	PF	EFF	AMP	PF	EFF	AMP
0.5			3.90			2.02			1.01
0.75			5.85			2.83			1.51
1			7.80			3.49			2.02
1.5			11.70			4.74			3.03
2						6.08			4.04
3						8.24			6.05
5						13.00			10.09
7.5						19.51			15.13
10						25.17			20.18
15						37.76			30.27
20						49.81			40.36
25						62.26			50.45
30						69.67			60.54
40						92.89			80.72
50						116.11			100.90
60						139.34			121.08
75						174.17			151.35
100						232.23			201.80
125						290.28			252.25
150						348.34			302.70
200						464.45			403.59

Table E-4: NEMA AC 60 Hz Motor Technical Data—Horsepower-Ampere-Voltage

| | | 50 Hz | | | 60 Hz |
| kW | IEC | 380 V | 380 V | 380 V | 380 V |
OUTPUT	FRAME	AMP	PF	EFF	AMP
0.18	63	0.7	0.68	0.61	0.659329
0.25	71	0.8	0.68	0.61	0.915735
0.37	71	1.2	0.72	0.68	1.14823
0.55	80	1.5	0.74	0.71	1.590527
0.75	80	1.9	0.78	0.74	1.974256
1.1	90	2.7	0.78	0.76	2.819376
1.5	90	3.5	0.82	0.8	3.474209
2.2	100	5.0	0.85	0.81	4.854978
3	100	6.7	0.85	0.82	6.539687
4	112	8.7	0.84	0.85	8.511974
5.5	132	11.5	0.85	0.86	11.43178
7.5	132	15.5	0.85	0.88	15.2345
11.5	160	23.5	0.85	0.88	23.35957
15.5	160	30.5	0.85	0.91	30.44668
19	180	37.5	0.85	0.91	37.32173
22.5	180	44.0	0.85	0.91	44.19679
30	200	59.0	0.83	0.93	59.0512
38	225	74.0	0.84	0.93	73.90773
48	225	92.0	0.85	0.94	91.27734
57	250	107.0	0.86	0.94	107.1315
75	280	142.0	0.86	0.94	140.9625
90	280	169.0	0.86	0.95	167.3744
110	315	204	0.86	0.95	204.5687
132	315	245	0.86	0.955	244.1972
160	315	294	0.87	0.96	291.0704
200	355	365	0.87	0.957	364.9785
250	355	455	0.87	0.96	454.7975
315	355	570	0.87	0.96	573.0448
355	355	630	0.89	0.96	631.2998
400	355	710	0.89	0.966	706.9055
450	400	770	0.9	0.966	786.4324

		50 Hz			60 Hz
kW	IEC	380 V	380 V	380 V	380 V
OUTPUT	FRAME	AMP	PF	EFF	AMP
500	400	870	0.9	0.967	872.9101
560	400	965	0.91	0.968	965.917
600	400	1070	0.88	0.968	1070.192

Table E-5: IEC 380 V, 50 Hz, and 60 Hz Three-Phase Motor Technical Data—Horsepower-Voltage-Ampere (full load ampere)

Appendix F   Horsepower and kilowatt calculations

Pump-motor horsepower calculations

Horsepower calculation—US units

Calculate motor horsepower for shipboard auxiliary equipment such as pumps.

By definition horsepower is equal to 550 ft-lb per second or 33,000 ft-lb per minute.

Pump Brake Horse Power (BHP) $= \dfrac{work}{time(\sec onds)x550}$

(Where Work = Force × distance)

Brake Horse Power for Pump $= \dfrac{GPMx\Pr essurehead(FT)xSpecifcGravity}{3960\,xefficiency}$

GPM = Pump capacity in gallon per minute.

SG = Specific Gravity (SG) of the fluid.

η =Efficiency of the pump

Pressure head (FT) = 2.31 × PSI (lbf pressure per square in)

	Typical specific gravity
Fresh Water	1.0
Salt Water	1.024
Lube Oil	0.85
Diesel Oil	0.85

Example 1 – Horsepower Calculation

A seawater service pump installed in the engine room at a 50° c ambient. Calculate the horsepower nameplate rating of the motor, when the pump capacity is 550 gallons at 300ft pressure head, at 89 percent pump efficiency (η), and motor efficiency (η) is 92 percent.

Step 1

Fluid Horse Power for Pump $= \dfrac{GPMx\Pr essure(Ft)xSpecificG\,ravity}{3960}$

Fluid Horse Power for Pump $= \dfrac{550\,x300\,x1.024}{3960} = 42.7$ HP

Step 2

Mechanical output of the pump = Fluid horse power/ pump efficiency

$= \dfrac{42.7}{0.89} = 48HP$

Step 3, Motor brake horsepower

Note that mechanical output of the pump is equal to the brake horsepower of the motor, which is 48 hp.

Step 4, Motor indicate horsepower (IHP)

Motor IHP is equal to motor brake horsepower divided by the motor efficiency.

$$MotorIHP = \frac{48}{0.92} = 52HP$$

Step 5, Motor nameplate horsepower

Motor nameplate horsepower is the rating of the motor is to be purchased.

For 52 IHP, the next available size motor is 60 hp.

Step 6, Special ambient consideration

In general the motors are rated for 40 °C ambient. Appropriate derating factor should be applied for using 40 °C ambient motor 50 °C application. Motor manufacturer's recommendation should be applied. For example if the derating factor is .95 the 60 hp nameplate rated motor, the motor is good for $(60 \times 0.95) = 57$ hp service.

Step 7

Motor rated at 60 hp is sufficient for this application at 50 °C.

Remarks

In this example the motor is selected from NEMA frame rated horsepower such as 30, 40, 50, 60, 75 etc. For IEC motor the IEC frame rating in kilowatts is used. Traditionally the US shipbuilding uses horsepower rated equipment and all supporting engineering documentations are also horsepower related. The IEC standard kilowatt system should be adapted. The next example is shown in kilowatt rating.

Example 2 - Kilowatt Power Calculation—SI units

Calculate motor kilowatt rating of a Seawater pump motor for shipboard application:

Kilowatt calculation for seawater pump motor

$$Kilowatt = \frac{Qx(H)xSG}{270x1.36x\eta}$$

Where

Q is the capacity of the pump in m³/hr, H is the pumping head in meter (m).
SG is the specific Gravity (SG of seawater 1.025), pump efficiency η is 70%.

Example 3 – Kilowatt Power Calculation

A seawater service pump installed in the engine room at a 50 °C ambient. Calculate the kilowatt rating of the motor, when the pump capacity is m³/hr at 100 m pressure head, at 80% pump efficiency (η), and motor efficiency (η) is 92%.

Step 1

$$MotorKilowatt = \frac{Qx(H)xSG}{270x1.36x\eta} = \frac{100x80x1.025}{270x1.36x0.8} = 27.9 \text{ kW}$$

Appendix G   AC/DC network analysis

Direct Current (DC) – Network analysis

The DC voltage source is with a positive and negative voltage terminal. Examples of DC voltage sources are:

— Batteries
— DC generators

Connecting an electrically conductive load to the terminal forces a current to flow. The polarity of the voltage source is annotated with + (positive) and – (negative). The arrow shows positive direction of current flow.

If the DC voltage is V_{dc} and the electrical resistance in the load is R, the current I is found by Ohm's law:

$I = V_{dc}/R$

The power dissipated in the resistance equals the power delivered by the power source:

$P = V_{dc} . I = V^2_{dc} / R = R . I^2$

Example: Signal-light bulb

A signal light is powered with a 12 V battery and 0.5 W light bulb.

The current in the bulb is

$I = Power/V_{dc} = 0.5/12 = 0.0416 A = 41.6 mA$

The resistance in the bulb is

$R_{light\ bulb} = (V_{dc})^2 /P = (12 \times 12) / 0.5 = 288\Omega$

AC – Alternating Current – Network analysis

The AC voltage source is bipolar and normally sinusoidal time varying. United States AC power source frequency rating is 60 Hz and in some other parts of the world the frequency is 50 Hz. IEC countries follow 50 Hz convention. The AC power supplies are mostly three phase and single phase. For some special applications, 400 Hz is used.

Examples of AC voltage sources are

— Generators (also called alternators)
— Static converters
— UPS, which are static converters supplied from batteries if supply voltage vanish, applied for power supply to critical equipment

For IEC AC voltages are sinusoidal time varying, 50 Hz voltages have a cycle time of 20 ms.

For US AC voltages are sinusoidal time varying, 60 Hz voltage have a cycle time of 16.6 ms.

$V(t) = V . SIN (2\pi f\ t) = V . SIN (\omega t)$

Here the $\omega = 2\pi f$ is the angular frequency, expressed in rad/s. If this voltage source is connected to a resistive load (ohmic load), as shown in Figure D19, a current will flow through the circuit. And assuming that there is no voltage drop in the voltage source, the current will be given by the resistance in the load:

$i(t) = v(t) / R = (\hat{V} / R) \cdot \sin(\omega t) = \hat{I} \cdot \sin(\omega t)$

From this it is seen, that the current is proportional to the voltage, and that the voltage and current are in phase, i.e. phase angle equals to zero. This is always true for a pure resistive load.

Since the voltages and currents are time varying, the power consumed by the resistor is also time varying:

$$p(t) = v(t) \cdot i(t) = \hat{V} \cdot \hat{I} \cdot \sin^2(2\pi f t) = \hat{V} \cdot \hat{I} \cdot \frac{1}{2}(1 - \cos(2\omega t))$$

Normally, the instantaneous power is less interesting than average power. This is found by integrating the power and averaging it over e.g. one cycle time, T:

$$P = \overline{p(t)} = \overline{v(t) \cdot i(t)} = \frac{1}{T}\int_0^T \hat{V} \cdot \hat{I} \cdot \sin^2(2\pi f t)dt = \frac{1}{2\pi}\hat{V} \cdot \hat{I}\int_0^{2\pi}\sin^2(\omega t)d\omega t = \frac{1}{2\pi}R \cdot \hat{I}^2\int_0^{2\pi}\sin^2(\omega t)d\omega t = R \cdot \frac{\hat{I}^2}{2}$$

Introducing the root mean square value for the current, or rms value:

$$I_{rms} = \sqrt{\frac{1}{2\pi}\hat{I}^2\int_0^{2\pi}\sin^2(\omega t)d\omega t} = \frac{\hat{I}}{\sqrt{2}}$$

one can see that the power equation simplifies to a form similar as for DC systems:

$$P = R \cdot I^2{}_{rms}$$

With similar definition of rms value for the voltage, it yields:

$$V_{rms} = R \cdot I_{rms} \text{, and } P = V_{rms} \cdot I_{rms}$$

Rms values for voltages and currents enable us to use well-known formulas from DC systems in AC systems, when average power is concerned.

Example 1

1. A lamp connected to a single-phase 120 V AC network

> A light bulb of 60 W is connected to a 120 V_{rms}, voltage.
>
> The current in the bulb is then I bulb = P / V = 60/120 = 0.5 A
>
> The resistance in the bulb is: R bulb = V / I = 120/ 0. 5 = 240 Ω

2. Three phase AC sources and resistive loads:

The power is in all larger installations generated as a three-phase voltage, because

— Three-phase distribution systems give a better utilization of the material in the generator itself, interconnecting cabling, and other distribution equipment and loads.
— Smooth torque in rotating electrical machines, i.e. motors and generators.
— Satisfactory start-up torque with proper design of rotating electrical machines.
— Smoother rectified voltage in static frequency converters.
— A three-phase voltage generator can be regarded as three single-phase voltage sources connected in a common point called the neutral point.

The voltages are called <u>symmetrical</u> if they all are sinusoidal with same amplitude and 120° phase shift.

$$v_a(t) = \sqrt{2}V_{n,rms} \cdot \sin(\omega t),$$

$$v_b(t) = \sqrt{2}V_{n,rms} \cdot \sin(\omega t - 120^\circ),$$

$$v_c(t) = \sqrt{2}V_{n,rms} \cdot \sin(\omega t - 240^\circ)$$

In analysis of voltages and currents, it is convenient to express the sinusoidal time varying variables by using complex notation, which for voltage means:

$$v_a(t) = \mathrm{Re}\left(\sqrt{2}V_{n,rms}e^{j\omega t}\right) = \sqrt{2}V_{n,rms}\,\mathrm{Re}\left(e^{j\omega t}\right) = \sqrt{2}V_{n,rms} \cdot \sin(\omega t)$$

$$v_b(t) = \mathrm{Re}\left(\sqrt{2}V_{n,rms}e^{j\omega t - \frac{2\pi}{3}}\right) = \sqrt{2}V_{n,rms}\,\mathrm{Re}\left(e^{j\omega t - \frac{2\pi}{3}}\right) = \sqrt{2}V_{n,rms} \cdot \sin(\omega t - 120^\circ)$$

$$v_c(t) = \mathrm{Re}\left(\sqrt{2}V_{n,rms}e^{j\omega t + \frac{2\pi}{3}}\right) = \sqrt{2}V_{n,rms}\,\mathrm{Re}\left(e^{j\omega t + \frac{2\pi}{3}}\right) = \sqrt{2}V_{n,rms} \cdot \sin(\omega t - 240^\circ)$$

The sinusoidal variables can be graphically presented in a complex vector diagram. The vectors are in electrotechnical terms annotated phasors, which per definition is represented as a vector with length equal to the rms value of the variable, as for the voltage is as follows:

$$V_a = V_{n,rms}e^{j\omega t},$$

$$V_b = V_{n,rms}e^{j\omega t - \frac{2\pi}{3}},$$

$$V_c = V_{n,rms}e^{j\omega t + \frac{2\pi}{3}}$$

The line-voltages (line to line voltages) is found as the voltage difference between two phases:

$$v_{ab}(t) = v_a(t) - v_b(t);$$

$$v_{bc}(t) = v_b(t) - v_c(t);$$

$$v_{ca}(t) = v_c(t) - v_a(t)$$

In the phasor diagram, the phasor of the line voltages vector difference between the phasors of the corresponding phases:

$$V_{ab} = V_a - V_b;$$

$$V_{bc} = V_b - V_c;$$

$$V_{ca} = V_c - V_a.$$

Hence, the rms value which is the length of the phasor, for all the line voltages is the same and:

$$V_{ll,rms} = \left|V_a - V_b\right| = 2 \cdot V_{n,rms} \cdot Cos(30) = \sqrt{3} \cdot V_{n,rms}$$

If connecting a <u>symmetrical</u> load, i.e. equal load, to a symmetrical three-phase voltage source as shown in Fig. D20, there will flow symmetrical currents. The currents in each of the load branches are the same in amplitude but phase shifted. If the loads are resistive, the currents in each of the branches are

---

$$i_a(t) = \sqrt{2}\frac{V_{n,rms}}{R} \cdot Sin(\omega t),$$

$$i_b(t) = \sqrt{2}\frac{V_{n,rms}}{R} \cdot Sin(\omega t - 120^\circ),$$

$$i_c(t) = \sqrt{2}\frac{V_{n,rms}}{R} \cdot Sin(\omega t - 240^\circ)$$

In phasor notation:

$$I_a = I_{rms}e^{j\omega t} = \frac{V_{n,rms}}{R}e^{j\omega t},$$

$$I_b = I_{rms}e^{j\omega t - \frac{2\pi}{3}} = \frac{V_{n,rms}}{R}e^{j\omega t - \frac{2\pi}{3}},$$

$$I_c = I_{rms}e^{j\omega t + \frac{2\pi}{3}} = \frac{V_{n,rms}}{R}e^{j\omega t + \frac{2\pi}{3}}$$

The power dissipated by the load is:

$$P = 3 \cdot R \cdot I_{rms}^2 = 3 \cdot R \cdot \frac{V_{n,rms}}{R} \cdot I_{rms} = 3 \cdot V_{n,rms} \cdot I_{rms} = \sqrt{3} \cdot V_{ll} \cdot I_{rms}$$

NOTE—In a three-phase system, the power transfer capability of a three-phase cable is $\sqrt{3} = 1.73\,2$, i.e. 73% above a one phase transmission, with only 50% more conductor material, which means that three-phase transmission is more cost efficient than one-phase.

Electrical Impedance

The electric network on a ship will typically consist of a set of alternating generators (three-phase voltage sources), a distribution system with switchboards and transformers, and loads. For electrical analysis of the network, the components are modeled by use of equivalent parameters, such as voltage, or current sources and impedances. In complex notation, the impedance of the circuit components are also expressed as complex values:

Resistors: $Z_R = R$, i.e. positive real.

Inductors: $Z_L = j\omega L = jX_L$, i.e. positive imaginary.

Capacitors: $Z_C = \frac{1}{j\omega C} = -jX_C$, i.e. negative imaginary.

Ohm's law can be applied also to phasors. The relation between the current phasor I through an impedance, and the corresponding voltage phasor V over the impedance is: $V = Z \cdot I$

The current through a resistor will be in phase with the voltage, since the resistor is a positive real. An inductor is a positive imaginary, hence the current will lag the voltage by 90°. For a capacitor, the current will lead the voltage by 90°.

For a combination of resistors and inductors, the current's phase angle will be leading between 0 and 90°, dependent on the R/L ratio. Similarly, the current's phase angle will be lagging between 0 and –90° in a resistor-capacitor combination.

In phasor notation, the following definitions of power related terms apply, per phase:

Apparent power (VA): $S = V \cdot I^* = P + jQ$

Power (W): $P = \text{Re}[S] = \text{Re}[V \cdot I]$

Reactive power (VAR): $Q = \text{Im}[S] = \text{Im}[V \cdot I]$

For a resistor, the apparent power equals to the power, since the voltage over the resistor and the current through it are in phase. This is also seen by calculating the apparent power:

$$S = V \cdot I^* = Z_R \cdot I \cdot I^* = R \cdot |I|^2 \text{, which is a real value.}$$

For an inductor with only inductance, and a capacitor with only capacitance, the power is similarly zero, and hence apparent power and reactive power are equal.

Defining the phase angle between voltage and current as φ, where positive phase angle means leading (capacitive) current and negative phase angle means lagging (inductive) current, it is seen that:

$$P = |S| \cdot \cos\varphi$$
$$Q = |S| \cdot \sin\varphi$$

The $\cos\varphi$ is called the power factor (PF) of the load. The power factor is 1.0 if the current is in phase with the voltage, and lower if the load is inductive or capacitive.

Voltage: 230 V rms, Current: 100 A rms.

Since reactive power is a measure of currents which load the network without contributing to usable power, but only gives losses in the distribution and generation system, it is usually desired to design and operate a system with highest possible power factor. If the load had been purely resistive, the power factor would be 1.0, but due to rotating machinery and transmission lines with inductances and capacitances, the power factor will in all practical applications be lower, normally lagging (inductive).

Example: Single-phase voltage source with inductive -resistive load

A one-phase, 230 V, 60 Hz voltage source is connected to an inductor with inductance L = 30 mH and resistance R = 2 Ω.

The impedance of the load is

$$Z = R + j\omega L = R + j \cdot 2\pi f \cdot L = (2 + j15.7)\Omega.$$

Selecting V= 230 V real, then the current in the circuit is

$$I = V / Z = [\frac{230V}{(2 + j15.7)\Omega}]A = (1.83 - j14.4) = 14.5e^{-j1.44}$$

which is a current of magnitude 14.5A, and it is lagging the voltage with a phase angle of 82.8°, which means that the power factor is Cos (82.8) = 0.125.

The apparent power of the load is

$$S = V \cdot I^* = 230 \cdot (1.83 + j14.4) = (421 + j3312)$$

The dissipated power is hence P = 421W and the reactive power, Q = 3312VAR.

Verifying the results by checking the power consumption in the resistor

$$P_R = R \cdot |I|^2 = 2 \cdot (14.5)^2 W = 421W.$$

The inductor contributes to current and reactive power, but the power dissipation, or losses, only occur in the resistor and results in heat generation.

The power factor is

$$PF = \cos\varphi = P/|S| = P/\sqrt{P^2 + Q^2} = 0.126$$

NOTE—Currents magnified 10x. Current is lagging voltage by 82.8°.

Example: Three-phase voltage source with inductive-resistive load

A three-phase, 400 V (line to line) 50 Hz voltage source is connected to a three-phase inductor with inductance L = 30 mH and resistance R = 2 Ω per phase.

Comparing with example D-3, it is seen that the loads will have the same voltage conditions, with a phase to neutral voltage of 230 V. Hence, in the three-phase system, the following applies:

Apparent power: $S = 3 \cdot V_n \cdot I^* = (1263 + j9936)VA$

$$|S| = 3 \cdot |V_n| \cdot |I| = \sqrt{3} \cdot |V_{ll}| \cdot |I| = 10016VA = 10.016kVA$$

Power: $P = \text{Re}[S] = 1263\,W$

Reactive power: $Q = \text{Im}[S] = 9936\ VAr$

The power factor is then $P/|S|=0.126$, which is the same as for the one-phase example D-.5, a fact that should not be surprising, since the three-phase system actually is a triple of the one-phase system. Therefore, analysis of symmetrical three-phase systems is usually carried out by only regarding one of the phases, since all quantities are equivalent, and phase shifted by 120° and 240° in the remaining two phases. Such models are called per phase representation of the three-phase system.

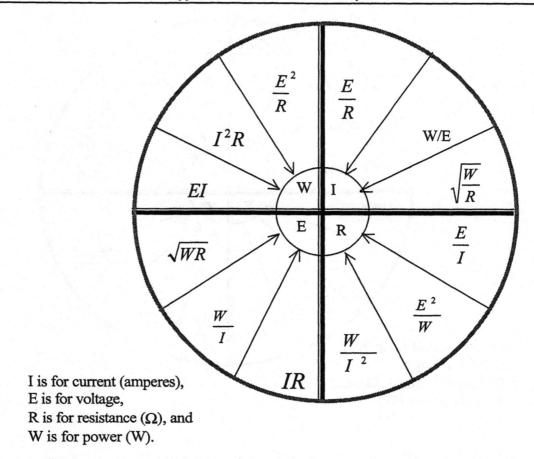

I is for current (amperes),
E is for voltage,
R is for resistance (Ω), and
W is for power (W).

Figure G-1: Ohmic Law for Resistance, Current, Voltage, and Power—DC System

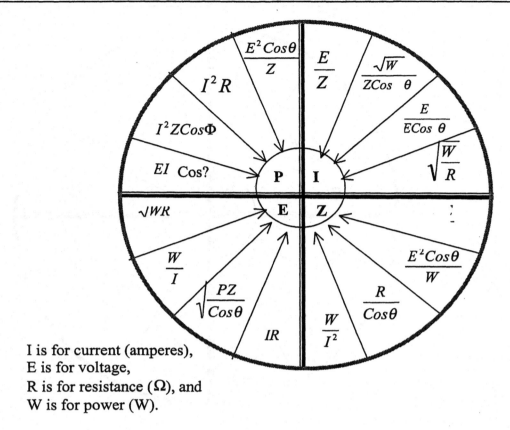

I is for current (amperes),
E is for voltage,
R is for resistance (Ω), and
W is for power (W).

Figure G-2: Ohmic Law for Resistance, Current, Voltage, and Power—AC System

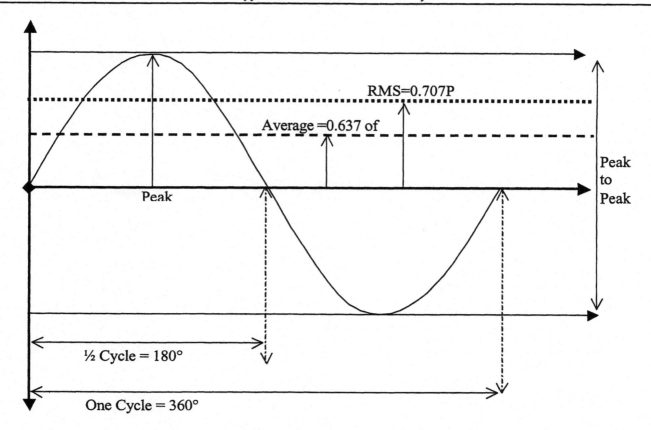

Figure G-3: Sinusoidal Wave Form

Element	Defining equation	Steady state behavior	Energy/Power
Resistance	$e = R\,i$	$R = e/i$	$(e)\,(R)$ or $(i^2\,R)$
Capacitance	$i = C\,de/dt$	$i = 0$ (open circuit)	$W(t) = \tfrac{1}{2}\,C\,e^2$
Inductance	$e = L\,di/dt$	$e = 0$ (short circuit)	$W(t) = \tfrac{1}{2}\,L\,i^2$

Table G-1: AC Voltage-Current Relationship for Combination of Resistance, Inductance and Capacitance

Appendix H   Ship Work Breakdown System (SWBS) for Navy ships (electrical only)

Navy shipboard systems are identified with unique SWBS sequence numbers. This table lists example SWBS numbers for the electrical system. For a more -detailed SWBS listing, refer to US Navy "Ship Work Breakdown Structure" Document number S9040-AA-IDX-010/SWBS.

SWBS #	SWBS Title	SWBS Topic
433	Announcing Systems	General
405	Antenna Requirements	Personnel Safety
405	Antenna Requirements	Installation Standards
405	Antenna Requirements	Material Standards
310	Auxiliary / Emergency Generator Set	Power Generation System Equipment
310	Auxiliary / Emergency Generator Set	Emergency Operation
310	Auxiliary / Emergency Generator Set	Generator Control Systems and Switchboard
310	Auxiliary / Emergency Generator Set	Over-voltage Protection
310	Auxiliary / Emergency Generator Set	General
310	Auxiliary / Emergency Generator Set	Generator Controls
310	Auxiliary / Emergency Generator Set	Generator Control Systems and Switchboard
310	Auxiliary / Emergency Generator Set	Generator Selection
502	Auxiliary Diesel Engines	General
500	Auxiliary Machinery	General
633	Cathodic Protection	General
410	Command and Control Systems	System Architecture
410	Command and Control Systems	General
651	Commissary and Messing Spaces	Equipment and Installation
664	Damage Control	General
415	Digital Data Communications	System Power
415	Digital Data Communications	General
529	Drainage and Ballasting Systems	Podded Propulsor Bilge Pumps
304	Electric Cable	Medium-Voltage Cable
304	Electric Cable	Fiber Optic Cable
304	Electric Cable	Application, Construction and Testing of Low-Smoke Cable
304	Electric Cable	Medium-Voltage Cable
302	Electric Motors and Associated Equipment	Motor Controller
302	Electric Motors and Associated Equipment	Construction
300	Electric Plant General	Ship Service Electric Plant Design
300	Electric Plant General	Standard Voltage and Frequency Ratings
300	Electric Plant General	Ground Detection
300	Electric Plant General	Power Quality
300	Electric Plant General	Ship Service Electric Plant Design

SWBS #	SWBS Title	SWBS Topic
300	Electric Plant General	Rating of Electrical Equipment and Machinery
300	Electric Plant General	Equipment Grounding and Bonding Techniques
235	Electric Propulsion and Central Power Plant	Propulsion Generators
235	Electric Propulsion and Central Power Plant	Main Switchboard
235	Electric Propulsion and Central Power Plant	General
235	Electric Propulsion and Central Power Plant	Power Quality and Harmonics
235	Electric Propulsion and Central Power Plant	Propulsion Generators and Motors
235	Electric Propulsion and Central Power Plant	Propulsion Motor Controls
235	Electric Propulsion and Central Power Plant	Main Switchboard
235	Electric Propulsion and Central Power Plant	Propulsion Generators and Motors
235	Electric Propulsion and Central Power Plant	Central Power Plant and Electric Propulsion Systems
235	Electric Propulsion and Central Power Plant	Propulsion Motor Drives (Power Conversion Equipment)
235	Electric Propulsion and Central Power Plant	Propulsion Motor Controls
235	Electric Propulsion and Central Power Plant	Central Power Plant and Electric Propulsion Systems
235	Electric Propulsion and Central Power Plant	Power Quality and Harmonics
235	Electric Propulsion and Central Power Plant	Central Power Plant and Electric Propulsion Systems
235	Electric Propulsion and Central Power Plant	Main Switchboard
235	Electric Propulsion and Central Power Plant	Propulsion Generators and Motors
235	Electric Propulsion and Central Power Plant	Motor Drives
235	Electric Propulsion and Central Power Plant	Power Quality and Harmonics
235	Electric Propulsion and Central Power Plant	Generator Prime Movers
235	Electric Propulsion and Central Power Plant	Main Switchboard
436	Electrical Alarm, Safety and Warning System	Fire and Smoke Detection System
436	Electrical Alarm, Safety and Warning System	Sprinkling Alarm System
305	Electrical and Electronic Designating and Marking	General
305	Electrical Designating and Marking	General
426	Electrical Navigation System	Gyrocompass System
426	Electrical Navigation System	Underwater Speed Log
406	Electromagnetic Interference	General
406	Electromagnetic Interference	Installation Standards
423	Electronic Navigation Systems, Radio	Design Standards
423	Electronic Navigation Systems, Radio	Equipment Standards
402	Electronic Systems	Installation Standards
440	Exterior Communications	General

SWBS #	SWBS Title	SWBS Topic
555	Fire Fighting Systems	AFFF Concentrate
400	General Standards for Electronics Systems	General
400	General Standards for Electronics Systems	EMI/RFI
400	General Standards for Electronics Systems	Cable Runs
400	General Standards for Electronics Systems	General
400	General Standards for Electronics Systems	Uninterruptible Power Supply (UPS)
400	General Standards For Electronics Systems	Design Standards
400	General Standards for Electronics Systems	Personnel Safety
400	General Standards for Electronics Systems	Installation Standards
512	Heating, Ventilation, and Air Condition Systems (HVAC)	General
512	Heating, Ventilation, and Air Condition Systems (HVAC)	Non-Sparking Fans
512	Heating, Ventilation, and Air Condition Systems (HVAC)	Battery Charging
512	Heating, Ventilation, and Air Condition Systems (HVAC)	Electric Heaters
512	Heating, Ventilation, and Air Condition Systems (HVAC)	Smoke Dampers
512	Heating, Ventilation, and Air Condition Systems (HVAC)	Fans
602	Hull Designating And Marking	Damage Control Classification of Material Condition
556	Hydraulic Systems	General
430	Interior Communication System	Circuit Designations
430	Interior Communication System	General
430	Interior Communication System	Wireless Communications
330	Lighting Systems	Low-level Lighting
330	Lighting Systems	Lighting Circuits and Equipment
330	Lighting Systems	Emergency Lighting System
330	Lighting Systems	Ice Lights
330	Lighting Systems	Weather deck Lighting
330	Lighting Systems	Lighting Calculations / Illumination Levels
252	Machinery Control Stations	Engineering Control Station
252	Machinery Control Stations	Equipment Enclosures
202	Machinery Plant Control and Monitoring System	EMI/RFI
202	Machinery Plant Control and Monitoring System	Main Diesel Engines
202	Machinery Plant Control and Monitoring System	General

SWBS #	SWBS Title	SWBS Topic
202	Machinery Plant Control and Monitoring System	MPCMS Engineering Workstation
202	Machinery Plant Control and Monitoring System	Online Electronic Manuals, Publications, and Drawings.
202	Machinery Plant Control and Monitoring System	MPCMS Workstation Printer
202	Machinery Plant Control and Monitoring System	Tank Level Indicating (TLI) System
202	Machinery Plant Control and Monitoring System	Trending and Analysis System
202	Machinery Plant Control and Monitoring System	Graphical User Interface Style Guide
202	Machinery Plant Control and Monitoring System	Sensors (Resistive Temperature Detector RTD)
202	Machinery Plant Control and Monitoring System	Emergency Manual Control
202	Machinery Plant Control and Monitoring System	Graphical User Interface (Video Terminal Display Graphic)
202	Machinery Plant Control and Monitoring System	General
202	Machinery Plant Control and Monitoring System	EMI/RFI
202	Machinery Plant Control and Monitoring System	Power
233	Main Diesel Engines	General
233	Main Diesel Engines	Starting System
233	Main Diesel Engines	Seating and Alignment
233	Main Diesel Engines	Lube Oil System
233	Main Diesel Engines	Conventional Cooling System
233	Main Diesel Engines	Fuel Oil System
233	Main Diesel Engines	General Governor and Trip Requirements
233	Main Diesel Engines	Emergency Shutdown
233	Main Diesel Engines	Blowers and Turbochargers
568	Maneuvering Thrusters	Design
568	Maneuvering Thrusters	Material and Component Requirements
568	Maneuvering Thrusters	Design
568	Maneuvering Thrusters	Vibration
568	Maneuvering Thrusters	Material and Component Criteria
422	Navigation Lights, Signal Lights and Searchlights	Navigation Lights and Panel
422	Navigation Lights, Signal Lights and Searchlights	Searchlights and Signal Lights
730	Noise, Vibration and Resilient Mounts	Resilient Mounts
493	Non-Combat Data Processing System	Programmable Logic Controllers (PLCs)

SWBS #	SWBS Title	SWBS Topic
661	Office Spaces	General
641	Officer Berthing Spaces	Commanding Officer
245	Podded Propulsors	Fire Fighting, Dewatering, Shafting, Propellers Seal. Bearing, Cooling, Lubrications, Steering, Vibrations
314	Power Conversion Equipment	General, Power Conversion Equipment, Motor Generator Set, Transformer
314	Power Conversion Equipment	S/S Motor Generator Control Systems and Switchboard
314	Power Conversion Equipment	S/S Motor Generator General
314	Power Conversion Equipment	General
314	Power Conversion Equipment	Semiconductor Converters, VFD Drives, UPS Systems, AC to DC Drives, Inverters DC to DC
314	Power Conversion Equipment	S/S Motor Generator General
314	Power Conversion Equipment	Semiconductor Converters, VFD Drives, UPS Systems, AC to DC Drives, Inverters DC to DC
314	Power Conversion Equipment	Semiconductor Converters, VFD Drives, UPS Systems, AC to DC Drives, Inverters DC to DC
314	Power Conversion Equipment	General
314	Power Conversion Equipment	Semiconductor Converters, VFD Drives, UPS Systems, AC to DC Drives, Inverters DC to DC S/S Motor Generator
314	Power Conversion Equipment	Semiconductor Converters, VFD Drives, UPS Systems, AC to DC Drives, Inverters DC to DC, S/S Motor Generator Control Systems and Switchboards
314	Power Conversion Equipment	Motor Generator Sets
314	Power Conversion Equipment	Semiconductor Converters, VFD Drives, UPS Systems, AC to DC Drives, Inverters DC to DC
314	Power Conversion Equipment	Semiconductor Converters, VFD Drives, UPS Systems, AC to DC Drives, Inverters DC to DC
314	Power Conversion Equipment	General
314	Power Conversion Equipment	S/S Motor Generator Control Systems and Switchboard
320	Power Distribution Systems	Electric Steering Systems
320	Power Distribution Systems	General Design
320	Power Distribution Systems	Shore Power Facilities
320	Power Distribution Systems	Electronic Equipment
320	Power Distribution Systems	Emergency Power Distribution System
320	Power Distribution Systems	Final Emergency Loads
320	Power Distribution Systems	Final Emergency Loads

SWBS #	SWBS Title	SWBS Topic
320	Power Distribution Systems	Receptacles, General Purpose
320	Power Distribution Systems	Shore Power Facilities
200	Propulsion Plant	Safety
303	Protective Devices for Electric Circuits	Protection of Power Generation and Distribution System Equipment
303	Protective Devices for Electric Circuits	Transformers
303	Protective Devices for Electric Circuits	Protection of Adjustable Speed Drive (ASD) System Equipment
303	Protective Devices for Electric Circuits	Protection of Power Generation and Distribution System Equipment
503	Pumps	Centrifugal Pumps
441	Radio System	General
441	Radio System	Equipment Standard
441	Radio System	Installation Standard
516	Refrigeration Plants and Equipment	Refrigeration Plants
516	Refrigeration Plants and Equipment	Components
404	RF Transmission Lines	Material Standards
404	RF Transmission Lines	Installation Standards
521	Seawater Service Systems	Cargo Hold Sprinkling
637	Sheathing	Ceilings Systems
111	Shell Plating, Shell framing, Structural Bulkheads, Decks and Platforms	Icebelt
432	Shipboard Telephone Systems	Features
432	Shipboard Telephone Systems	Power
760	Small Arms, Ammunition, Pyrotechnics and Stowage	General
760	Small Arms, Ammunition, Pyrotechnics and Stowage	Physical Security
760	Small Arms, Ammunition, Pyrotechnics and Stowage	Design Guidance
671	Special Stowages	General
671	Special Stowages	General
313	Storage Batteries and Servicing Facilities	Battery Charging
313	Storage Batteries and Servicing Facilities	Battery Application
313	Storage Batteries and Servicing Facilities	Battery Charging
451	Surface Search Radar	Personnel Safety
451	Surface Search Radar	Personnel Safety
324	Switchgear	Ship Service, Emergency and Auxiliary Switchboards Design, Controls, Performance, Application, Installation and Testing

SWBS #	SWBS Title	SWBS Topic
324	Switchgear	Ship Service, Emergency and Auxiliary Switchboards Design, Controls, Performance, Application, Installation and Testing
324	Switchgear	Motor Control Centers (MCC Group Control)
324	Switchgear	Ship Service, Emergency and Auxiliary Switchboards Design, Controls, Performance, Application, Installation and Testing
324	Switchgear	Motor Control Centers (MCC Group Control)
443	Visual and Audible Systems	Signaling Whistle
443	Visual and Audible Systems	Bells
443	Visual and Audible Systems	Controls

Table H- 1: Example SWBS Numbers, Electrical System

Appendix I   Bibliography

I.1 List of Regulatory body abbreviations specific to shipbuilding

ABS – American Bureau of Shipping – Rules for Building Steel Vessels

DNV – Det Norske Veritas

SOLAS – Safety of Life at Sea – International Maritime Organization (IMO) regulation

USCG – United States Coast Guard regulations

UL – Underwriter's Laboratory

IEC – International Electro-technical Commission

I.2 ANSI standards

ANSI/ISA-RP 12.6-1995, Wiring Practices for Hazardous (Classified) Locations—Instrumentation, Part I: Intrinsic Safety.

ANSI/NEMA FU1-1986, Low-Voltage Cartridge Fuses.

ANSI/NEMA ICS 1-1993, Industrial Control and Systems General Requirements.

ANSI/NEMA MG 1-1993, Motors and Generators.

ANSI 27.20, Switchboard and Bus Duct Standard.

ANSI C37.04-C37.010, Medium and High Voltage Circuit Breaker Requirements.

ANSI C37.13, Low Voltage Power Circuit Breaker.

ANSI/NEMA 250-1991, Enclosures for Electrical Equipment (1000 Volts Maximum).

ANSI/UL 913-1988, Standard for Intrinsically Safe Apparatus and Associated Apparatus for Use in Class II, and III, Division I, Hazardous Locations.

I.3 ASTM standards

ASTM D229-96, Standard Test Methods for Rigid Sheet and Plate Materials Used for Electrical Insulation.

ASTM E662-97, Standard Test Method for Specific Optical Density of Smoke Generated by Solid Materials.

ASTM F1003-86 (R1992), Standard Specification for Searchlights on Motor Lifeboats.

ASTM F1166-95a, Standard Practice for Human Engineering Design for Marine Systems, Equipment and Facilities.

ASTM G23-96, Standard Practice for Operating Light-Exposure Apparatus (Carbon-Arc Type) With and Without Water for Exposure of Nonmetallic Materials.

I.4 Canadian standards

CSA C22.1-1994, Canadian Electrical Code Part 1/Seventeenth Edition Safety Standard for Electrical Installations.

CSA C22.2 No. 38-1995, Thermoplastic Insulated Wires and Cables; (Gen. Instr. 1).

I.5 IEC standards

IEC 60056, High-Voltage Alternating-Current Circuit Breakers, 1987, (Including Amendment 1, 1992, Amendment 2, 1995, and Amendment 3, 1996).

IEC 60079, Electrical Apparatus for Explosive Gas Atmospheres.

IEC 60092-350- Electrical Installations in Ships.

IEC 60363, Short-Circuit Current Evaluation with Special Regard to Rated Short-Circuit Capacity of Circuit Breakers in Installations in Ships, 1972.

IEC 60-529, Degrees of protection provided by enclosures (IP Code) 1989.

IEC 60533, Electromagnetic Compatibility of Electrical and Electronic Installations in Ships, 1977.

IEC 60947-2, Low-Voltage Switchgear and Control gear, Part 2: Circuit Breakers, 1989 (Incl Amendment 1, 1992; Amendment 2, 1993).

I.6 IEEE standards

IEEE Std 4™-1995, IEEE Standard Techniques for High-Voltage Testing.

IEEE Std 43™-1991, IEEE Recommended Practice for Testing Insulation Resistance of Rotating Machinery.

IEEE Std 45™-1998, Recommended Practice for Electrical Installations on Shipboard.

IEEE Std 45™-2002, Recommended Practice for Electrical Installations on Shipboard.

IEEE Std 74™-1974, IEEE Standard for Test Code for Industrial Control (600 Volts or Less).

IEEE Std 112™-1996, IEEE Standard Test Procedure for Poly-phase Induction Motors and Generators.

IEEE Std 115™-1995, IEEE Guide: Test Procedures for Synchronous Machines, Part I—Acceptance and Performance Testing, Part II—Test Procedures and Parameter Determination for Dynamic Analysis.

IEEE Std 432™-1992, IEEE Guide for Insulation Maintenance for Rotating Electric Machinery.

IEEE Std 444™-1992, Standard Practices and Requirements for Thyristor Converters for Motor Drives, Part 1—Converters for DC Motor Armature Supplies.

IEEE Std 519™-1992, IEEE Recommended Practices and Requirements for Harmonic Control in Electric Power Systems.

IEEE Std 576™-1989, IEEE Recommended Practice for Installation, Termination, and Testing of Insulated Power Cable as Used in the Petroleum and Chemical Industry.

IEEE Std 1202™-1996, IEEE Standard for Flame Testing of Cables for Use in Cable Tray in Industrial and Commercial Occupancies.

IEEE Std C37.20.2™-1993, IEEE Standard for Metal-Clad and Station-Type Cubicle Switchgear.

IES RP-12-1998, Recommended Practice for Marine Lighting.

I.7 Military standards (USA)

MIL-C-17G, Cables, Radio Frequency, Flexible and Semi-rigid, General Specification for.

MIL-C-17587, Low Voltage Circuit Breaker.

MIL-C-17361, Molded Case Circuit Breaker.

MIL-HDBK-299-October 1994, Cable Comparison Handbook—Data Pertaining to Electric Shipboard Cable.

MIL-S-16036, Power Switchgear.

MIL-C- 24640A, Cable, Electrical, Lightweight for Shipboard Use, General Specification for.

MIL-C-24643 A, Cable and Cord, Electrical, Low Smoke, for Shipboard Use, General Specification for.

DDS-300-2, Fault Current Calculation for Alternating Current.

DDS-311-3, Ship Service Power System, Application and Coordination of Protective Devices.

DDS-9620-2, Electric Cable Voltage Drop Calculations, 1 November 1963.

I.8 NEMA standards

NEMA AB1-1975, Molded Case Circuit Breakers.

NEMA SG3-1975, Low-Voltage Power Circuit Breakers.

Appendix J   Cross references of standards

Cross References - IEEE Std 45-2002, USCG CFR-46, ABS-2002, and SOLAS for Shipboard Electrical Installation

Title	IEEE Std 45-2002 (Clauses)	USCG CFR-46 Regulation Section #	ABS-2002 Rule Section #	SOLAS 2001 Regulation-Chapter and section #
Plan and Data			4-8-1-5	
Alarms—Fire Detection and Sprinkler	28			II-2-12
Ambient Temperature		111.01-15	4-8-3-1.17	
Batter Types	22.4			
Battery Charging	22.10	111.15-25,30	4-8-3-5.9	
Battery Installation	22.6.2	112.55	4-8-4-5.3	
Battery Rating	22.9			
Battery Ventilation	22.7	111.15-10		
Cable – General	23			
Cable – Propulsion	25.12, 31.14			
Cable Ampacity	24.8			
Cable Ampacity Table—Medium Voltage	Tables 27, 28, 29			
Cable Ampacity Table—Distribution, Control, and Signal	Table 25			
Cable Application	24			II-1-D-5
Cable Generator		111.12-9		
Cable Splicing	25.11	111.60-19		
Circuit Breaker Application	8.6			
Circuit Designations	Annex B			
Communications—External	27			
Communications—Means of	26.7.1			
Continuity of Power—Uninterruptible Power Supply (UPS)	9.8			
Control Hierarchy	9.5.2			
Control System Design	9.3			II-1-6
Control System—Environmental Conditions	9.20			
Control System Programming	9.17			
Coordination of Protective Devices		111.51		
Controls, Start, Stop, and Shutdown Conditions	9.5.3			
Demand Factors	5.4			

Title	IEEE Std 45-2002 (Clauses)	USCG CFR-46 Regulation Section #	ABS-2002 Rule Section #	SOLAS 2001 Regulation-Chapter and section #
Electric Propulsion—Drive	31.8		4-8-3-7, 4-8-5-5.17.9	II-1-D-41
Electric Propulsion—Drive Transformer	31.6		4-8-3-7	
Electric Propulsion—Switchboard	31.9			II-1-D-41
Electric Propulsion—Generator	31.5		4-8-5-5.5.1	II-1-D-41-5
Electric Propulsion—Podded Propulsion	31.13			
Electric Propulsion—Power Management	31.11			
Electric Propulsion—Power Quality	31.3.2			
Electric Propulsion—Prime Movers	31.4			
Electric Propulsion—Propulsion Control	31.10			
Electric Propulsion—Propulsion Motor	31.7			
Emergency Generator	6.2	112.5, 112.50		II-1-D-41-44
Emergency Power Distribution	6.4	112.05	4-8-2.5	II-1-D-41-44
Emergency Stops	11.7	111.103	4-8-2-11.9	
Enclosure Types	Annex C	111.01-9	Tables 1A, 1B, 2	
Engine Order Telegraph	26.2			
Fault Current—AC System	5.9.2			
Feeder—Generator			4-8-2-7.7.2	
Feeder—Transformer			4-8-2-7.7.3	
Feeder—Steering Gear			4-8-2-7.11	
Fiber Optics Cable		111.60-6		
Final Emergency Circuits	6.8			
General Alarm	26.5	113.25		
General Alarm—Power Supply	26.5.5			III-6.4.2, III/50, II42.43
Generator Protective Relay Control	7.5			
Generator Selection and Size	7.4.2		4-8-2-3.1, 3 &5	
Governors	7.3.4			
Ground Detection	5.9.7	111.05, 111.30-25(e)(1)		
Grounding Points	5.9.6.2			
Hazardous Area—Approved Equipment	33.6	111.105		

Title	IEEE Std 45-2002 (Clauses)	USCG CFR-46 Regulation Section #	ABS-2002 Rule Section #	SOLAS 2001 Regulation-Chapter and section #
Hazardous Area Classification for Types of Ship	33.3, Annex A	111.105	4-8-4-27	
Individual and Multiple Motor Protection	5.4.4			
Intrinsically Safe Equipment		111.105-11	4-8-4-27.7	
Lighting—Navigation	21.2			
Lighting—Emergency	20.10			
Lighting—Searchlight	20.9			
Location of Overcurrent Protection	5.9.5.7			
Low-Voltage Switchboards—Description	8.3			
Medium-Voltage Switchboards—Description	8.4			
Minimum Switchboard Equipment—AC Generator	7.6			
Motor Branch Circuits	5.9.5.3			
Motor Duty Rating	14.2	111.25-15		
Motor Enclosure	14.1			
Motor Limits of Temperature Rise	13.13			
Motor Locked Rotor Kilovoltampere	13.6			
Motor Overload Test	13.2 0			
Motor—AC Controller	11.1.2			
Motor—Steering Gear	14.3			
Motor—Undervoltage Protection	11.1.1	111.70-3(f) & (g)		
Protection—Generator		111.21-11	4-8-2-9.11	
Protection—Feeder		111.50-3	4-8-2-9.13	
Protection—Motor Circuit		111.7	4-8-2-9.17	
Protection—Transformer		111-.20-15	4-8-2-9.19	
Power Quality and Harmonics	4.6			
Power Supply—Gyro Compass	30.3			
Power Supply—External Communications	27.4			
Prime Mover	7.3	111-.21-1		
Public Address/General Announcing/Talkback	26.7.2		4-8-2-11.5.4	
Rating and Setting of distribution circuit Protection devices	5.9.5.8			
Rectifier—Ambient Temperature	10.20.7			

Title	IEEE Std 45-2002 (Clauses)	USCG CFR-46 Regulation Section #	ABS-2002 Rule Section #	SOLAS 2001 Regulation-Chapter and section #
Safety System			4-9-2.9	
Ship Test—Generator Set	34.3		4-8-3-3.15	
Ship Test—Motor and Controllers	34.5			
Ship Test—Switchboards	34.4			
Ship Test—Voltage Drop	34.13			
Shore Power	8.9.4 and 17.7		4-8-2-11	
Starter—Magnetic	10.7	111.70-3		
Starter—Medium Voltage	10.9			
Starter—Solid State	10.8			
Starter Overload Relays and ambient temperature	11.1.4			
Starters—Manual	10.6			
Steering—Controls	32.6			
Steering—Feeder Circuit	32.5.2			
Steering—General	32.5.1			
Switchboard—Installation and Location	8.2			
Switchboard—Controls		111.30-25	4-8-3-5	
Switchboard—Emergency		111.30-29	4-8-2.5.13	
Synchronizing Control	7.5.5			
Temporary Emergency Circuits	6.7			
Transformer Temperature Rise	12.6	111.20-5		
Transformer—Feeder		111.20-15		
Transformer Type, Number, and Rating	12.3	111.2	4-8-3-7, 4-8-5-3.7.5	
Voltage Drop	5.5			
Voltage Regulation	7.4.7			
Watertight Doors	29	111.97		
Whistle Control	21.5	113.65		

Table J-1: Cross References - IEEE Std 45-2002, USCG CFR-46, ABS-2002, and SOLAS for Shipboard Electrical Installation

Appendix K   Cross reference of IEEE Std 45-1998 and IEEE Std 45-2002 Clauses

The sections in this Handbook have been arranged for better understanding of intersystem and intrasystem functionality of ship board electrical systems, particularly the power generation, distribution, controls, and communications. The following table provides cross-references from the clauses in IEEE Std 45-2002 to the clauses in IEEE Std 45-1998 with brief descriptions of major changes. Additional details of the changes are provided in appropriate sections in this Handbook. Note that Clause 8 in IEEE Std 45-1998 cable manufacturing was removed from IEEE Std 45-2002 and published in a stand-alone standard, IEEE Std 1580-2001.

Table K-1: IEEE Std 45-1998 and IEEE Std 45-2002 Cross-References of Clauses

IEEE Std 45-2002 Clause # and title	IEEE Std 45-1998 Clause # and title	Remarks
1 – Overview	1 – Overview	IEEE Std 45-2002, 1.7: Added with caution and precaution for mixing and matching equipment designed to different national and international standards. It is recommended that coordination of different equipment design and testing standards, construction ratings, installation methods, and performance be carefully analyzed.
2 – References	2 – References	
3 – Definitions	3 – Definitions	IEEE Std 45-1998, 3.7: NEMA enclosures were supported by equivalent IEC enclosure types. IEEE Std 45-2002, 3.7: NEMA enclosure types are given without IEC enclosure type equivalency as such equivalency I different enclosure types are not correct. IEEE Std 45-2002, Tables C1 through C4: Added NEMA enclosure types, IEC enclosure types and approximate equivalent between these two types of enclosures, emphasizing the fact that they not exact equivalent. The definition of NEMA enclosure type is completely different from the definition of IEC enclosure types.
4 – Power system characteristics	4 – Power system characteristics	IEEE Std 45-2002, 4.4: Introduces system voltage levels of 11 000 V and 13 800 V.
5 – Power system design	11 – Distribution	IEEE Std 45-2002, 5.8.3: Added IEC 1.5 mm^2 conductor to the 15 AWG branch circuit conductor requirement. IEEE Std 45-2002, 5.9.6: Gave better understanding of continuous insulation monitoring system.
6 – Emergency power system	26 – Emergency Power System	
7 – Electric power generation	5 – Electric power generation	Generator Insulation Class F added to the Class B requirement to consider initial cost versus service life requirement. See IEEE Std 45-2002, 7.4.3
8 – Switchboards	7 – Switchboards	IEEE Std 45-2002, 8.4: Added medium-voltage switchboards (601 V to 38 kV for ANSI and 1010 V to 35 kV for IEC).
9 – Control systems	37 – Control systems	
10 – Control apparatus	17 – Control apparatus	
11 – Control applications	18 – Control applications	
12 – Transformers	34 – Transformers	
13 – Motors	14 – Motors	
14 – Motor application—general	15 – Motor applications	
15 – Brakes	19 – Brakes	
16 – Magnetic friction clutches	21 – Magnetic friction clutches	

IEEE Std 45-2002 Clause # and title	IEEE Std 45-1998 Clause # and title	Remarks
17 – Distribution equipment	13 – Distribution equipment	
18 – Heating equipment	22 – Heating equipment	
19 – Galley equipment	32 – Galley equipment	
20 – Lighting equipment	23 – Lighting equipment	
21 – Navigation lights and signals	24 – Navigation lights and signals	
22 – Storage batteries	6 – Storage batteries	
23 – Cable types for installation on shipboard		IEEE Std 45-2002, new Clause 23 is for shipboard cable installation requirements.
24 – Cable application	9 – Cable application	IEEE Std 45-2002, Tables 27, 28, 29: Added for medium-voltage cable ampacities.
25 – Cable installation	10 – Cable installation	IEEE Std 45-2002, 25.12: Added requirements for propulsion cables.
26 – Interior communications systems	27 – Interior communications systems	
27 – Exterior communication and navigation systems	31 – Exterior communication and navigation systems	
28 – Fire detection, alarm, and sprinkler systems	28 – Fire detection, alarm, and sprinkler systems	
29 – Watertight and fire door equipment	16 – Watertight and fire door equipment	
30 – Gyro compass systems	29 – Gyro compass systems	
31 – Electric propulsion and maneuvering system	36 – Electric propulsion and maneuvering system	IEEE Std 45-2002, 31.12–31.22: Check additions, such as Failure Mode and Effect Analysis (FMEA), podded propulsion and propulsion cables
32 – Steering systems	30 – Steering gear	
33 – Hazardous locations, installations, and equipment	33 – Hazardous locations, installations, and equipment	
34 – Ship tests	38 – Ship tests	
35 – Spare parts		
Annex A, Hazardous locations	Annex A, Hazardous locations	
Annex B, Circuit designations	Annex B, Circuit designations	

IEEE Std 45-2002 Clause # and title	IEEE Std 45-1998 Clause # and title	Remarks
Annex C, NEMA and IEC enclosures		Table C.1: Gives NEMA enclosure classification. Table C.3: Gives IEC, IP enclosure classification. Tables C.2: Gives approximate conversion between NEMA enclosure and IEC enclosure. In IEEE Std 45-1998, NEMA and IEC enclosure ratings were given side by side throughout the standard. However, it was found that the comparative listing was somewhat misleading. It was decided to not show NEMA and IEC enclosure types throughout the standard. The user must establish the appropriate enclosure type for either standard and seek approval of the certifying agency that has the authority of certification.

NOTES

1—IEC requirements were added throughout IEEE Std 45-2002 to harmonize with internationally accepted standards.

2—SI units of measurement were added to the US units throughout IEEE Std 45-2002 to harmonize with internationally accepted standards.

Appendix L   IEEE cable manufacturing requirements comparison

Table L-1: Comparison of Cable Construction—IEEE Std 45-1998, IEEE Std 45-2002, and IEEE Std 1580-2001

Conductor	IEEE Std 45-1998, 8.1		IEEE Std 45-2002	IEEE Std 1580-2001, 5.1.1	
	The conductors should be of soft annealed copper wire. All conductors should be tinned or alloy coated where necessary to ensure compatibility with primary insulation.		None	The conductors should be of soft annealed copper wire. All conductors should be tinned or alloy coated where necessary to ensure compatibility with primary insulation.	
Insulation materials	IEEE Std 45-1998, 8.2		IEEE Std 45-2002, Table 25	IEEE Std 1580-2001, 5.3.1	
	Insulation-type designation	Maximum conductor temperature		Insulation-type designation	Maximum conductor temperature
	T – Poly vinyl chloride	75 °C (167 °F)	T (75 °C)	T (PVC) – Polyvinylchloride	75 °C
	T/N – Polyvinylchloride/nylon	90 °C (194 °F)	T/N (90 °C)	T/N (PVC/polyamide) – Polyvinylchloride/nylon	90 °C
	E – Ethylene propylene rubber	90 °C (194 °F)	E (90 °C)	E (EPR) – Ethylene propylene rubber	90 °C
	X – Cross-linked polyethylene	90 °C (194 °F)	X (90 °C)	X (XLPE) – Cross-linked polyethylene	90 °C
	LSE – Low-smoke, halogen-free ethylene propylene rubber	90 °C (194 °F)	LSE (90 °C)	LSE (LSEPR) – Low-smoke, halogen-free ethylene propylene rubber	90 °C
	LSX – Low-smoke, halogen-free cross-linked polyethylene	90 °C (194 °F)	LSX (90 °C)	LSX (LSXLPO) – Low-smoke, halogen-free cross-linked polyolefin	90 °C
	S – Silicone rubber	100 °C (212 °F)	S (100 °C)	S (Silicone) – Silicone rubber	100 °C
	P – Cross-linked polyolefin	100 °C (212 °F)	P (100 °C)	P (XLPO) – Cross-linked polyolefin	100 °C
Cable jacket (sheath)	IEEE Std 45-1998, 8.9		IEEE Std 45-2002	IEEE Std 1580-2001, 5.12.1	
	The jacket should be thermoplastic Type T or TPO, thermosetting (Type CP, N or L) complying with the requirements of Table 8-9 or Table 8-10. The manufacturer should perform type tests and periodic testing to ensure jacket materials meet these requirements.		None	The jacket should be thermoplastic [Type T (PVC) or TPO (TPPO)], thermosetting [Type CP (CSPE), CPE, N (PCP), or L (XLPO)] complying with the requirements of Table 16 or Table 17. The manufacturer should perform type tests and periodic testing to ensure jacket materials meet these requirements. The temperature rating of a jacket shall be not less than 15 °C lower than the temperature rating of the insulation.	
	IEEE Std 45-1998, Table 8-11		IEEE Std 45-2002	IEEE Std 1580-2001, Table 18, which is same as IEEE Std 45-1998, Table 8-11 (except unit)	

	Calculated diameter of cable under jacket (in)	Jacket thickness (mils)	None	Calculated diameter of cable under jacket (mm)	Jacket thickness minimum average (mm)
	0–0.425	45		0–10.79	1.14
	0.426–0.700	60		10.80–17.78	1.52
	0.701–1.500	80		17.79–38.10	2.03
	1.501–2.50	110		38.11–63.50	2.79
	2.51 and larger	140		63.51 and larger	3.56
Metal braid armor	**IEEE Std 45-1998, Clauses 8–10**		**IEEE Std 45-2002**	**IEEE Std 1580-2001, 5.19.1.5**	
	The wire should be commercial bronze, aluminum, or tin-coated copper.		None	Aluminum (A), bronze (B), tinned copper (T), continuous corrugated metal (CWCMC)	

Appendix M   Codes and standards (US national and international)

— NEC® – National Electric Code® (United States)
— ABS – American Bureau of Shipping's Rules for Building Steel Vessels
— IEC – International Electrotechnical Commission
— IEEE Std 45-2002, Recommended Practice for Electrical Installation on Shipboard
— IEEE Std 1580-2001, Commercial Cable Manufacturing Standard for Shipboard Application
— IMO-SOLAS – International Maritime Organization's Safety of Life at Sea (SOLAS is throughout the Handbook)
— US Navy Military Specifications related to ship construction only
— DNV – Det Norske Veritas (Norway)
— UL – Underwriter's Laboratory (United States)
— NEMA – National Electrical Manufacturer's Association (United States)
— IEEE/ASTM SI 10-1997, Standard for Use of the International System of Units (SI): The Modern Metric System
— USCG CRF-46 – US Coast Guard Code of Federal Register #46
— BV – Bureau of Varitas (France Classification Society)
— GL – Germanischer Lloyd (Germany Classification Society)
— Lloyd's (British Classification Society)
— ANSI – American National Standards Institute
— ASTM – American Society for Testing Material
— NVIC –Navigation and Vessel Inspection Circular (United States Coast Guard)

Appendix N   Standard US and IEC Voltages for Shipboard Application (see IEEE Std 45-2002, Table 3, in 3.2)

Table N-1: US and IEC Shipboard Power Generation and Distribution Levels at 50 Hz and 60 Hz

AC voltage generation 60 Hz and 50 Hz (V)	AC voltage distribution 60 Hz and 50 Hz (V)	Remarks
120	115	Mostly US applications
208	200	
230	220	Mostly IEC application and some US commercial applications
240	230	
380	350	
400	380	50 Hz
450	440	Mostly military applications
480	460	Mostly commercial applications
600	575	
690	660	Mostly IEC applications
2400	2300	50 Hz
3300	3150	60 Hz and 50 Hz
4160	4000	60 Hz and 50 Hz
6600	6000	50 Hz
11 000	10 600	60 Hz and 50 Hz
13 800	13 200	60 Hz and 50 Hz

Appendix O   Miscellaneous

Harmonization of national and international shipbuilding standards

NEMA and IEC equipment—mixing and matching installation (refer to IEEE Std 45-2002, 1.7) (typical example)

Shipboard installation of IEC motors, with IEEE cables, and NEMA switchboards, i.e., the mixing and matching of equipment designed to different standards, must be carefully considered. The IEC motors are kilowatt rated, whereas NEMA motors are horsepower rated. IEC equipment is manufactured with SI units. The US equipment is manufactured mostly with US units. The cable termination at the IEC motor pothead is in millimeters, an SI unit. The cable gland at the IEC motor pothead is also in millimeters. The IEEE cables are in inches, a US unit, and the cable terminating lugs in the NEMA (ANSI) switchboard are also in inches. The intermixing of SI and US unit equipment is a major engineering challenge which often demands additional design considerations.

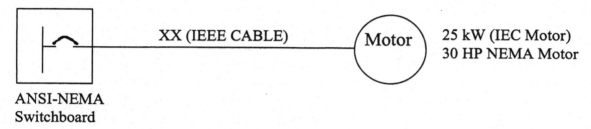

Figure O-1: IEC Motor with IEEE Std 45 Cable and NEMA Switchboard

NOTES

1—The diameter of the stuffing gland at the IEC motor pothead is in millimeters, an SI unit. The IEEE Std 45 cable is in inches, a US unit. However, the cable diameter (in) matches the motor pothead cable entry hole (mm) so that the termination is watertight or drip proof as required.
2—The IEEE Std 45 cable terminating lug holes for stud connection inside the motor pothead are also in inches, but must be suitable for matching the stud, which is measured in millimeters.
3—The space inside the IEC motor pothead may not be sufficient for lugs sized in inches, a situation which often requires additional engineering.

Shipboard underway vibration: recommendations for electric equipment

In IEEE Std 45-1998, the shipboard electrical equipment vibration withstanding requirements were in US units and SI units. In IEEE Std 45-2002, all units were changed to SI units. Both versions are provided for easy reference.

IEEE Std 45-1998, 1.6.1.d, Vibration of vessel underway—electrical equipment

1) Vibration freq range of 5-50 Hz with velocity amplitude of 20 mm/sec
2) Peak accelerations due to ship motion in a seaway of ± 0.6 g for ship's exceeding 90 m in length and ± 1.0 g for smaller ships, with duration of 5 s to 10 s.

NOTES

1—g is acceleration of a free fall (or acceleration due to gravity), which is 9.80.7m/ sec^2 in SI unit and 32.2 ft/s^2 in US unit
2—1.0 g = 9.8 m/s^2
3—0.6 g =5.9 m/s^2

IEEE Std 45-2002, 1.5.1, Vibration of vessel underway—electrical equipment (extract)

d) Vibration of a vessel underway: Electrical equipment should be constructed to withstand at least the following:
   1) Vibration frequency range of 5–50 Hz with a velocity amplitude of 20 mm/s.
   2) Peak accelerations due to ship motion in a seaway of $\pm$ 5.9 m/sec^2 for ships exceeding 90 m in length, and $\pm$ 9.8 m/sec^2 for smaller ships, with a duration of 5–10 s.

Index